1617 919 430

Land and Agrarian Reform in Zimbabwe

4/4
One week loan
Benthyciad un wythnos
WITHDRAWN
AF759258

This book is a product of CODESRIA National Working Group on Zimbabwe

Land and Agrarian Reform in Zimbabwe

Beyond White-Settler Capitalism

Edited by

Sam Moyo

Walter Chambati

African Institute for Agrarian Studies
HARARE

CODESRIA

Council for the Development of Social Science Research in Africa
DAKAR

333.316891 LAN

© CODESRIA 2013

Council for the Development of Social Science Research in Africa
Avenue Cheikh Anta Diop, Angle Canal IV
BP 3304 Dakar, CP 18524, Senegal
Website: www.codesria.org

ISBN: 978-2-86978-553-3

All rights reserved. No part of this publication may be reproduced or transmitted in any form or by any means, electronic or mechanical, including photocopy, recording or any information storage or retrieval system without prior permission from CODESRIA.

Typesetting: Alpha Ousmane Dia
Cover Design: Ibrahima Fofana

Distributed in Africa by CODESRIA
Distributed elsewhere by African Books Collective, Oxford, UK
Website: www.africanbookscollective.com

The Council for the Development of Social Science Research in Africa (CODESRIA) is an independent organisation whose principal objectives are to facilitate research, promote research-based publishing and create multiple forums geared towards the exchange of views and information among African researchers. All these are aimed at reducing the fragmentation of research in the continent through the creation of thematic research networks that cut across linguistic and regional boundaries.

CODESRIA publishes *Africa Development*, the longest standing Africa based social science journal; *Afrika Zamani*, a journal of history; the *African Sociological Review*; the *African Journal of International Affairs*; *Africa Review of Books* and the *Journal of Higher Education in Africa*. The Council also co-publishes the *Africa Media Review*; *Identity, Culture and Politics: An Afro-Asian Dialogue*; *The African Anthropologist* and the *Afro-Arab Selections for Social Sciences*. The results of its research and other activities are also disseminated through its Working Paper Series, Green Book Series, Monograph Series, Book Series, Policy Briefs and the CODESRIA Bulletin. Select CODESRIA publications are also accessible online at www.codesria.org.

CODESRIA would like to express its gratitude to the Swedish International Development Cooperation Agency (SIDA/SAREC), the International Development Research Centre (IDRC), the Ford Foundation, the MacArthur Foundation, the Carnegie Corporation, the Norwegian Agency for Development Cooperation (NORAD), the Danish Agency for International Development (DANIDA), the French Ministry of Cooperation, the United Nations Development Programme (UNDP), the Netherlands Ministry of Foreign Affairs, the Rockefeller Foundation, FINIDA, the Canadian International Development Agency (CIDA), the Open Society Foundations (OSFs), TrustAfrica, UN/UNICEF, the African Capacity Building Foundation (ACBF) and the Government of Senegal for supporting its research, training and publication programmes.

CARDIFF UNIVERSITY
24 MAR 2014
PRIFYSGOL CAERDYDD

Contents

List of Abbreviations

A1	small-scale family farms
A2	smaller-sized capitalist farms
AIAS	African Institute for Agrarian Studies
ARDA	Agricultural and Rural Development Authority
AREX	Department of Agricultural Research and Extension
ASPEF	Agricultural Sector Productivity Enhancement Facility
BIPPA	Bilateral Investment Protection and Promotion Agreement
CA	Communal Area
CFU	Commercial Farmers Union
CSC	Cold Storage Commission
CSO	Central Statistical Office
DCC	District Coordination Committee
DTZ	Development Trust of Zimbabwe
ESAP	Economic Structural Adjustment Programme
FAO	Food and Agriculture Organization (United Nations)
FCTZ	Farm Community Trust of Zimbabwe
FTLRP	Fast Track Land Reform Programme
GAPWUZ	General Plantation and Agriculture Workers Union of Zimbabwe
GMB	Grain Marketing Board
GoZ	Government of Zimbabwe
IMF	International Monetary Fund
ISI	Import Substitution Industrialisation
JAG	Justice for Agriculture
LSCF	Large-Scale Commercial Farms
MAEMI	Ministry of Agricultural Engineering, Mechanisation and Irrigation
MAMID	Ministry of Agriculture, Mechanisation and Irrigation Development
MDC	Movement for Democratic Change
MLLR	Ministry of Lands and Rural Resettlement
NCA	National Constitutional Assembly

NGO	Non Governmental Organisation
NRA	New Resettlement Area
PDL	Poverty Datum Line
RBZ	Reserve Bank of Zimbabwe
RDC	Rural District Council
SADC	Southern African Development Community
SSCF	Small-Scale Commercial Farms
UNDP	United Nations Development Programme
USAID	United States of America Aid
ZANU-PF	Zimbabwe African National Union-Patriotic Front
ZCTU	Zimbabwe Congress of Trade Unions
ZIPRA	Zimbabwe People's Revolutionary Army
ZNLWVA	Zimbabwe National Liberation War Veterans Association
ZRP	Zimbabwe Republic Police

List of Tables, Figures and Boxes

Tables

Figure

Boxes

Acknowledgements

We are grateful for the assistance that many people afforded us during this book project. Invaluable comments were received from various Zimbabwean researchers and activists at a workshop convened by the African Institute for Agrarian Studies (AIAS) in December 2009 in Harare. We received extensive feedback on some chapters from participants in the annual Agrarian Studies Summer School between 2009 and 2012 in Harare, Dar es Salaam and Grahamstown (Rhodes University). In this regard, special thanks are due to Issa Shivji and Ngwanza Kamata (University of Dar es Salaam), Dzodzi Tsikata (University of Ghana), Praveen Jha (Jawaharlal Nehru University), Fred Hendricks (Rhodes University), Lungisile Ntsebeza (University of Cape Town) and Marcelo Rosa and Antonadia Borges (University of Brasilia) for their critical comments.

Shepherd Siziba and Godfrey Mahofa of the Department of Agricultural Economics and Extension at the University of Zimbabwe assisted us with some of the statistical analyses. We also acknowledge the research assistance provided by AIAS interns, including Steven Mberi, Charity Dangwa and Tatenda Matengu; and the survey enumeration and data entry support provided by several undergraduate students from the University of Zimbabwe and Midlands State University. Hundreds of land reform beneficiaries and other informants gave their valuable time to respond to our interviews. We also thank Lisa María Noudéhou for the editorial support she provided.

We are also grateful to various sponsors. The Council for the Development of Social Science Research in Africa (CODESRIA) funded the convening of the National Working Group (NWG) on Zimbabwe and some follow-up field work. The baseline field surveys which provided the bulk of the empirical material for most of the chapters were generously funded by the Royal Norwegian Embassy in Harare and the W.K. Kellogg Foundation. The Canadian International Development Agency (CIDA), Trust Africa and the Open Society Initiative for Southern Africa (OSISA) funded the workshops wherein the research was debated.

As usual, the views expressed in each of the chapters are those of the authors and not those of the institutions they are affiliated to, the people mentioned above, AIAS as an institution, or the sponsors.

Notes on Contributors

Walter Chambati is a researcher at the African Institute for Agrarian Studies (AIAS) in Harare and was Future Agriculture's Consortium Research Fellow for 2011. He received a BSc. (Hons) in Agricultural Economics from the University of Zimbabwe and a Masters in Public and Development Management from the University of Witwatersrand. His research interests are in rural labour issues and agricultural development in Africa and he is studying for a doctorate at the School of Public and Development Management, University of Witwatersrand, focusing on agrarian labour changes after the land reform programme in Zimbabwe.

Tendai Chari is a media analyst and a media studies lecturer in the Department of Communication, School of Human and Social Sciences, University of Venda, South Africa. His research interests are political communication, media ethics, media and development and media policy.

Louis Masuko is a research associate at the African Institute for Agrarian Studies. He was a lecturer in economics and research with the Institute of Development Studies at the University of Zimbabwe until 2011. He is editor of *Economic Policy Reforms and Meso-Scale Rural Market Changes in Zimbabwe: The Case of Shamva* (1998) and has published extensively on development issues in Zimbabwe.

Sam Moyo is Executive Director of the African Institute for Agrarian Studies (AIAS), Harare and former President of the Council for the Development of Social Research in Africa (CODESRIA, 2008–11). He has taught at the Zimbabwe Institute of Development Studies, University of Zimbabwe and has served on the boards of various research institutes and non-governmental organizations, including ZERO. He is currently professor at Fort Hare University and Editor of Agrarian South: Journal of Political Economy (Sage India). He has published widely in academic journals and is the author and editor of several books, including: The land question in Zimbabwe (SAPES, 1995), Land reform under structural adjustment in Zimbabwe (Nordiska Afrikainstitutet, 2000), Reclaiming the land (Zed Books, 2005), African land questions, agrarian transitions and the state (CODESRIA, 2008); Land and sustainable development in Africa (Zed Books, 2008), Reclaiming the nation (Pluto Press, 2011) and The Agrarian Question in the Neoliberal Era (Pambazuka, 2011).

Tendai Murisa recently completed his PhD at Rhodes University. He wrote the paper included in this volume when he was a fulltime research fellow at the African Institute for Agrarian Studies in Harare, Zimbabwe. He is now a programme specialist at the Dakar-based Trust Africa, responsible for the coordination of the Pan-African Agriculture Advocacy project.

Ndabezinhle Nyoni is a research fellow at the African Institute for Agrarian Studies (AIAS). He received a BSc Agriculture (Hons) and an MPhil in Animal Science from the University of Zimbabwe. His research interests are international politics and agricultural development.

Zvakanyorwa Wilbert Sadomba is a researcher with the Centre for Applied Social Sciences (CASS) at the University of Zimbabwe. He is a rural development expert and a veteran of Zimbabwe's liberation war. He has published several journal articles and two books: *War Veterans in Zimbabwe's Land Occupations: Complexities of a liberation movement in an African postcolonial settler society* (2008) and *War Veterans in Zimbabwe's Revolution: Challenging neo-colonialism & settler & international capital* (2011). His research interest is to study the effect of African liberation movements on current societies.

Paris Yeros is Adjunct Professor International Economics at the Federal University of ABC, São Paulo, Brazil' and a research associate of AIAS. He is co-editor of Reclaiming the Land (2005) and Reclaiming the Na¬tion (2011).

1

Introduction: Roots of the Fast Track Land Reform in Zimbabwe

Sam Moyo and Walter Chambati

Introduction

The Fast Track Land Reform Programme (FTLRP) implemented during the 2000s in Zimbabwe represents the only instance of radical redistributive land reforms since the end of the Cold War. It reversed the racially-skewed agrarian structure and discriminatory land tenures inherited from colonial rule, whereby over 6,000 large-scale white farmers and a few foreign and nationally- owned agro-industrial estates controlled most of the prime land, water resources and bio-reserves, while relegating the majority of the population to marginal lands and cheap-labour services. The land reform also radicalised the state towards a nationalist, introverted accumulation strategy, against a broad array of unilateral Western sanctions. Indeed, Zimbabwe's Fast Track Land Reform, in its social and political dynamics, must be compared to the leading land reforms of the twentieth century, which include those of Mexico, Russia, China, Japan, South Korea, Taiwan, Cuba and Mozambique.

Zimbabwe's land reform may be unique in the post-Cold War period, but the social and economic conditions which brought it about are not. In the last three decades, the whole of the South has been subjected to a neoliberal regime of accumulation under the leadership of monopoly-finance capital, which reversed many of the socio-economic gains obtained in the aftermath of decolonisation, spawning a new generation of land movements (Moyo and Yeros 2005a.) The successful redistribution of wealth ushered in by the FTLRP, therefore, places Zimbabwe at the forefront of an emergent nationalism in the South, which is still poorly understood (Moyo and Yeros 2011). Even among 'progressive' circles, the debate remains fixed only on the resurgence of left politics in Latin America under similar socio-economic conditions (e.g., Sader 2011), although none of these cases has progressed to the point of a radical redistribution of wealth, as in Zimbabwe. The only other major case of substantial redistribution and economic re-orientation of recent years, though under very different political conditions, has been that of Cuba (Altieri and Funes-Monzote 2012).

The fact that the Zimbabwe case has not been recognised as a vanguard nationalism has much to do with the 'intellectual structural adjustment' which has accompanied neoliberalism (Moyo and Yeros 2005b) and a hostile media campaign (see Chari, Chapter 8 in this volume). This has entailed dubious theories of 'neopatrimonialism', which reduce African politics and the state to endemic 'corruption', 'patronage' and 'tribalism' (see de Grassi 2008; Olukoshi 2011; Mustapha 2002; Mkandawire 2012), while overstating the virtues of neoliberal good governance. Under this racist repertoire, it has been impossible to see class politics, mass mobilisation and resistance, let alone believe that something progressive can occur in Africa. Such intellectual trends have also been found among some scholars who espouse Marxism, who have similarly ignored the advances made in Zimbabwe by reproducing the right-wing repertoire of 'corruption' and deploying tactics to suppress or distort the evidence, especially via their control of publication outlets.

Yet, it should come as no surprise that there was a re-radicalisation of nationalism in Zimbabwe. Land reform was a central demand of the nationalist movement and the armed liberation struggle, which more broadly sought popular sovereignty and majority rule, including control over land, natural resources and the economy (Mkandawire 2001), despite its flirtation with socialist ideals (see Saul 2005; Mandaza 1986). Accordingly, the FTLRP is most likely to be remembered in Zimbabwe and Africa as the culmination of the anti-colonial struggle, despite the liberal democratic deficit (Mamdani 2008) and economic policy contradictions that accompanied it.

It is often pointed out that structural change was suspended during the first two decades of independence. Indeed, land redistribution was compromised by the Lancaster House Constitution of 1979, which protected settler political power and proscribed land expropriation, leading to limited land redistribution, using market-based land acquisition mechanisms (Moyo 1995). Even this market initiative hibernated in the 1990s when the Economic Structural Adjustment Programme (ESAP) was adopted and as land concentration and foreign land ownership expanded, reinforcing historical land alienation (Moyo 2000).

Yet, another constraint was the simultaneous transition in neighbouring South Africa, which had posed security threats in the region as a whole and further mitigated radical politics through 'destabilisation'. Even as the apartheid regime unravelled in 1994 and neoliberal restructuring spread throughout Africa, a new phase of foreign land alienation, led by South African white farmers, intensified in the sub-region as a whole. This converged in the 2000s with extensive foreign land grabbing in non-settler Africa, as the crisis of monopoly-finance capitalism unfolded (Moyo 2011b).

The general trends on the continent have been adverse for radical politics, but they have not been able to stem the tide completely. Notably, the return of the land question as an issue of national sovereignty in Zimbabwe has rowed against the current of foreign land-grabbing and the intensification of export-oriented large-scale farming. The Zimbabwe case has also challenged the received wisdom in international land reform debates, which predicted the replacement of popular and state- sponsored land reforms by market approaches, due to the political and institutional constrictions imposed by neoliberalism (Borras 2006; Rosset et al 2006). Even within South Africa, the so- called 'willing-seller-willing-buyer' approach has since been questioned by that state (Republic of South Africa 2005).

Before we delve further into the issues of how the causes, process and outcome of Zimbabwe's Fast Track Land Reform has been interpreted, it is important to understand its historical background and the challenges that it poses for agrarian transformation, democratic politics and national development.

Accumulation by dispossession in former settler-colonial Africa

Since colonisation in the late-nineteenth century, Zimbabwe was integrated subordinately into the world capitalist system, mainly through the settler-colonial mode of political rule and social relations of production, based largely on unequal and repressive agrarian relations which defined the character of the state. Land dispossession and extra-economic regulation and taxes turned Zimbabwe into a labour reserve economy (Amin 1972) reliant on cheap domestic and foreign migrant labour (Arrighi 1973), while repressing the peasantry and small-scale rural industry and commerce, without creating full-scale proletarianisation (Bush and Cliffe 1984; Weiner 1988; Yeros 2002). Large-Scale Commercial Farms (LSCF) were consolidated by discriminatory subsidies to the white farmers and a few black Small-Scale Commercial Farmers (SSCF) (Moyana 2002).

Although the Zimbabwean economy at independence was relatively semi-industrialised, it remained in a disarticulated pattern of accumulation. Agriculture contributed about 20 per cent of gross domestic product (GDP) and 40 per cent of the exports and employed 70 per cent of the workforce (Stoneman and Cliffe 1989; Yeros 2002). The essentially bi-modal agrarian structure involving large-scale farms and marginal small capitalist and peasant farms was tied to industrial interests wholly-owned by the local white minority and foreign capital. Individual white settlers held average land sizes of 2,000 hectares, while foreign and domestic agro-industrial estates (plantations) held average landholdings well above 5,000 hectares. In the 1950s, the settler-colonial state sought to turn the peasantry into full-time agricultural, industrial and mining workers disconnected from the land (Yeros 2002), while allowing for 'subsistence-based social reproduction' of the black migrant labourers and their families in the labour reserves. But the top-down methods of this policy and the limited scope for wage increases doomed it to failure.

Rhodesia's strategy for agricultural transformation from the 1960s also entailed accumulation through large-scale estate farming structures supported by the state (e.g., through irrigation facilities, dams, rural electrification and other infrastructures) to expand exports and reduce sugar and wheat imports (see Moyo 2011b). Such private estates were largely owned by South African and British transnational corporations, as well as by domestic white-owned agribusiness conglomerates, involving pioneer white family landowners, some of whom held mining concessions. The estates also created outgrower farms among whites, including new Mauritian and South African immigrants, with average landholdings of 217 hectares (European Union 2007). Around 1971, the privately-owned Mkwasine Estate created black sugar outgrowers with 10 hectares each (Moyo 2011b). During the 1970s, state- owned farm estates were created, including through the Agricultural and Rural Development Authority, formerly called the Tribal Trust Lands Development Authority, the Cold Storage Commission and other parastatals.

However, the narrow import-substitution and export-led development strategy involving mining and agriculture continued to serve mainly the consumption and investment interests of the white minority and the mother land (Stoneman 1988), while repressing indigenous capitalist classes (West 2002) and limiting wage-goods consumption. The Import Substitution Industrialisation (ISI) began to falter by the early 1970s after a final boom in the late 1960s (Arrighi et al 2010).

Zimbabwe's 'sub-type' of ('semi-peripheral') neocolonialism, derived from white-settler colonial capitalism, entailed perennial contradictions between introverted and extroverted capitalist accumulation strategies (Moyo and Yeros 2005a) and the organisation of labour through both 'direct' and 'indirect' power over indigenous populations and institutionalised racial segregation. These social relations of production induced the 'sale' of cheap and 'semi-bonded' black labour among a growing landless population (Yeros 2002), restricting the peasantry's social reproduction and accumulation from below. Petty-commodity production in the Communal Areas (CA) and especially unwaged female labour, subsidised the social reproduction of male labour-power on mines and farms. Neither a settled industrial proletariat nor a viable peasantry was established. A workforce in motion, which can best be conceptualised as a semi-proletariat, was established instead (ibid). This labour force straddled communal lands, white farms, mines and industrial workplaces, aggregating peasant-worker households, differentiated by gender and ethno-regional divisions.

These structural parameters established the basis of an enduring process of super-exploitation of the semi-proletariat, as well as severe economic and political contradictions. These class and racial features of the agrarian and national question defined the dynamics of the liberation movement and its visions of land and agrarian reform. After independence, such aspirations were fuelled by mounting grievances over the failures of the market- based land reform and the social crisis generated by structural adjustment policies by the mid-1990s (Moyo 2000). Furthermore, since Zimbabwe, like most African countries, remained largely an agrarian society, with the majority of the population being dependent on low- intensity agricultural techniques, its agrarian question continued to require reforms which transform productivity and diversify employment opportunities in the face of rising unemployment and food insecurity.

The prospects and challenges of Zimbabwe's land and agrarian reform

Agrarian reform requires restructuring the role of dominant agrarian classes (such as land owners, merchant capital and agribusinesses), particularly with regard to their super-exploitation of labour, as well as of the extroverted state, towards improving the social relations of agricultural production and the productivity of land and labour. The overall goal is to enhance agriculture's contribution to national accumulation and economic sovereignty, while the immediate purpose is to 'create the conditions for a rise in [agricultural] productivity, such that [the] raw materials and wage–goods needs of a growing manufacturing sector can be met, while labour is released' (Patnaik 2003). Since 'national' commodity and capital flows intersect with and are shaped by, global processes of production and markets, involving transnational capital (Bernstein 2002), the regulation of trade and industrial policy are critical mechanisms which shape the agrarian transition (Chang 2009). Mal-adjustment and mal-integration into the world economy, entrenched by the neoliberal policies adopted in Africa since the 1980s need to be reversed to realise development (Amin 1990; Mkandawire et al 1999).

Land reform is a necessary dimension of agrarian reform in settler societies, but it is not on its own a sufficient condition for national development (Moyo and Yeros 2005a). Agrarian reform requires state-facilitated land redistribution and support to build the productive and social capabilities of small producers (Evans 2009). Re-orienting agricultural production to the home market, broadening the consumption of wage and industrial goods and services and promoting cooperativism is essential to increasing intersectoral linkages and agricultural productivity growth, in synergy with rising domestic wages and laying the basis of a deeper process of democratic transformation (Moyo and Yeros 2005a). For small and middle-sized farms to realize their employment and productive potential, there has to be a synergy between state interventions in industrial, trade and rural development policies, so as to enhance the potential multiplier effects of agriculture and to increase aggregate demand. A basic related concern is how to address food security, which in Africa was increasingly undermined by neoliberal economic policies and manipulation by dominant monopoly capital (Mafeje 2003) and the underdevelopment of irrigation resources. The recent world food price crises hurt Africa the most, largely due to its weak food production trajectory and dependence on food imports (FAO 2010).

However, for land reform to be redistributive, it need not transfer all the landowners' land to the non-landed poor, but a substantive amount enough to alter agrarian relations and state power (see Borras 2005). Diminishing the influence of agrarian elites through land reforms (as happened in China, Korea and Taiwan) is necessary to remove conservative pressures on development policy-making, that is, the economically powerful and politically reactionary landed classes that monopolise land, force the population to be landless labourers and reduce incentives among landowners and workers to invest (Evans 2009). Accumulation from below is thus a key objective of redistributive land and agrarian reform.

Zimbabwe's historical processes of accumulation by dispossession of land and labour power eschewed accumulation from below and undermined social reproduction (see Arrighi et al 2010), while post-independence policies failed to resolve the national questions of broad-based development, social inclusion and national integration, including substantive democratisation (Moyo and Yeros 2009). Zimbabwe's agrarian question thus concerned the transformation of the dominant settler-colonial agrarian relations towards a racially and socially equitable structure of access to economic and natural resources by the majority of the peoples towards building a diversified economy.

Agrarian reforms in former settler-Africa were expected to re-orient state intervention towards broader-based rural development and to redress entrenched racially-unequal political and economic power relations. This required the reversal of the territorial and spatial segregation of society maintained through unequal control of land and retrogressive agrarian labour relations, so as to free labour and uplift the lives of working people, straddling between rural and urban areas. Land reform implies restructuring the distribution of land ownership towards a more democratic agrarian structure in order to promote social, economic and political transformation, which creates security of land tenure for all (ANC Conference 2007).

Zimbabwe's land reform experience since 1980 has entailed three different approaches to the acquisition and reallocation of land (see Moyo, Chapter 2). Until 1990, it was a market-based and state-managed land redistribution process, which primarily targeted landlessness, 'overcrowding' and land shortages among the rural and urban poor and unemployed (Moyo 1995). However, some political leaders lobbied for the 'de-racialisation of commercial farming' (Nkomo 2001) and a few black middle-class actors and political elites secured land on the market (Moyo 1995). Meanwhile, various localised illegal land occupations were mobilised to gain access to land (Moyana 1987; Tshuma 1997). This was followed by an unsuccessful attempt between the state and landowners to negotiate land transfers during the ESAP regime, which began in 1990, alongside limited attempts at land expropriation by the state between 1993 and 1997. In 1998, Zimbabwe's revised Land Reform Programme sought to redistribute 5 of the 12 million hectares owned by white famers, most of which were deemed to be underutilised and near congested Communal Areas and to tackle multiple farm and foreign land ownership (Moyo 2000).

The land and agrarian reform strategy became explicitly bi-focal in 1990 when it sought to redistribute land to small-scale family farms (called A1 schemes) and to smaller-sized capitalist farmers called A2 schemes (GoZ 1998),[1] reflecting the growing influence of the domestic black petty-bourgeoisie on policy-making. *A priori* therefore, the outcome expected was a differentiated pattern of access to land, driven by the belief that it was desirable to maintain some large-scale farms which, it was believed, could use agrarian resources 'efficiently' (ibid). This policy proposed secondary redistributive mechanisms targeting the new larger landholdings, including progressive land taxes, lease rental fees and charges for the use of infrastructures left on the redistributed land, to further the redistribution agenda through revenue transfers. A share-equity transfer approach to the indigenisation of business (including of large agro-industrial estates and conservancies) was also debated in the 1990s, but this was gradually adopted from the early 2000s. The popular expectation was that such shares would be distributed proportionately among a broad base of beneficiaries, including neighbouring communities.

Whereas the land reform process was initially technocratic and conservative in terms of liberal market criteria, it became radicalised through the use of land expropriation on an extensive scale from 1997. In 2000, the FTLRP, which was fuelled by extensive land occupations, mobilised and led by the veterans of the liberation war (see Sadomba and Masuko, Chapters 3 and 4), entailed state land expropriation (see Moyo, Chapter 2), leading to intense conflicts over land, as well as over the numerous elections held from 2000. By 2010, the FTLRP had redistributed over 10 million hectares of Zimbabwe's prime agricultural lands, which in 2000 were held by over 4,000 white Large-Scale Commercial Farmers (LSCFs) and about 200 black LSCFs, to over 145,000 peasant families and over 20,000 small and middle-scale capitalist farmers, while retaining some large-scale farms, agro-industrial estates and conservancies (see Moyo, Chapter 2; Moyo 2011a, 2011b).

This process reconstituted Zimbabwe's agrarian structure (Moyo 2004), altering the nature of property rights and labour relations based on new forms of access to resources and markets. Land redistribution also increased access to farm infrastructures and natural resources which were previously dominated by a racial minority and partially reconstituted gender and ethno-regional dimensions of land and resource control, although it also led to some exclusion, especially of farm workers. This outcome has generated new forms of social differentiation (Moyo and Yeros 2005b), although wider access to land created a new basis for the social legitimacy of land tenure relations, despite the persistence of some land tenure insecurities and demands for the re-institution of private property relations in agricultural land.

As we had observed, before the FTLRP (Moyo 2000; Yeros 2002) and during its early phases (Moyo and Yeros 2005b), the 'illegal' land occupations and official land reforms were reinforcing an incipient process of re-peasantisation and enabling broader- based small-scale capitalist farming. We argued that this was creating prospects for accumulation from below, while generally modifying the semi-proletarianisation process and the conditions of labour exploitation. Contrary to the alleged process of 'de-agrarianisation' (Bryceson 1996), re-peasantisation led by radical movements was found to have a wider incidence in the South (Moyo and Yeros 2005a).

The overall outcome of the FTLRP has essentially been the constitution of a qualitatively different tri-modal agrarian structure, in which various classes compete for control over land, labour, water and natural resources, as well as for access to state support and influence over the regulation of markets (see Moyo and Nyoni, Chapter 6). This pits production for domestic needs against a resurgent export orientation, as capital reconstitutes itself and redefines its role, in a context of growing diversification towards the East. Radical state interventions have pervaded the last decade, in the context of sanctions and hyper-inflation. Various economic contradictions emerged, leading to a renewed liberalisation from mid-2008, although this has in turn introduced new contradictions arising from unequal access to and exploitative, commodity and financial markets (see Chapter 6).

Nonetheless, the structural reforms of the FTLRP promise a deeper form of substantive democracy, in a society previously pervaded by deep racial and class inequities, than the limited demands focused on good governance reforms. True, the immediate conflicts over the land reform process and around the elections held within a context of 'regime change' politics during the 2000s led to the narrowing of democratic space and a human rights deficit (Moyo and Yeros 2009). Despite these notable casualties during the period (Mamdani 2008), land reform itself has created the social and economic foundation and potential for broad-based development and democratisation. New forms of rural mobilisation are promising in this regard (see Masuko and Murisa, Chapters 4 and 7).

Moreover, critical deficiencies in the agrarian outcome, such as low levels of agricultural land and labour productivity, the security of land tenure for the various classes of small producers and the weak protection of peasant markets, reflect continuing struggles over development and democratisation. Also in question is the character of Zimbabwe's re-integration into global circuits.

The structural adjustment of intellectual discourse: some methodological notes

It remains imperative to engage with the dominant interpretations of the Fast Track Land Reform, as these continue to deflect the debate from addressing the requirements of realizing further social progress, while enabling aggressive 'regime change' politics, now justified through the 'right to protect' (R2P) on humanitarian grounds. As we have noted, the inability of most analysts to come to terms with the causes, the process and the outcome of the land reform is related to a profound process of 'intellectual structural adjustment' which has been underway since the 1990s.

To be sure, dubious intellectual positions always existed in and on Zimbabwe, but in the 1990s they germinated with a number of new tendencies. Specifically, these have included: a peculiar mix of liberalism and Weberianism peddled by American political science, especially via the notion of 'neopatrimonialism'; a rudderless culturalist theory of 'identity politics', whose post-structuralism has managed to replicate with great success the settler-colonial obsession with fragmented cultures; and, not least, an escapist 'left' critique, which has often sought refuge in pseudo-Gramscian theories of 'hegemony', whereby patrimonialism and culturalism substitute for class analysis. Indeed, some 'Marxists' succumbed to similar imperialistic and anti-nationalist impulses, to the effect of silencing class analyses which demonstrate the progressive nature of the land reform (see, for instance, Moore 2004).

More recently, some have openly acknowledged the redistributive outcome of the FTLRP, but have redirected the debate into an essentially liberal perspective on 'livelihoods' (e.g., Scoones et al 2010; Scoones et al forthcoming). This approach eschews the interrogation of class formation processes and exploitative relations of production (especially in the emerging labour relations) and the continued extraction of surplus value (particularly from peasants) through exchange relations driven by monopoly-finance capital. The critical role of state intervention in the overall outcome is also visibly downplayed by its liberal-populist orientation. As its main audience is the donor industry, sanctions are not featured in this narrative.

Overall, this is fed by the persistence of inappropriate theoretical perspectives applied to the analysis of the state and society in Africa and the general diversion of scholarly attention away from Marxian political economy. Thus, there is a systematic neglect of the continent's subordinate relations to monopoly-finance capital, as well as empirical analyses of class formation, political alliances, emergent social movements under the current crisis and the implications for state intervention and development. Furthermore, the nature and sources of external intervention have been reduced to a celebration of a narrow humanism, together with a condemnation of especially the East, rather than a careful understanding of the strategic economic and security changes underway and the new forms of resistance to the new assault on national sovereignty.

In the specific case of Zimbabwe, a new revisionist historiography (e.g., Ranger 2004; Alexander 2006; Raftopoulos and Phimister 2004) has downplayed the long historical legacy of settler-colonial relations of production and their ideological power structures. Such historiography has absorbed the legacy of technocratic thought and planning associated with large-scale farming and production practices and its corollary view of blacks mainly as cheap-wage labourers with limited social aspirations and capabilities. Furthermore, struggles over 'property rights' are reduced to abstract concerns with formal procedures of land tenure (e.g., Matondi 2011; Cross 2012), rather than conceptualising land tenure as a social relation of production shaped by historically specific class relations and the attitude of dominant forces to political and economic sovereignty (Moyo 2007).

The dominant Africanist writings, largely by self-proclaimed Zimbabweanists, but also the 'structurally adjusted' Zimbabwean intellectuals, have been able to offer only a peripheral critique which calls for limited social and political reforms. In the past, they had claimed to give 'voice' to various forms and sources of national resistances (proto-nationalists, nationalists, cultural nationalists, ethno-regional protests and local ZAPU nationalists), but restricted their notions of 'resistance' around identity-based organisation (e.g. Alexander et al 2000), rather than elucidate the dynamics of class-based social formations and political alliances.

This is the historical intellectual tendency which has now eclectically disaggregated the nationalist resurgence of the late 1990s from its variety of inter-connected social elements, so as to privilege the agency of a narrow civil society (read NGO) mobilisation supported by foreign donors. As neoliberalism gained ascendency, the dominant intellectual current had concluded that no alternative political economy and no radical agency against the settler-colonial edifice could be envisioned. Radical nationalism, now treated narrowly as 'patriotic history' (Ranger 2004), has been rendered as a patrimonial, politically self-serving exercise of rent-seeking by black capitalists associated to the ruling party and as an anathema to the ongoing progress of a 'globalised world'. Consequently, the analysis of agrarian politics overlooks actual class struggles over land, labour and markets, including the ways in which new farmers have organised themselves (see Chapters 4, 5 and 7). In this schema, all nationalist forms of social mobilisation and their redistributive outcomes are foreclosed by conceptual fiat.

Indeed, debates on the FTLRP were hinged around contested interpretations of the motives and political party alignments of Zimbabwe's land occupation movements, to the neglect of political economy processes which led to the gradual radicalisation of the land reform policy. The dominant literature has alleged that the FTLRP was centrally directed by ZANU-PF 'elites' and the state (for a rendition of this view, see Marongwe 2011 and Zamchiya 2011) and that, as such, it could only represent the narrow motives of an authoritarian nationalism seeking to retain receding electoral support, while marginalising farm workers who had voted against it (Raftopolous 2003, 2009; Scarnecchia et al 2008). Aspects of the violence which occurred were selectively highlighted, deducing 'disorder' and the wanton abrogation of the rule of law (Hammar et al 2003; Hellum and Derman 2004). Rather than systematically examine the nature of violence that did occur (Moyo and Yeros 2007, 2009), the entire land occupation movement itself was vilified by equating it with extreme human rights violations (Johnson 2009). This volume debates the actual FTLRP process extensively.

Even the structural transformation that has occurred in Zimbabwe since 2000 has been subjected to either prolonged silence or abstract contestations, despite the growing volume of empirical evidence. Among 'progressives', the tendency has been to deny radical or progressive change (Saul 2005; Raftopoulos 2010), or to accept it with reservations (Bernstein 2002, forthcoming), in order to uphold a narrowly-based, essentially liberal narrative of authoritarianism and despotism.

Most liberation war veterans regard the FTLRP process as nothing short of a 'Land Revolution',[2] while some of them qualify this to reflect a 'completion of the national democratic revolution' (see Sadomba 2008 and McCandless 2011, quoting interviewees). This view is in stark contrast to caricatures of the notion of 'land revolution' (Hammar et al 2003) or the more benign counter-positioning of the process as part of a liberal democratic 'revolution' (Madhuku 2004). Indeed, some elements in the Movement for Democratic Change (MDC) claimed credit for instigating the radical land reforms, arguing that the FTLRP would never have happened without their agency of challenging ZANU-PF politically (see Gwisai 2002). The evidence points to a variety of local and regional processes of decentralised socio-political mobilisation (see Masuko, Chapter 4) which shaped the dynamics of the 'revolution'. Its direction entailed compromises within a two-pronged land allocation strategy seeking to accommodate peasants and small capitalist farmers, as well as negotiating normalisation with sections of transnational capital (e.g., some agro-industrial estates).

Instead, it is alleged that the FTLRP only benefited a few cronies (Scarnecchia et al 2008), reducing the Fast Track struggles for land to an elite affair dominated by ZANU-PF (Marongwe 2011; Zamchiya 2011). The language of 'land grabbing' was used to create a moral and political equivalence between the restitutive appropriation of colonially- dispossessed lands through the state for land redistribution and externally- inspired land grabs. Only 'elite capture' could be envisioned through the dominant neopatrimonialist lenses. Although some elites grabbed land during the FTLRP (Moyo et al 2009; Scoones et al 2010), by overstating the agency of the domestic political elite, the class and wider social dynamics of struggles and competing claims over land, between a range of elites and the landless, vis-à-vis foreign and white domestic capital, were overlooked (Moyo 2011b). Few have examined how the staggered shifts in the FTLRP land allocation processes that characterised the 2000s evolved through various contradictions (see Moyo, Chapter 2).

By 2005, the debate had shifted to claims that agricultural production had collapsed and that food insecurity had reached levels of humanitarian crisis (Richardson 2005; Campbell 2008; Bond 2008). It was assumed that the FTLRP outcome was an unstructured replacement of 'commercial farming' with 'subsistence farming' and that this could not feed Zimbabwe's limited agro-industrial structures (Tupy 2007). Others continue to purvey a strange pessimism on the prospects for agricultural recovery (Dekker and Kinsey 2011; Cliffe et al 2011), despite evidence to the contrary and at the expense of understanding the social and structural basis of the agrarian change that is underway (e.g., as proposed by Moyo 2011c; Chambati 2011).

Moreover, there has been a failure to examine the nature and sources of the radical state intervention towards agrarian reform and its evolution over time during the 2000s (see Moyo and Nyoni, Chapter 6) and thus to decipher the contradictions entailed in the face of a highly polarised political landscape and a capital strike, shaped by external isolation and economic sanctions. Thus, the reconfiguration of domestic agrarian markets and struggles over these, in relation to changing forms of state intervention, in the context of a gradual re-orientation of critical commodity and financial markets to the East, have been overlooked. This oversight leads many analysts to ignore the struggles during the decade between the state and capital over domestic production and supply of agrarian inputs and to miss the wider economic contradictions which shaped the agrarian reforms and their distributional outcome. Few have examined the more recent sources of pressures to indigenise the economy, in relation to the imperatives for agricultural finance, beyond speculating over the likelihood of elite capture.

Consequently, numerous neoliberal visions intended to 'rationalise' the FTLRP landholdings and agrarian policies have been proffered to: re-privatise land; reinstate large-scale export oriented farming; integrate farming into agribusiness; and deepen trade liberalisation (Richardson 2005; Tupy 2007; Adam Smith International 2007; UNDP 2010; USAID 2010). Fortuitously, calls for the 'reversal' of the FTLRP land redistribution were dealt a political blow in 2008 when the Inclusive Government of Zimbabwe signed against this, calling instead for improved land tenure security, wider access to land, including for women and for measures to improve land utilisation (For the GPA 2008, see Inclusive GoZ 2009).

Much of this ungrounded discourse against the FTLRP is a consequence of its over-reliance on media renditions of the FTLRP process and outcomes,[3] and related reports issued by western funded NGOs (e.g., JAG/ RAU 2008; HRW 2004), Commercial Farmers unions (see CFU 2003) and a few MDC aligned 'public intellectuals', who are routinely pitted against ZANU-PF 'patriotic intellectuals' (scc Tendi 2010). Such research has mostly focused on the 'victims' of the land reform (e.g., former land owners and farm workers), to the neglect of the views and experiences of key actors within and protagonists of the FTLRP (such as the land occupiers, land beneficiaries, former and new farm workers and war veterans), as well as of local and central state officials, local NGOs engaged with the FTLRP and the key actors in agribusiness. Official reports and policies are often ignored,[4] while empirical researchers who go contrary to the dominant narrative are labelled as defenders of the FTLRP (e.g., Cliffe et al 2011; see also Hungwe 2011; Chiweshe 2012). Moreover, the limited empirical analysis pursued has tended to focus on spotting the 'elite' beneficiaries, many of whose unequal access to land is not the real dispute, although what is in question is the definition of such elites and the extent to which they benefited. Such 'elites' are not treated as a differentiated class of people, whose dynamic growth or demise in the process of class formation is worth examining. Beneficiaries employed by the state and war veterans are all assumed to be state and/or ZANU-PF 'elites', let alone assessing their class status and/or rank in such institutions. The many private sector executives who got land are not readily identified as elite beneficiaries, unless they can be tagged as ZANU-PF members. Thus, diverse social groups are lumped into the category of elites, despite their social differentiation in terms of labour relations, assets and access to finance, let alone their varied positions in the political hierarchy and economy (see Moyo et al 2009). Such 'elites' are all deemed to have used political connections and/or corruption to gain access to land and inputs.

Furthermore, in such narratives, only high-profile formal policy statements in the media are considered worthy of note in examining the role of state intervention, to the exclusion of policy shifts induced through informal structures and planned or spontaneous struggles over land and agrarian resources (Moyo 2011a). Thus, the actual substance and politics of agrarian policy-making since 2000 is understudied, leading to a failure to identify its defensive and proactive dimensions, as well as the nuanced and shifting attitude of the state towards capital (which was not fully ousted). The ideological and political influences of big business in relation to the wider race-based intra-capitalist relations and conflicts, including within the ZANU-PF leadership and among other black capitalists, is rarely examined to gauge the reaction of capital (and donors), including the mechanisms they deployed against the FTLRP. The capacity and planning deficiencies of the state are not commented upon, since all the contradictions of the FTLRP are explained as instances of neo-patrimonial corruption and ZANU-PF patronage, within an 'authoritarian state', led by atavistic nationalists, supported by violent liberation war veterans.

Yet the origins, processes and outcome of the FTLRP cannot be explained away as being driven merely by a monolithic neopatrimonialist agenda. For, the FTLRP experience raises wider and fundamental questions regarding the 'best practise' of implementing redistributive land reform under a neoliberal policy regime, while sustaining electoral democracy. In Zimbabwe's historically specific context, grievances over settler-colonial dispossession and wider race-class inequalities and the limits to reform imposed by political compromise and market reforms after 1980, shaped and sustained land struggles which culminated in the FTLRP experiences. The political power struggles that ensued and the mobilisation of radical land movements, as well as the authoritarian posture of the state towards a new political opposition from 1999, reflected wider social struggles, which provoked confrontations with the judiciary, capital and the non-state media, as well as with the western donors and related NGOs, most of which were opposed to radical land reform.

Outline of the chapters

This volume presents the findings of eight scholars, who have undertaken sustained empirical work on Zimbabwe's Fast Track Land Reform Programme (FTLRP) since 2000, in order to redress the conceptual malaise and the empirical deficiencies of the debate on FTLRP. The chapters examine the nature and implementation of the FTLRP, the social forces which mobilised the land occupations, the transformation of the labour regime during the FTLRP, the patterns of agrarian change that have emerged, the organisation of new farmers at the local level and the media representations of the FTLRP. A concluding chapter distils the key findings.

Research by these scholars entailed local field surveys undertaken between 2000 and 2011 and a national baseline survey in 2005 and 2006, which were organised by the African Institute for Agrarian Studies (AIAS) (see AIAS Baseline Survey 2007),[5] and numerous follow up field visits undertaken during 2007 and 2011 by the authors. The authors examined multiple local and national sources of secondary data over the decade and undertook interviews with various actors.

Most of the authors have pursued class-oriented ethnographic analyses of a wide range of actors involved 'within' and 'outside' the FTLRP and agrarian reform processes, as well as wider political economy analyses of the process and outcome. The studies build on earlier efforts to trace the social basis of the FTRLP and the social differentiation that has emerged within the newly redistributed areas and related Communal Areas and nationally (see Moyo and Yeros 2005b; Moyo et al 2009),[6] and seek to understand the character of land and agrarian change since 2000, within its historical context. Various social and economic factors which characterise the new landholders, the agrarian structure and related struggles in relation to control of land, natural resources, production assets and control of labour, as well as their access to markets and the role of capital and the state were explored.

In the second chapter, Sam Moyo identifies the phased manner in which the FTLRP process was implemented over a decade since 2000, in relation to various social struggles and legal changes and examines the redistributive outcome of the FTLRP. His results reinforce earlier projections that both re-peasantisation and the growth of small-scale capitalist producers have widened the prospects for accumulation from below, despite the new class struggles observed (Moyo and Yeros 2005b). He argues that the reform has triggered wider progressive changes in access to land for farming, natural resources and various non-farm work and self-employment 'opportunities', as well as various symbolic benefits, including the recognition of communities' rights and place in society. Indeed, the FTLRP unravelled the wider roots of unequal power relations, including the territorial segregation inherited from settler-colonialism. The extent to which some economic and political elites benefited more from the FTLRP process and some foreign-owned agro-industrial estates and conservancies were retained, are demarcated. This highlights the persistence of land ownership inequalities in class, gender and ethno-regional terms, which are also reflected in the differentiated access among beneficiaries to farming resources, services and infrastructure and varied capacities to hire labour. This foments new social differentiation processes and agrarian struggles (as discussed in Chapter 6).

Wilbert Sadomba, a war veteran and scholar who participated in the land occupations, traces the role of Zimbabwe's liberation war veterans in the FTLRP process, beginning with their re-organising in the 1990s. He examines how they were mobilised into a militant organisation which confronted the state, demanding welfare benefits, recognition and land redistribution, as well as their eventual leadership of the 'land revolution', involving peasants, farm workers and other urban homeless people. He argues that failing neoliberal economic policies, increasing authoritarian rule and heightened external interventions in the country led to the mobilisation into two formidable movements pitting sections of civil society NGOs and trade unions and the political opposition (the MDC) against the land occupation and wider liberation movement. The nationwide land revolution challenged elitist state tendencies by transforming the localised peasant struggles into a national class struggle with critical ramifications for the structure of state power, leading the state to institute the formal dimension of the 'fast track' land reform programme, in order to simultaneously co-opt the 'land revolution' and suppress political opposition. He contends that the state failed to co-opt the land movement, despite military style and retributive violence, hence the stalemate which led to the sharing of power through a Government of National Unity (GNU).

Louis Masuko, also a war veteran and scholar who participated in the land occupations, extends this debate, arguing that suggestions that the land occupation movement was a top down movement conceived of and directed by a beleaguered state and that it was a one- issue movement, are empirically unfounded. He shows how the land movement was deeply rooted and argues that its long-term social, political and economic sustainability is guaranteed by this origin. Using the Nyabira-Mazowe War Veterans Association as a case study, he conceptualises how groups of land seekers were mobilised and organised for collective action, using strategies that addressed individual and collective deprivation. He argues that beyond land ownership, the land movement has created new social structures which place new demands on the agrarian policy agenda, including land tenure security and state support for improved land use.

Walter Chambati traces how the FTLRP substantially reformed labour relations due to the restructuring of settler-colonial agrarian structures. He argues that the dominant literature misses this point because of its blindness to the redistributive outcome of the FTLRP and inadequate examination of the labour processes within the emerging agricultural production structures, as well as due to a conservative orientation towards protecting the poorly remunerated former LSCF workers' jobs and an overestimation of formal job losses. He elucidates the new and diverse sources of rural employment that have emerged, including the shortage of wage-labour, as more people work for themselves on their own farms. Wage-labour continues to be provided mostly on a part-time basis to various classes of farms and through new forms of agrarian labour supply, including the increased supply of independently organised specialist labour to larger farms by skilled workers. He also highlights the persistence of labour-tenancy on some farms and the differentiated patterns of labour hiring which are associated with the unequal command over larger land sizes and access to farming resources. He suggests that the FTLRP could improve the social relations of labour by redistributing more land to former farm workers and improving state interventions to protect labour and support social reproduction.

Sam Moyo and Ndaba Nyoni trace the phased introduction of agrarian reform policies instituted on the heels of the equally phased redistributive land reform implemented since 2000 and examine the nature of change in some of Zimbabwe's agrarian relations, as well as their contradictions. They argue that by broadening the agricultural producer base, the sources of food production and the consumption base have been extensively expanded, creating wider prospects for social reproduction, despite the failure to realise past levels of output in some commodities. This highlights the new inequities in access to land and farm input and output markets based on class and regional differences and a growing focus on export-oriented production, although wider sections of rural society have attained food security and realise new sources of income and self-employment. The chapter also explores the re-composition of agricultural outputs and questions dominant behaviourist approaches which focus on new farmers to explain output decline, providing evidence on the structural limitations imposed on the new producers, including declining inputs and credit supplies, droughts and the rise of exploitative but diverse forms of contract farming. The role of agro-industrial capital and its gradually increased dominance in reconfigured domestic markets, as well as the increased influence of Asian capital, is also explored in the context of the new agrarian politics and persistent sanctions. The chapter reflects on the uneven prospects for accumulation and social reproduction within the new tri-modal agrarian structure, highlighting the wider social and structural changes which place the question of national and economic sovereignty and democratisation on the agenda.

Tendai Murisa argues that the Fast Track Land Reform Programme led to a fundamental re-structuring of the agrarian landscape beyond just the physical repartitioning of large-scale farms into smaller A1 farms and medium- sized A2 farms. It altered previously existing social relations of ownership, access and utilisation of land, by relocating people from diverse areas and backgrounds. The chapter demonstrates the manner in which the land beneficiaries are forging new social relations of production, involving 'strangers' from diverse social and ethnic backgrounds, including from customary land tenure, urban and former Large Scale Commercial Farming areas. The case studies presented discuss the factors which shape the emergence of Local Farmer Groups (LFGs) and their contribution, th rough a variety of mobilisation strategies, to their social reproduction.

Tendai Chari argues that the FTLRP was an enduring issue in the media over the past decade, although the attention paid to it was not matched by adequate factual information. He examines how the international, locally owned private and state-owned media 'framed' the land and agrarian reform issue and contributed towards shaping public opinion and perceptions about them, using samples of news articles, interviews with key informants and document analysis. The framing of the FTLRP, he finds, was circumscribed by ideological, partisan and commercial motives, limiting the media's ability to broaden and deepen public understanding of land reform and its ramifications for Zimbabwean politics. Whereas the state-owned media accentuated the necessity of redistributing land to correct historical imbalances, they routinely downplayed the negative aspects of the Fast Track Land Reform, some of which were public knowledge. The private domestic media and international media converged to downplay the West's imperialist project in Zimbabwe, while foregrounding neoliberal discourses (such as human rights, democracy, rule of law and good governance and 'commercial farming') as the only basis on which to measure success and ignoring the redistributive outcome. The selective voicing of different actors, self-serving and 'simulated' evidence of failure or success of the FTLRP and sweeping generalisations, undermined critical analyses of the FTLRP. He thus queries neoliberal conceptualisations of the media as 'public spheres' or 'watchdogs', especially in fragile societies grappling with unresolved post-colonial national questions.

The concluding chapter by Sam Moyo and Paris Yeros synthesises the key lessons arising from Zimbabwe's FTLRP and agrarian reform based on the chapters in this volume and their previous work, highlighting key gaps in the dominant narrative and the scope for further research. Their chapter argues that the case of Zimbabwe presents distinct and innovative characteristics when compared to other cases of radicalisation, reform and resistance. The process of mobilisation includes a number of key elements, namely the multi-class, decentralised and anti-bureaucratic character of the movement, its adherence to radical nationalism, its capacity to articulate grievances across the rural-urban divide and also the radicalisation of its petty-bourgeois components. The process of reform and resistance has entailed the deliberate creation of a tri-modal agrarian structure to accommodate and balance the interests of various domestic classes, the progressive restructuring of labour relations and agrarian markets, the continuing pressures for radical reforms, through the indigenisation of mining and other sectors and the rise of extensive, albeit relatively weak, producer cooperative structures. A final element has been the emergence of a non-aligned foreign policy (despite its official naming as a 'Look East Policy'), largely in response to political isolation and sanctions from the West and in search for an alternative accumulation strategy, in the context of the growing East-West scramble for resources in the region and emerging global security re-arrangements.

The concluding chapter also highlights some of the resonances between the Zimbabwean land struggles and those on the continent, as well as in the South in general, arguing that there are some convergences and divergences worthy of intellectual attention. Thus, the chapter calls for greater endogenous empirical research which overcomes the pre-occupation with failed interpretations of the nature of the state and agency in Africa.

Notes

1. The A1 model targeted landless and poor families, providing land use permits on small plots for residence cropping and common grazing, while the A2 scheme targeted new 'commercial' farmers, providing larger individual plots on long-term lease to beneficiaries supposedly with farming skills and/or resources (including for hiring managers).
2. Personal Communication by John Gwitira to Sam Moyo in Harare, June 2000.
3. These depend on reports from the media (*e.g. Financial Gazette*; *The Independent*; *The Standard*) and the commercial farmers' unions (Commercial Farmers' Union [CFU]; Justice for Agriculture [JAG]) and some NGOs [Human Rights Forum etc.]).
4. A few studies use official GoZ audit reports (Buka 2002; Utete 2003) and local land allocation distribution data, to which they otherwise give little credence (*e.g.* Zimbabwe Institute 2007).
5. The baseline survey districts were of varied agro-ecological potential with diverse mixed farming systems, ethno-regional and local political situations. Questionnaire surveys, key informant checklists and interviews, focus group discussions, field observations and other opportunistic open interviews were used in both the A1 and A2 settlement schemes. These covered 2,089 households, with 79 per cent of these being A1 settlers, as well as 761 farm workers. This captured data on land tenure, socio-economic characteristics, production, markets, labour and farmer organisational issues. This data is cited as AIAS (2007), while its analysis was reported in Moyo et al (2009). Most of the authors undertook further case studies on their own and/or through the AIAS and some did work for the government's land audits (Buka 2002; Utete 2003) and participated in studies organised by the European Commission in 2004 and the World Bank in 2006.
6. See Moyo and Sukume 2004; Chambati and Moyo 2004; World Bank 2006; Chambati 2009; Moyo et al 2009; Moyo 2001; Yeros 2002; Helliker 2008; Murisa 2008; Moyo 2006; Masuko et al 2008; Moyo et al 2009; post-graduate theses by Sadomba 2008, Murisa 2010, Chambati 2011.

References

Adam Smith International, 2007, '100 days: An agenda for Government and Donors in a New Zimbabwe', *Paper from Adam Smith International's Fragile States and Post Conflict Series*, July 2007.

AIAS Baseline Survey, 2007, 'Inter-district household and whole farm survey', Harare: African Institute for Agrarian Studies (AIAS).

Alexander, J., 2006, *The Unsettled Land: State-Making and the Politics of Land in Zimbabwe 1893-2003*, Harare: James Currey and Weaver Press.

Alexander, J., J. McGregor and T. Ranger, 2000, *Violence and Memory: One Hundred Years in the 'Dark Forests' of Matabeleland*, Oxford, Portsmouth, Cape Town and Harare: James Currey, Heinemann, David Philip and Weaver Press.

Altieri, M.A. and F.R. Funes-Monzote, 2012, 'The Paradox of Cuban Agriculture', *Monthly Review*, Vol. 63, No. 8, pp. 23–33.

Amin, S., 1972, *Neocolonialism in West Africa*, Harmondsworth: Penguin.

Amin, S., 1990, *Maldevelopment: Anatomy of a Global Failure*, London: Zed Books.

ANC Conference, 2007, 'ANC 52nd National Conference 2007– Resolutions', ANC Conference held at the University of Limpopo in Polokwane, 16–20 December 2007.

Arrighi, G., 1973, 'The Political Economy of Rhodesia', in G. Arrighi and J.S. Saul, eds., *Essays in the Political Economy of Africa*, New York and London: Monthly Review Press. pp. 336–77.

Arrighi, G., Aschoff, N. and Scully, B., 2010, 'Accumulation by Dispossession and its Limits: The Southern Africa Paradigm Revisited', *St Comp Int Dev*, Vol. 45, pp. 410-438.

Bernstein, H., 2002, 'Agrarian Reform after Developmentalism?', Presentation at the Conference on Agrarian Reform and Rural Development, *Taking Stock*, Social Research Centre of the American University in Cairo, 14-15 October 2001.

Bernstein, H., 2012, 'Zimbabwe's Land Reform: Myths and Realities – by Ian Scoones, Nelson Marongwe, Blasio Mavedzenge, Jacob Mahenehene, Felix Murimbarimba and Chrispen Sukume', *Journal of Agrarian Change*, Vol. 12, pp. 170–173.

Bond, P., 2008, 'Response to Lessons of Zimbabwe'. (http://links.org.au/node/815/9693) 8 June 2011.

Borras, S.M. Jr., 2005, 'Can redistributive reform be achieved via market-based land transfer schemes? Lessons and evidence from the Philippines', *Journal of Development Studies*, Vol. 41, No. 1, pp. 90–134.

Borras, S.M. Jr., 2006, 'The Philippine land reform in comparative perspective: some conceptual and methodological implications', Journal of Agrarian Change, Vol. 6, No. 1, pp. 69–101.

Bryceson, D., 1996, 'Deagrarianization and rural employment in sub-Saharan Africa: A sectoral perspective', *World Development*, Vol. 24, No. 1, pp. 97-111.

Buka, F., 2002, A preliminary audit report of the Land Reform Programme. Unpublished report, Harare: Government of Zimbabwe.

Bush, R. and Cliffe, L., 1984, 'Agrarian Policy in Labour Migrant Societies: Reform or Transformation in Zimbabwe?', *Review of African Political Economy*, Vol. 29, pp. 77-94.

Campbell, H., 2008, 'Mamdani, Mugabe and the African Scholarly Community.' (http://www.pambazuka.org/en/category/features/52845). 8 June 2011.

CFU, 2003, 'An overview of commercial Agriculture in Zimbabwe', 11 June 2003.

Chambati, W., 2009, 'Land reform and changing agrarian labour processes in Zimbabwe', MA Thesis, Johannesburg: University of Witwatersrand.

Chambati, W., 2011, 'Restructuring of agrarian labour relations after Fast Track Land Reform in Zimbabwe', *Journal of Peasant Studies*, Vol. 38, No. 5, pp. 1047-1068.

Chambati, W. and Moyo, S., 2004, 'Land Reform and the Political Economy of Agricultural Labour', Harare: African Institute for Agrarian Studies. Mimeo.

Chang, H., 2009, 'How to "do" a developmental state: political, organisational and human resource requirements for the developmental state', in O. Edigheji, ed., *Constructing a democratic developmental state in South Africa: potential and challenges*, Cape Town: HSRC Press. pp. 82–96.

Chiweshe, M., 2012, 'Farm level institutions in emergent communities in post fast track Zimbabwe: case of Mazowe district', PhD. Thesis, Grahamstown: Rhodes University.

Cliffe, L., Alexander, J., Cousins, B. and Gaidzanwa, R., 2011, 'An overview of Fast Track Land Reform in Zimbabwe', editorial introduction, *Journal of Peasant Studies*, Vol. 38, No. 5, pp. 907-938.

Cross, E., 2012, 'The issue of property rights in Africa'. (http://eddiecross.africanherd.com). 15 January 2012.

De Grassi, A., 2008, '"Neo-patrimonialism" and Agricultural Development in Africa: Contributions and Limitations of a Contested Concept', *African Studies Review*, Vol. 51, No. 3, pp. 107–133.

Dekker, M. and Kinsey, B., 2011, 'Contextualizing Zimbabwe's land reform: long-term observations from the first generation', *Journal of Peasant Studies*, Vol. 38, No. 5, pp. 995-1019.

European Union, 2007, 'Zimbabwean adaptation strategy to the European Union', Sugar regime reform final report, Brussels: European Union.

Evans, P. B., 2009, 'Constructing a twenty-first century developmental state: potentialities and pitfalls', in O. Edigheji, ed., *Constructing a democratic developmental state in South Africa: potential and challenges*, Cape Town: HSRC Press, pp. 37–58.

FAO, 2010, 'Food Outlook: Global Market Analysis', Rome: Food and Agriculture Organization of the United Nations, World Food Programme.

GoZ, 1998, 'Land Reform and Resettlement Programme Phase II: A Policy Framework and Project Document', Harare: Government Printers.

GoZ, 2001, *Fast Track Land Reform Programme*, Harare: Government Printers (Draft 2000).

Gwisai, M., 2002, *Revolutionaries, resistance and crisis in Zimbabwe: anti-neoliberal struggles in periphery capitalism*, Harare: International Socialist Organisation/Socialist Worker -Zimbabwe.

Hammar, H., Raftopoulos, B. and Jensen, S., eds., 2003, *Zimbabwe's unfinished business: rethinking land, state and nation in the context of crisis*, Harare: Weaver Press.

Helliker, K., 2008, 'Dancing on the same spot: NGOs', in S. Moyo, K. Helliker and T. Murisa, eds., *Contested terrain: land reform and civil society in contemporary Zimbabwe*, Pietermaritzburg: S&S Publishers, pp. 239–274.

Hellum, A. and Derman, B., 2004, 'Land reform and human rights in contemporary Zimbabwe: balancing individual and social justice through an integrated human rights framework', *World Development*, Vol. 32, No. 1, pp. 1785–1805.

Human Rights Watch (HRW), 2004, 'The politics of food assistance in Zimbabwe', A Human Rights Watch Briefing Paper, Human Rights Watch.

Hungwe, M., 2011, 'In Search of "Community" in Zimbabwe's Fast Track Resettlement Area of Mazowe District', PhD Thesis, Lund, Sweden: Lund University.

Inclusive GoZ, 2009, *Global Political Agreement (GPA) between the Zimbabwe African National Union-Patriotic Front (Zanu-PF) and the two Movement for Democratic Change (MDC) formations, on resolving the challenges facing Zimbabwe*, Harare: Government of Zimbabwe, September 2008.

JAG/RAU, 2008, 'Land, retribution and elections: post-election violence on Zimbabwe's remaining farms 2008', A report prepared by the Justice for Agriculture (JAG) Trust and the Research and Advocacy Unit, May 2008.

Johnson, D., 2009, 'Mamdani, Moyo and "Deep Thinkers of Zimbabwe"'. (http://www.pambazuka.org/en/category/comment/54039), 12 February 2009.

Madhuku, L., 2004, 'Law, Politics and the Land Reform Process', in S. Masiyiwa, ed., *Post Independence Land Reform in Zimbabwe: Controversies and Impact on the Economy*, Harare: Friedrich Ebert Stiftung and Institute of Development Studies. pp. 124-146.

Mafeje, A., 2003, 'The Agrarian Question, Access to Land and Peasant Responses in Sub-Saharan Africa, UNRISD Civil Society and Social Movements Programme Paper Number 6 May 2003, Geneva: United Nations Research Institute for Social Development.

Mamdani, M., 2008, 'Lessons of Zimbabwe', London Book Review.

Mandaza, I., 1986, *The State and Politics in a Post-White Settler Colonial Situation in Zimbabwe: The Political Economy of Transition, 1980–1986*, Dakar: CODESRIA.

Marongwe, N., 2011, 'Who was allocated Fast Track land and what did they do with it? Selection of A2 farmers in Goromonzi District, Zimbabwe and its impacts on agricultural production', *Journal of Peasant Studies*, Vol. 38, No. 5, pp. 1069-1092.

Masuko, L., 2008, 'War veterans and the Re-emergence of Housing Cooperative', in S. Moyo, K. Helliker and T. Murisa, eds., *Contested Terrain: Land Reform and Civil Society in Contemporary Zimbabwe*, Pietermaritzburg: S&S Publishers.

Matondi, P., ed., 2011, *Inside the political economy of redistributive land and agrarian reforms in Mazoe, Shamva and Mangwe districts in Zimbabwe*, Harare: Ruzivo Trust.

McCandless, E., 2011, *Polarisation and Transformation in Zimbabwe: Social Movements, Strategy Dilemmas and Change*, Lanham: Lexington Books.

Mkandawire, T. and Soludo, C., 1999, *Our Continent Our Future: African Perspectives on Structural Adjustment,* Dakar: CODESRIA.

Mkandawire, T., 2001, 'Thinking about Developmental States in Africa', *Cambridge Journal of Economics,* Vol. 25, No. 3, pp. 289–314.

Mkandawire, T., 2012, *Neopatrimonialism and the political economy of economic performance in Africa: Critical reflections,* London School of Economics.

Moore, D., 2004, 'Marxism and Marxist intellectuals in Schizophrenic Zimbabwe: How many Rights for Zimbabwe's left? A comment', *Historical Materialism,* Vol. 12, No. 4, pp. 405-25.

Moyana, H.V., 1987, *The Victory of Chief Rekayi Tangwena,* Harare: Longman Publishers.

Moyana, H.V., 2002, *The political economy of land in Zimbabwe,* Gweru: Mambo Press.

Moyo, S., 1995, *The land question in Zimbabwe,* Harare: SAPES Books.

Moyo, S., 2000, *Land reform under structural adjustment in Zimbabwe: land use change in Mashonaland provinces,* Uppsala: Nordic Africa Institute.

Moyo, S., 2001, 'The Land Occupation Movement and Democratization in Zimbabwe: Contradictions of Neoliberalism', *Millennium Journal of International Studies,* Vol. 30, No. 2, pp. 311-330.

Moyo, S., 2004, 'The overall impact of the Fast Track Land Reform Programme', AIAS monograph, Harare: African Institute for Agrarian Studies.

Moyo, S., 2007, 'Emerging Land Tenure Issues in Zimbabwe', AIAS Monograph, Harare: African Institute for Agrarian Studies.

Moyo, S., 2011a, 'Three decades of agrarian reform in Zimbabwe', *Journal of Peasant Studies,* Vol. 38, No. 3, pp. 493-531.

Moyo, S., 2011b, 'Land concentration and accumulation after redistributive reform in post-settler Zimbabwe', *Review of African Political Economy,* Vol. 38, No. 128, pp. 257-276.

Moyo, S., 2011c, 'Changing agrarian relations after redistributive land reform in Zimbabwe', *The Journal of Peasant Studies,* Vol. 38, No. 5, pp. 907–934.

Moyo, S., Chambati, W., Murisa, T., Siziba, D., Dangwa, C., Mujeyi, K., and Nyoni, N., 2009, *Fast Track Land Reform baseline survey in Zimbabwe: trends and tendencies, 2005/06,* Harare: African Institute for Agrarian Studies (AIAS).

Moyo, S., Helliker, K. and Murisa, T., eds., 2008, *Contested terrain: land reform and civil society in contemporary Zimbabwe,* Pietermaritzburg, South Africa: S&S publishers.

Moyo, S. and Sukume, C., 2004, 'Farm Sizes, decongestion and land use: Implications of the Fast Track Land Redistribution Programme in Zimbabwe', Harare: AIAS Mimeo.

Moyo, S. and Yeros, P., 2005a, 'Introduction', in S. Moyo and P. Yeros, eds., *Reclaiming the land: the resurgence of rural movements in Africa, Asia and Latin America,* London: Zed Books. pp. 1–7.

Moyo, S. and Yeros, P., 2005b, 'Land occupations and land reform in Zimbabwe: towards the National Democratic Revolution', in S. Moyo and P. Yeros, eds., *Reclaiming the land: the resurgence of rural movements in Africa, Asia and Latin America,* London: Zed Books. pp. 165–208.

Moyo, S. and Yeros, P., 2007, 'The Radicalised State: Zimbabwe's Interrupted Revolution', *Review of African Political Economy,* Vol. 111, pp. 103-121.

Moyo, S. and Yeros, P., 2009, 'Zimbabwe ten years on: results and prospects'. (http://www.pambazuka.org/en/category/features/54037). February.

Moyo, S. and Yeros, P., 2011, 'After Zimbabwe: State, Nation and Region in Africa', in S. Moyo and P. Yeros, eds., *Reclaiming the Nation: The Return of the National Question in Africa, Asia and Latin America,* London: Pluto Press.

Murisa, T., 2010, 'Emerging forms of social organisation and agency in the newly resettled areas of Zimbabwe: the cases of Goromonzi and Zvimba districts', PhD. Thesis, Grahamstown: Rhodes University.

Nkomo, M.J., 2001, *The story of my life,* Harare: Sapes Books.

Olukoshi, B., 2011, 'Democratic Governance and Accountability in Africa: In search of a Workable Framework', Discussion paper, no. *4,* Uppsala: Nordic Africa Institute.

Patnaik, U., 2003, 'Global capitalism, deflation and agrarian crisis in developing countries', *UNRISD* Social Policy Paper Series.

Raftopoulos, B., 2003, 'The State in Crisis: Authoritarian Nationalism, Selective Citizenship and Distortions of Democracy in Zimbabwe', in A. Hammar, B. Raftopoulos and S. Jensen, eds., *Zimbabwe's Unfinished Business: Rethinking Land, State and Nation in the Context of Crisis,* Harare: Weaver Press.

Raftopoulos, B. and Phimister, I., 2004, 'Zimbabwe Now: The Political Economy of Crisis and Coercion', *Historical Materialism,* Vol. 12, No. 4, pp. 355-82.

Raftopoulos, B., 2009, 'The crisis in Zimbabwe', in B. Raftopoulos and A. Mlambo, eds., *A history from the pre-colonial period to 2008: becoming Zimbabwean,* Harare: Weaver Press. pp. 201–32.

Raftopolous, B., 2010, 'The Global Political Agreement as a "Passive Revolution": Notes on Contemporary Politics in Zimbabwe', *The Round Table,* Vol. 99, No. 411, pp. 705-718.

Ranger, T., 2004, 'Nationalist Historiography, Patriotic History and the History of the Nation: The Struggle over the Past in Zimbabwe', *Journal of Southern African Studies,* Vol. 30, pp. 215-234.

Republic of South Africa, 2005, 'Report of the National Land Summit', *Nastec,* Johannesburg: Land Affairs, 27-30 July 2005.

Richardson, C., 2005, 'The loss of property rights and the collapse of Zimbabwe', *Cato Journal,* Vol. 25, No. 2, pp. 541–565.

Rosset, P. and M. Courville, 2006, *Promised Land: Competing Visions of Agrarian Reform,* Oakland, CA: Food First Books.

Sader, E., 2011, *The New Mole: Paths of the Latin American Left,* London: Verso.

Sadomba, W., 2008, 'War veterans in Zimbabwe: complexities of a liberation movement in an African post-colonial settler society', PhD. Thesis, Netherlands: Wageningen University.

Saul, J.S., 2005, *The Next Liberation Struggle: Capitalism, Socialism and Democracy in Southern Africa,* Between The Lines, London Toronto: UKZN Press; Scottville: Monthly Review Press; New York: The Merlin Press.

Scarnecchia, T. and 33 other scholars, 2008, 'Response to lessons of Zimbabwe'. (http://www.lrb.co.uk/v31/n01/letters.html). December 2008.

Scoones, I., Marongwe, N., Mavedzenge, B., Murimbarimba, F., Mahenehene, J. and C. Sukume, 2010, *Zimbabwe's Land Reform: Myths and Realities,* Harare: Weaver Press and Oxford: James Currey.

Scoones, I., Marongwe, N., Mavedzenge, B., Murimbarimba, F., Mahenehene, J. and Sukume, C., 'Livelihoods after Land Reform in Zimbabwe: Understanding Processes of Rural Differentiation', *Journal of Agrarian Change.*

Stoneman, C., 1988, *Zimbabwe's prospects: issues of race, class, state and capital in Southern Africa,* London: Macmillan.

Stoneman, C. and L. Cliffe, 1989, *Zimbabwe: Politics, Economics and Society,* New York: Pinter Publishers.

Tendi, M., 2010, *Making History in Mugabe's Zimbabwe: -Politics, Intellectuals and the Media,* Oxford: Peter Lang Publishers.

Tshuma, L., 1997, *A matter of (in) justice, law, state and the agrarian question in Zimbabwe,* Harare: SAPES Books.

Tupy, M.L., 2007, 'A four-step recovery plan for Zimbabwe', CATO Institute. (http://www.cato.org/pub_display.php?pub_id¼8191). 9 October 2008.

UNDP (United Nations Development Programme), 2010, *Human Development Report: The Real Wealth of Nations: Pathways to Human Development,* New York: UNDP.

USAID, 2010, 'Zimbabwe Agricultural Sector Market Study', Washington D.C.: Weidemann Associates, Inc.

Utete Report, 2003, 'Report of the Presidential Land Review Committee', under the chairmanship of Dr Charles M.B. Utete, Vols. 1 and 2: Main report to His Excellency the President of the Republic of Zimbabwe, Presidential Land Review Committee (PLRC), Harare: Government Printers.

Weiner, D., 1988, 'Land and Agricultural Development', in C. Stoneman, ed., *Zimbabwe's Prospects: Issues of Race, Class and Capital in Southern Africa,* London: Macmillan., pp, 63-89.

West, Michael O., 2002, *The Rise of an African Middle Class: colonial Zimbabwe, 1898-1965,* Bloomington: Indiana University Press.

World Bank, 2006, *Agricultural Growth and Land Reform in Zimbabwe: Assessment and Recovery Options,* Report No. 31699-ZW, Harare: World Bank.

Yeros, P., 2002, 'The political economy of civilisation: peasant-workers in Zimbabwe and the neo-colonial world', PhD. Thesis, University of London.

Zamchiya, P., 2011, 'A synopsis of land and agrarian change in Chipinge District, Zimbabwe', *Journal of Peasant Studies,* Vol. 38, No. 5, pp. 1093-1122.

Zimbabwe Institute, 2007, *Understanding the impact of the Fast Track Land Reform Resettlement Programme,* Cape Town: Zimbabwe Institute.

2

Land Reform and Redistribution in Zimbabwe Since 1980[1]

Sam Moyo

Introduction

Although it is increasingly recognised that Zimbabwe's Fast Track Land Reform Programme (FTLRP), initiated in 2000, was redistributive (Moyo et al 2009; Scoones et al 2010), few studies have examined the qualitative character of this outcome and its prospects for progressive social and political transformation in a largely agrarian society. Most critics of the FTLRP (e.g., Cliffe et al 2011) continue to underplay the significance of the settler-colonial roots of Zimbabwe's land question and its exacerbation under neoliberal rule after independence, in fomenting the social and political crisis which provoked the popular reclamation of land (see Moyo and Yeros 2005, 2007a) and in shaping the transformational character of the FTLRP (Moyo 2011a, 2011c). Focusing on the narrowly defined 'human rights transgressions' that accompanied the FTLRP, using abstracted neoliberal good governance norms, the critics miss the important rolethat broad-based social mobilisation played in shaping state action towards accommodating a wide array of land demands.

Land reform was meant to redress historical settler-colonial land dispossession and the related racial and foreign domination, as well as the class-based agrarian inequalities which minority rule promoted. Post-independence land reforms sought to alter the resultant repressive social relations of production and reproduction, through broadening access to land and promoting peasant productivity. Implicitly, land reform would free labour from the exploitative tenancy relationship used by large-scale farmers to compel the landless to work for low wages. Local government reforms were instituted to undo the territorial, administrative and social segregation of Communal Areas from former Large-Scale Commercial Farming areas, free the movement of people, goods and services and consummate local and national sovereignty over alienated territory in a unitary system of government (Mtizwa-Mangiza 1991). By eliminating the enclaves of unequal political power and economic domination, land reform would also promote democratic land administration (Shivji et al 1998).

Land reform policy in Zimbabwe was implemented in three phases characterised by critical shifts in economic policy and performance and by changes in the electoral political circumstances. Between 1980 and 1989, land reform was based on state-led purchases of land on the market and its allocation to selected beneficiaries, in the context of heterodox economic policies, which enabled increased public spending on social services and peasant agriculture. From 1990, neoliberal policies restricted state interventions in markets, in general and restricted social welfare subsidies. Furthermore, land redistribution slowed down, despite the adoption of land expropriation laws. In the third phase, an escalating social crisis, which culminated in extreme political polarisation by 1997, saw the land redistribution programme shift towards land expropriation, leading to extensive land redistribution and increased state interventions in the economy, alongside bitterly contested elections.

Initially the Government of Zimbabwe's (GoZ) land reform was narrowly conceived to address displacement, landlessness and 'overcrowding'.[2] From 1985, a wider range of indigenous classes of people agitated for access to land (Moyo 1995) and land reform policy gradually accommodated middle- class demands for land (GoZ 1998). Official land reform programmes were always accompanied by varying degrees of popular ('illegal') land occupations throughout the countryside (Moyo 2001; Tshuma 1997). The state repressed these from 1985 (Herbst 1990), despite their being encouraged by some political party and state officials and by war veterans (Moyo 2001). In 2000, however, it condoned the nationally coordinated land occupations led by war veterans. The post-colonial political dynamics of class formation, including the resuscitation of the peasantry and the rise of an aspirant petty-bourgeoisie, in the face of the consolidation of settler and foreign capital, alongside the demise of working classes, animated land reform policy shifts.

Dominant discourses on land reform in Zimbabwe, from 1980, tended to argue against large-scale land redistribution on the grounds that it would disrupt agricultural productivity, while assuming that peasant agriculture was unproductive (Whitesun Foundation 1983), despite the extensive under-utilisation of land among Large-Scale 'Commercial' Farmers (LSCF) (Weiner et al 1985). Economic policy emphasised industrial development and urban employment within a minimalist rural development strategy which promoted agricultural productivity through improving peasant markets (World Bank, 1982; GoZ 1986; Moyo 1986). Land redistribution was gradually considered relevant only to competent black farmers, notwithstanding the wider livelihood requirements of smallholder farmers (Moyo 2000). When the Economic Structural Adjustment Programme (ESAP) was adopted in 1990, the land policy was re-oriented to support export led agricultural growth, based mainly on large-scale farming (Moyo 2000), while the building of domestic grain reserves for food security was discouraged (Moyo 2011d).

During the 1990s, however, political demands to de-racialise large-scale commercial farming to achieve equitable capitalist opportunity and political stability increasingly shaped policy (GoZ 1998). The critique against land reform soon turned to the alleged prevalence of elite capture in the programme, contrary to the evidence from official evaluations of the resettlement programme (Cusworth and Walker 1988). Donor- led land reform prescription soon emphasised poverty reduction and sustainable livelihoods approaches which 'make markets work for the poor' (DFID 1997 [2000]). Thus, when the FTLRP was initiated, some scholars considered it unwarranted, allegedly because it pursued the partisan political and land interests of ZANU-PF elites, whose 'nationalism' and legitimacy needed refuelling (Raftopoulos 2003).

The underlying assumption of this critique is that the FTLRP land occupations and state land expropriations and allocations were centrally controlled by ZANU-PF and state functionaries, who are perceived to be a monolithic political entity representing mainly petty-bourgeois interests. Local social forces and diverse class interests were allegedly not active in the politics and implementation of the FTLRP (see Sadomba and Masuko, Chapters 3 and 4 for alternative views). Instead, the whole land reform was deemed a 'chaotic and often violent', 'racially motivated land seizure' and 'politically vindictive land grab', which violated the legitimate land rights of white farmers (see Willems 2005). Putatively, the FTLRP only benefited President Mugabe's cronies and destroyed agriculture (Bond 2005), while the statutory land user rights provided to the beneficiaries by the state are allegedly insecure, leading to ubiquitous land disputes (Vudzijena 2007). Agricultural land has apparently been transformed into 'dead capital' (Mhishi 2007: 9; Richardson 2005; Robertson 2011), limiting the supply of credit. Without empirical substantiation, the critics vilified the entire land movement (Johnson 2009).

Consequently, white farmers have demanded that either their land be returned to them or they be compensated for the land at 'market rates'. It is claimed, furthermore, that the land tenure of FTLRP beneficiaries cannot be secure without such compensation, while the recovery of agricultural investment and re-engagement with and funding from donors can only occur when the 'contested land' is resolved (see JAG 2006). This perspective eschews the demands for the restitution of indigenous land rights in relation to colonial loss, contemporary social needs and the direct action to be taken by various classes to repossess land (Hunzvi 2000[3]), despite the state support they received. Not surprisingly, the scope of land redistribution has been understated even if it is now grudgingly recognised (see Cliffe et al 2011).

This chapter explores the social and structural distributional outcomes of the FTLRP based on extensive empirical research (see Chapter 1 on the sources used). It first summarises the evolution of land reform policy before and after the FTLRP, touching briefly on the social forces which mobilised for the radical reform (as elaborated upon by Sadomba and Masuko, Chapters 3 and 4). It then examines the extent to which the programme was redistributive and elaborates on the social transformations it evoked in terms of race, gender and other forms of identity and the recognition this purveys. The chapter finally assesses the class formation processes emerging from the new land ownership structure, briefly noting the changing agrarian land-labour relations (for details on labour relations see Chambati, Chapter 5). This chapter is also intended to set the stage for later assessments of the wider agrarian changes that have ensued.

Market-based land reform and its contradictions

The use of the market mechanism to redistribute land from 1980 to 1999 meant that landowners defined the land available for resettlement and central government reactively chose the land to acquire. The UK government contributed matching funds for 'resettlement' during the 1980s, within the logic of an 'aid' project, rather than as reparations for colonial land losses. This approach limited the amount, quality and location of the land acquired in relation to social needs and organised demand (Moyo 1995). Land prices rose in response to the growth of both private and public demand for land in the markets. Notably, some of the acquired land had been 'illegally' occupied by peasants, in the 'Accelerated Land Resettlement Programme' (see Tshuma 1997), setting a precedence of 'regularising' popularly occupied land, although, from 1984, 'squatters' were often violently evicted by the police and white farmers.

Constitutional restrictions on land expropriation were partially removed in 1990 and the Land Acquisition Act (1992, Chapter 20: 10) enabled the state to expropriate land for redistribution, although this was sparingly used between 1993 and 1997, partly due to successful litigation by landowners (Moyo 1999). While land reform had not fully challenged the legitimacy of exclusive private freehold property, which was generally regarded as a legally superior form of land tenure compared to customary land rights (Shivji et al 1998), the state allocated permissory land rights and tradable leasehold rights to the beneficiaries of land reform.

By 1999, approximately 3.4 million hectares had been redistributed, reducing the white commercial farming area to 11 million hectares, or approximately 35 per cent of the total agricultural land, most of which was 'prime' land. About 70,000 families were resettled, far short of the official targeted 162,000 families and others on official 'waiting lists' for land (Moyo 1999). The beneficiaries mainly included 'the landless, poor and overcrowded rural people', various 'disadvantaged groups' and some 'competent' farmers (Moyo 1995). By 1990, black professionals, entrepreneurs and political elites had also gained a limited amount of private and state-acquired land through the Commercial Farm Settlement Scheme (Moyo 1999). The extension of peasant farming on marginal grazing lands and increased 'squatting' stretched local government regulatory capacities, while rural social differentiation deepened (Moyo 2000).

Meanwhile, the adoption of the ESAP in 1990 led, not only to the slowing down of land redistribution, but to the expansion of land markets to foreigners and aspiring black commercial farmers, leading to increased private land subdivisions and consolidations (see Rugube et al 2003). Land concentration and foreign land ownership escalated, including in regions previously regarded as agro-ecologically marginal (Moyo 2000). The acquisition by blacks of large-scale farms grew to about 15 per cent of the LSCF areas. Intra-capitalist competition for land escalated through the 'indigenisation' ideology, while some blacks were co-opted by capital into large-scale farming.

The ESAP also had the effect of increasing the scale and sources of demand for land. Labour retrenchments led to increased illegal occupations of farming land (see Yeros 2002). Increased demographic pressure in Communal Areas fuelled more organised land occupations and natural resource poaching (Moyo 2000). In 1997, war veterans agitated for increased pensions and militantly demanded that President Mugabe and the state 'get on' with land expropriation. The state designated 4.1 million hectares for compulsory acquisition in 1997 and, in 1998, land 'invasions' by war veterans backed this policy (see Sadomba, Chapter 3). Thus, the structural and social contradictions of the ESAP and limited land redistribution fuelled the mobilisation of radical land reclamation movements, which influenced state expropriation, while most formal civil society organisations stood aside (Moyo 2001).

The radicalisation of the official land reform policy was also stoked by disagreements between the GoZ and the UK government over financing land acquisition, with the latter denying any colonial obligation (Short 1997). The Land Donors Conference of 1998 mobilised donors around a poverty-oriented land reform programme which prescribed limited and gradual land redistribution using 'market-assisted' approaches, rather than land expropriation (World Bank 1999). Also from 1999, the Zimbabwe Congress of Trade Unions (ZCTU) 'suddenly' created an externally-funded political opposition party (the Movement for Democratic Change), which was allied to various western- funded NGOs and the Commercial Farmers Union (CFU) and which exposed radical land reform. Political polarisation spiked over constitutional reform, pitting the ZANU-PF alliance against the MDC alliance over presidential powers and land expropriations, with the CFU playing a critical role in mobilising farm workers against the Draft Constitution.

These events dovetailed into wider political developments, surrounding campaigns for the Referendum on the Draft Constitution in February 2000, the mass occupations of land led by war veterans from March 2000 and the parliamentary election campaign of June 2000. External relations with the West deteriorated as its support for the political opposition grew and economic sanctions were escalated, allegedly in response to state- sponsored violence and abrogation of the rule of law.

The Fast Track Land Redistribution Policy and its implementation

Implementing the FTLRP entailed numerous actions phased over the decade, involving 'illegal land occupation' and formal state expropriations in relation to the changing balance of forces, pitting increasing popular demand for land against landowner litigations (see Table 2.1). The first phase of the FTLRP was dominated by extensive popular and 'illegal' land occupations and the mass designation of over 3,000 farm properties for expropriation. In March 2001, a law was enacted to protect 'illegal occupiers' on the farms from a barrage of litigations by white farmers and from a Constitutional judgement requiring the restoration of land to farmers. Land occupations initially entailed both violence in the takeover of land and negotiations over its sharing ('co-existence') between occupiers and white farmers.

The state gradually gained control over the land reform process from the 'illegal occupiers' through creating District Land Committees (DLCs), which involved the government bureaucracy, security agents, ZANU-PF members, war veterans and other social formations (chiefs, farmers associations, etc). The formal FTLRP policy document was issued in April 2001 (GoZ 2001). From late 2000, the FTLRP was increasingly dominated by formal land allocations to the 'illegal occupiers' and many others who applied for land through the DLCs (see Moyo 2005). This change entailed adjustments to the numbers of people accommodated and the sizes of their plots allocated, as well as converting some A1 allocations to A2 allocations. In April 2001, the white farmers negotiated the transfer of one million hectares under the proposed Zimbabwe Joint Resettlement Initiative (ZIJRI 2001), but the deal collapsed over various disagreements. By 2004, the bureaucracy had formalised two FTLRP land allocation schemes (A1 and A2),[4] initiated various land law reforms and recomposed the judiciary to facilitate expropriations and thwart landowner litigation. Land allocations by 2002 accommodated more of the petty-bourgeoisie. Surges of land allocations were, however, also shaped by the political mobilisation that accompanied the various elections and the related escalation of western sanctions from 2001, as dirigiste economic policies were instituted.

Table 2.1: Periodisation of the Fast Track Land Reform Programme 2000-2012

Land Reform Phase (Period)	Key Political and Economic Events	Key Land Policy Events	Key Land Transfer Processes
March 2000 to June 2001 Revolutionary situation	-Feb 2000 Referendum -Mass land occupations -Parliamentary election -Judicial reforms	-Constitutional Judgment -FTLRP policy paper -Land Occupiers Act -DLCs established -ZIJRI negotiation	-Popular occupations -Mass A1 allocations -Mass expropriations -Mass land litigations
July-2001 to December 2003 Rationalisation of land reform	-Heterodoxy policy -Sanctions escalated -Presidential election	-Land audits -A2 scheme designed -'Corrections' exercise	-A2 allocations -A1/A2 land disputes -Expanded A1 allocations
January 2004 to June 2008 Bureaucratisation of land reform	Deep economic crisis -Parliamentary elections -Unified elections (2008)	- Expropriation laws -'Re-planning' processes -Leases designed	-A2 allocations -A1 allocations -Estates contested
July 2008 to December 2011 Residual land redistribution	-Inclusive Government formed -Economic liberalisation -New sanctions -Indigenisation policy -New foreign investment	-Negotiating land audit -Leases negotiated	-More A2 land allocations -Conservancies indigenised -Some land occupiers ac-commodated

Source: Compiled by author (see Moyo 2011d, 2007)

From mid-2001 to 2003, the second phase of the FTLRP entailed a re-design of the A2 scheme, which by 2003 had allocated 8,000 plots to beneficiaries (Utete 2003). The District Land Committees increasingly sidelined war veterans in the formal land transfer process (see Sadomba 2008), while illegal land occupations were sporadic and less formally condoned. The GoZ also sought to rationalise the FTLRP process by instituting two land audits and embarking on a 'correction' exercise (Buka 2002; Utete 2003). A third phase of the FTLRP saw both A1 and A2 land allocations increase twice above the original target of redistributing 5 million hectares. This increase constituted a response to mobilisations for land around the 2002 presidential election. By 2008, the A2 land allocations had surpassed 16,000 beneficiaries.

By mid-2008, when political accommodation and compromise over economic liberalisation were reached, the land allocations had tapered off. This fourth phase of the FTLRP entailed residual land allocations, including that of more A2 plots.

The FTLRP allocation process was not free of corruption. In particular, the A2 scheme entailed jumping application queues to gain better capitalised plots (e.g., with irrigation facilities) and 'whole farms' and unsubstantiated plans to justify access to larger-sized plots, using 'pseudo-legal' administrative loopholes. Some individuals corruptly gained access to multiple farms, while others got plots which are above the recommended A2 farm size, although their area represents a relatively small proportion of the overall redistributed lands (Buka 2002; Utete 2003). Some political elites, state officials and private citizens with connections gained temporary, but free, use of underutilised state farms and some of the state acquired, but unallocated, lands (e.g., for seasonal cropping, grazing). Some land beneficiaries 'grabbed' the moveable properties of the former farmers using 'illegal' means.[5]

However, below 20 per cent of the A1 land beneficiaries secured land through 'illegal occupations' and these were later 'regularised' by the GoZ (see Moyo et al 2009). The rest had applied for the land through various formal channels. Beyond the 'official' land beneficiaries, many people used 'informal' methods to gain access to sub-plots from official beneficiaries, who 'shared' their land with relatives, friends and neighbours, or even rented land out to tenants (Moyo et al 2009: 34). Thus, contrary to common assumptions about the FTLRP land beneficiaries, many more than the officially recognised beneficiaries were resettled formally and informally.

FTLRP policy contradictions and adjustments

The FTLRP land acquisition and allocation policy and its implementation procedures often seemed inconsistent. Agro-industrial estates were not sub-divided despite persistent demand for their land, while the share of A2 land allocations rose beyond expectation. The pace and regional distribution of land transfers was unevenly spread out over the decade, as a result of provincially differentiated mobilisations of the land occupations and formal expropriation. Provincial land reform managers often weighed local economic imperatives against social pressures for land and protected some white farmers to sustain local supplies (e.g., of milk) and jobs. Moreover, the demand for A2 land grew as the economic crisis escalated and certainty over the veracity of the FTLRP process grew among the middle classes. This demand led to various adjustments to the acquisition and allocation policy to accommodate expansive claims and their belated expression.

Moreover, land acquisition was actively opposed by white farmers in the courts on the ground, often involving some farm workers and at times the bribing of officials to stall land expropriations. Approximately 700 white farmers had successfully negotiated for their land not to be expropriated by 2007, but this was whittled down to 300 by 2010 as demand persisted. Meanwhile, the political opposition and some NGOs consistently vilified the FTLRP process, leading to escalating animosities. High profile media vilification and incessant litigation often sparked emotional and precipitous seizures of some LSCF farms.[6]

The key dilemma facing the policy elites by 2002 was that over 50,000 people had applied for A2 plots and only 8,000 had been offered land by 2003 (Utete 2003), while some within ZANU-PF and FTLRP critics insisted that the A2 scheme should only allocate land to people 'with means'. Few A2 applicants could prove they had investible resources. Some applicants felt excluded, alleging this was because they lacked connections to power and indeed such corruption did occur. Many simply lacked the resources and patience to persistently bid for land as officials faced land expropriation bottlenecks (Interviews: Centenary 2005[7]). When the 99-year lease document was first issued in 2006, only some with 'whole farms' got leases, ostensibly because these did not require survey and as the FTLRP process faced legal bottlenecks to confirm the state's ownership of expropriated lands. This disparity caused widespread disaffection and pressures for land allocations.

However, the main FTLRP implementation contradictions and policy ambiguities revolved around the bi-modal land and agrarian reform strategy, which, although formally adopted in 1990, was only implemented more generally from 2001, when the A2 scheme began. Therefore, class-based differentiation in access to redistributed land arose *a priori* from the deliberate policy design of addressing landlessness and racial imbalances in agrarian capitalist accumulation (i.e., 'black empowerment'). Initially, the policy had proposed to allocate 20 per cent or one million hectares of the targeted land to A2 farmers on a sliding scale depending on agro-ecological potential, while expecting most wealthy blacks to buy farms on the market (GoZ 1998). This scheme was intended to create medium-scale black 'commercial' farmers, who would pay progressive land taxes and lease fees charged according to the farm infrastructures availed with the land, hence the ideal of selecting those 'with means'. This secondary redistribution mechanism was never fully levied on A2 farmers, who pleaded poverty. By 2010, the A2 farmers had been allocated close to 40 per cent of the redistributed land, with over 20,000 having benefited compared to over 150,000 A1 beneficiaries (GoZ MLRR 2010).

During 2001, the A2 maximum farm size policy was adjusted to cater for three categories of A2 beneficiaries: small-scale (averaging 50 hectares in natural regions 1 and 2), medium-scale (averaging 200 to 600 hectares in natural region 3) and large-scale A2 farms (averaging 1,000 hectares in natural regions 4 and 5) (GoZ 2001). In order to accommodate demand, however, most of the A2 allocations were within the set maximum sizes, with the majority getting less than 100 hectares each. About 3,000 A2 farmers got more than 300 hectares each (see Table 2.1 and Figure 2.2), ostensibly because the terrain, soils and agro-ecological potential were poor (Sukume and Moyo 2003). Influence played a critical role in the selection of the beneficiaries on some of the A2 and A1 farms.

Zimbabwe's land reform policy was always ambiguous about the redistribution of agro-industrial estates, wildlife conservancies and forest plantations. The FTLRP was silent about creating new privately or publicly owned estates. Since some estates were not located 'near congested communal areas', partly because they had been established in remote and previously 'uninhabitable' areas and were buffered by white outgrower farms, they were less accessible to 'illegal' land occupations. However, most of the estate land acquisitions were more reactive to popular land occupations than state- planned. Until 2002, there was also official prevarication over the expropriation of foreign-owned farms covered by Bilateral Investment Protection and Promotion Agreements (BIPPAs), although many of them were eventually redistributed because they had been extensively occupied illegally. The logic of 'economies of scale' was mainly used by policymakers to justify retaining estates, alleging their superior (micro-economic) efficiency compared to small farms (see Sukume and Moyo 2003; Moyo 2011b). The agro-industrial estates were also considered critical to export, employment and agro-industrial growth by the state (GoZ 1998; Utete 2003).

From 2001, the GoZ was discouraging 'illegal settlements' in the estates and conservancies which it had not already expropriated, despite such lands being recommended for expropriation by provincial authorities. At the beginning of 2003, government officials and 'stakeholders' proposed FTLRP policies for forest estates and wildlife conservancies (GoZ 2004). These policies sought to redistribute shareholdings to black investors ('indigenisation'), rather than subdividing the lands and thereby to 'save' employment-intensive and specialised farms which produced 'strategic' needs (e.g., seed, citrus, dairy and timber). This step formalised the 1998 decision to limit the expropriation of estates (Moyo 1999). Thus, many estates which had been listed for expropriation during the FTLRP, including those 'illegally' occupied, were not appropriated by the state until 2008. While 39 middle-class people got shares in the seven conservancies during 2010, the equity share-holdings of agro-industrial estates were being negotiated by 2011. Following social pressures for such shares to be more broadly distributed, 'Community Equity Share Trusts' were being negotiated in 2011, as was being done for large mines.

Another contentious aspect of the FTLRP policy concerned the limited allocations of land to farm workers. By 2010, they comprised below 10 per cent of the official beneficiaries (see Chambati, Chapter 5). In fact, many farm workers were not merely victims of the FTLRP, but active agents who sought land through land occupations (Sadomba 2008), applying for land in local official channels, refusing to vacate farm compounds and 'squatting' on redistributed lands (AIAS 2007). Moreover, many farm workers were also mobilised against the land acquisitions by white farmers. However, by 2001 the GoZ policy was only able to prevent A2 farmers from evicting them from farm compounds and compelling such residents to provide labour, partially undermining the erstwhile farm labour-tenancy relationship. This was too late for about 45,000 former farm workers who had been physically displaced (Chambati and Magaramombe 2008). By 2004, however, the GoZ was encouraging A1 and A2 farmers to provide former farm workers with small 'subsistence plots' of about one acre per family (ibid), while in A1 areas this had happened, oblivious of official policy.

This policy contradiction essentially reflected the evolving class contest between farm workers and the new A2 capitalist employers. However, farm workers were also being mobilised by the competing political parties for votes. This constituency had tended to be influenced by the white farmers during the constitutional referendum and elections until 2002, as many farm workers were isolated in the LSCF compounds and did not have adequate social networks and political connections to the Communal Areas.

Persistent landlessness, however, also led to popular challenges of the official attempts to stabilise the FTLRP process and to normalise relations with capital. The retention of large-scale farms and agro-estates was met with active resistance by poor rural people, former farm workers, provincial elites and some land allocation officials (see Moyo 2011b). Land concentration was seen as continuing to exclude landless people and elites who aspired to gain land, while the economic crisis led various social forces in different localities to lobby for more land redistribution, leading to belated sub-divisions of parts of the estates which were 'illegally settled'. Some local authorities 'formally allocated' estate land to 'beneficiaries', contradicting the central government's evolving investment policy, while mobilising local grievances over foreign land ownership and exclusion (Moyo 2011b). The Development Trust of Zimbabwe (DTZ), whose 386,000 hectares were spared from expropriation because it is owned by indigenous people, remained 'illegally occupied' for a while, until the central government 'mediated' the dispute, leading the DTZ to cede over 60,000 hectares to settlers (GoZ 2009).

The Forest-Based Land Reform Policy of 2004, however, sustained its moratorium on 'illegal' land occupations of the state-owned Forest Commission of Zimbabwe (FCZ) and actually evicted occupiers. Similarly, most of the conservancies were still occupied by peasants by 2009 and they resisted government evictions, as also occurred in public parks such as Gonarezhou Park in southeast Zimbabwe (GoZ 2009). Local and central government officials were soon at loggerheads over plans to redistribute shareholdings under the Indigenisation Policy. Another dispute also arose between the GoZ and some former landowners over the ownership of wildlife itself, with the latter claiming compensation for them as private property (if their land was expropriated), while state officials considered them and other natural resources on LSCFs as public property (GoZ 2004).

By 2011, the policy conflict over agro-estates and conservancies reflected an intra-elite class struggle for access to shares coloured by ethno-regional sentiment (see Moyo 2011b), while the landless people who occupied them illegally were often pawns in such struggles. Moreover, the scale of publicly-owned agro-industrial estates which was retained was extensive enough to deprive many landless people of access to land and autonomous 'livelihoods'. The distributive value of retaining parastatal estates only began to materialise in 2011, when the foreign and domestic partnership over the ARDA estates started producing ethanol in Chisumbanje (Moyo 2011b), while promoting irrigated outgrower plots among neighbouring Communal Area families.

Much of the critique of the FTLRP land allocation process correctly identifies the government's failure to eradicate some multiple farm holdings and to sufficiently include former farmer workers, but it has failed to comprehend the complexities of its implementation. Few protested the limited redistribution of remaining estates and conservancies, although more recently there has arisen a correct query over the allocation of conservancy shareholdings, mainly to a few elites. However, the dominant narrative on the FTLRP has not been adequately informed by empirical data to comment substantively on its varied redistributive qualities.

Land redistribution and the reformation of property rights

An extensive land redistribution outcome

Of the 15 million hectares of land which in 1980 were controlled by about 6,000 white farmers, over 13 million had by 2009 been formally transferred to over 240,000 families of largely rural origin with widely differentiated landholding sizes within the A1 and A2 land redistribution schemes in various agro-ecological regions and provinces (Moyo et al 2009; Sukume and Moyo 2003). The FTLRP phase alone officially benefited 168,671 families on 9.2 million hectares, while we estimate that at least 20 per cent more families out of the official beneficiary families also have access to the redistributed land (Moyo 2011a).

Those families which acquired land through the A1 scheme hold an average 20 hectares of land each, including access to common grazing areas, while the peasantry now holds 70 per cent of all the agricultural land. The A1 allocations averaged 5 hectares of arable land in the wetter regions and 10 arable hectares in the drier regions, while access to grazing land per beneficiary varies between 7 and 60 hectares in wetter regions and 20 to 200 hectares in the drier areas. By 2010, the FTLRP had benefited over 22,000 new small-scale, medium-scale and large-scale capitalists with relatively larger plots averaging about 100 hectares under the A2 scheme and these beneficiaries grew in number from about 8,000 and 16,000 families in 2003 and 2008 respectively. While the white outgrowers around the sugar, coffee, tea and forestry agro-industrial areas were eliminated, redistribution substantially increased the number of smaller black outgrowers.

Table 2.2: Structure of agricultural landholdings and f arms, 1980 to 2010

Farm categories	Farms/households (000's)						Area held (000 ha)						Average Farm size (ha)		
	1980		2000		2010		1980*		2000*		2010*		1980	2000	2010
	No	%	No	%	No	%	ha	%	ha	%	ha	%			
Peasantry															
Communal	700	98	1,050	92	1,100	81.3	16,400	49.2	16,400	50.1	16,400	49.9	23	16	15
Old resettlement	-		75	7	75	5.5	-		3,667	11.2	3,667	11.2		49	49
A1	-				145.8	10.8	-		-		5,759	17.5			40
Sub-total	***700***	***98***	***1,125***	***99***	***1,321***	***97.6***	***16,400***	***49.2***	***20,067***	***61.3***	***25,826***	***78.6***	***23***	***18***	***20***
Middle farms															
Old SSCF	8.5	1.2	8.5	0.8	8.5	0.6	1,400	4.2	1,400	4.3	1,400	4.3	165	165	165
Small A2	-				22.7	1.7					3,000	9.1			133.9
Sub-total	***8.5***	***1.2***	***8.5***	***0.8***	***31.2***	***2.3***	***1,400***	***4.2***	***1,400***	***4.3***	***4,400***	***13.4***	***165***	***165***	***142***
Large farms															
Large A2	-		-		0.217	0.02					508.9	1.6			2,345
Black LSCF			0.956	0.08	0.956	0.07			530.6	1.6	530.6	1.6		555.0	555
White LSCF	5.4	0.8	4	0.4	0.198	0.01	13,000	39	8,161	24.9	117.4	0.4	2407	2,040	593
Sub-total	***5.4***	***0.8***	***4.956***	***0.4***	***1.371***	***0.1***	***13,000***	***39***	***8,691.6***	***26.6***	***1,156.9***	***3.5***	***2,407***	***1,754***	***844***
Agro-Estates															
Corporate	0.049	0.01	0.049	0.004	0.02	0.001	1,084	3.3	1,084	3.3	806.4	2.5	22,122	22,122	40,320
Parastatal	0.126	0.02	.126	0.01	0.106	0.01	379.2	1.1	379.2	1.2	295.5	0.9	3,010	3,010	2,788
Conservancy	0.008	0.001	0.008	0.001	0.008	0.001	958	2.9	958	2.9	247	0.8	119,750	119,750	30,875
Institution	0.113	0.02	0.113	0.01	0.113	0.01	145.7	0.4	145.7	0.5	145.7	0.4	1,289	1,289	1,289
Sub-total	***0.296***	***0.05***	***0.296***	***0.02***	***0.247***	***0.02***	***2,567***	***7.7***	***2,567***	***7.9***	***1,494.6***	***4.5***	***8,672***	***8,672***	***6,051***
Total	***714.2***	***100***	***1,138.8***	***100***	***1,354***	***100***	***33,367***	***100***	***32,726***	***100***	***32,878***	***100***	***46,7***	***28,7***	***24,3***

Sources: Compiled by S. Moyo and N. Nyoni from Moyo and Yeros (2005), GoZ MLRR (2009), GoZ 2009 data. Total hectarages do not tally due to rounding off, while some agricultural land is now urban residential and there is some missing data.

Over 18 per cent of Zimbabwe's 39 million hectares of land (including parks and forests) remained as state land. There was a decrease in agricultural land area owned by the state farming agency (ARDA) from 121,964 hectares to 115,601 hectares, despite the increase in its farms from 19 in 1999 to 24 in 2010. The area of large agro-industrial estates also declined marginally (Table 2.2). This persistent land concentration means control of water, wildlife and woodlands resources is also concentrated, ostensibly to preserve large-scale, specialised and integrated enterprises, to meet the state's agro-industrial development agenda (Moyo 2011b).

Nonetheless, about 13 per cent of Zimbabwe's entire agricultural land is now held by a range of middle-scale farmers (on A2 and Small-Scale Commercial Farms [SSCF]), while over 82 per cent is held by small farm producers (in the Communal Areas, in A1 areas and in informal settlements) and below 5 per cent is held by large farms and estates. This state of affairs stands in stark contrast to the pre-1980 and pre-1999 situations, when agricultural lands were predominantly held by the LSCFs and large agro-industrial estates, leading to the formation of a new tri-modal agrarian structure (see Chapter 6).

The FTLRP's redistribution process has led to a 'net transfer of wealth and power' from a racial minority of landed persons to various classes of black people, including mostly the previously landless and land-poor classes and a substantial number of low-income wage-earning and unemployed workers, as well as various categories of the petty-bourgeoisie. Most of the beneficiaries of FTLRP came from rural areas, being largely peasants from the Communal Areas, with a few coming from the farm worker populations in LSCF areas (Moyo et al 2009), while about 25 per cent of them were from urban areas (Moyo et al 2009). The latter settled mainly in peri-urban areas, reflecting the social pressure for land among lower-income working peoples and the organised demands of the petty-bourgeoisie. Fewer than 25 per cent of beneficiaries continued to be formally employed, largely in urban areas, while over 70 per cent of the land beneficiaries were unemployed people mainly from rural areas. These findings confirm the shift towards increased urban demand for land in the context of declining formal employment and wages under structural adjustment in the 1990s (Moyo 2000; Yeros 2002), as well as in response to the deepening economic crises during the 2000s.

Since the FTLRP did not redistribute most private and public estate lands, popular demand for such land has been widespread, with the GoZ reporting that over 100,000 people were still on its waiting lists for land redistribution (GoZ MLRR 2010). In addition to this pressure, unfulfilled demand for residential land in urban and rural areas, including among agricultural workers, is extensive. Persistent land concentration and class-based differentiation in access to land has generated a variety of new land struggles (as elaborated upon below and in Chapter 6).

Reformation of agricultural property rights and land tenure relations

The land tenure system was reformed by extinguishing most private property rights in agricultural land and broadening the effective occupation and use or ownership of the redistributed land through socially differentiated forms of land tenure provided to the A1 and A2 land beneficiaries. The latter get 99-year lease contracts providing land use rights to individual landholders. The A1 beneficiaries, on the other hand, receive statutory permits to occupy and use land in perpetuity as a family land right, which includes sub-plots to establish a homestead and for cropping and access to grazing woodlands used communally by a number of families.

The A1 permit provides similar forms of land rights to those provided under the 'customary tenure' system in Zimbabwe's Communal Areas, but their legal status differs as the state directly owns such land and controls the land allocation process using criteria which transcend those defined by memberships to given 'communities', despite the involvement of traditional leaders in nominating some of the land beneficiaries. The A1 land tenure relationship is thus a vertical legal and social relationship between the state and the families, which is complemented by elements of customary land administration practise, including empowering traditional leaders to enforce compliance with recommended land use and the management of natural resources and adjudication over land disputes, such as inheritance. However, the land permit tenure provided to resettlement beneficiaries before 2000 tended to have limited legal enforceability with regard to intra-family land rights at succession and, in particular, women's land rights were considered weak and open to abuse (Shivji et al 1998; Tshuma 1997).

It has recently been argued that the A1 land tenure conditions are insecure because they can easily be evicted by the state and they face numerous land disputes (World Bank 2009). In practise, around 20 per cent of the A1 and A2 farmers reported facing tenure insecurities, especially during the early years of the FTLRP when the GoZ was 're-planning' the land allocations and evicting or relocating some 'unlawful occupiers', including converting some A1 landholdings into A2 schemes. Existent land conflicts were focused on disputes over boundaries, competing claims to land allocations, the rights to use common natural resources and over access to 'inherited' on-farm infrastructures. Competing claims over allocation were more common in the peri-urban and higher potential agro-ecological districts. Neighbours, local authorities and former land owners were the key sources of such conflicts (Moyo et al 2009). These conflicts reflect the GoZ's land administration deficiencies, rather than the form of land tenure per se.

By 2006, about 16 per cent of the beneficiaries had been threatened with eviction once or more, particularly among the A1 families in better agro-ecological potential regions and peri-urban districts such as Goromonzi, where 31 per cent of the beneficiaries had faced eviction (Moyo et al 2009). Most eviction threats were, however, successfully resisted. But about 9 per cent of the land beneficiaries said they kept Communal Area homes because they feared eviction at some future point (Moyo et al 2009; Scoones et al 2010). Altogether, a minority (17 per cent) of the beneficiaries felt that their current forms of land tenure were too vague and that they needed formal documents to secure their rights. About 21 per cent of the beneficiaries reported encountering problems with access to credit because they did not have adequate land tenure documents and the majority of these were A2 landholders (Moyo et al 2009).

Permissory and customary tenures in A1 and Communal Areas, respectively, continue to be contested by some formal civil society actors (see Zimbabwe Institute 2007) and international agencies (UNDP 2008), who advocate for the greater individuation and tradability of land rights, ostensibly to enable their use as collateral for credit. About 30 per cent of the A1 beneficiaries represent a new generation of farmers. These farmers have, on average, higher levels of education, formal work experience and urban connections when compared to the rest of the peasantry. They demand more formal land rights, a limited role for traditional leaders in land administration and specified land inheritance procedures (Mhondoro Field discussions[8]). The extension of traditional leadership into newly redistributed areas has heightened such concerns, despite the fact that these new land user rights are derived from the state and not through custom. The GoZ argues against land tradability for fear of renewed land concentration, while the administrative requirements of registering the permit tenure is considered beyond its current capacity (Midzi and Jowa 2007).

The A2 scheme land beneficiaries received land 'offer letters', mostly as individuals rather than as married couples and these are expected to receive leasehold contracts (Utete 2003) which are legally recognised as a record of their 'real land use right'. By 2010, hardly 1,000 A2 farmers had received the lease contracts (Midzi and Jowa 2007). The lease covenants require beneficiaries to institute basic farm developments, minimum land utilisation and recommended natural resource management practises, as was the case with pre-2000 leases. Until 2009, the lease had required the beneficiaries to allocate 20 per cent of their land to growing food grains or to sell 20 per cent of their cattle to a parastatal agency (the Cold Storage Commission), but this was resisted by new landholders and eventually abandoned. The lessees are expected to pay rental fees, but the state has not been collecting these because beneficiaries resisted it, claiming that it is not affordable, while many refused to 'pay for repossessed land'. Failure to enforce this redistribution mechanism represents an inequity which is only recently being addressed through the introduction of a land tax (GoZ, MoF 2009 Finance Bill).

Some have argued that the main source of land tenure insecurity concerning the A2 leasehold is the limited land administration capacity to sustain effective land records, land registration and to survey subdivisions, as this weak capacity limits the enforceability of A2 land rights (Midzi and Jowa 2007). Some consider the slow issuance of leases to reflect the reluctance of the GoZ to secure lease tenure so as to retain political influence over the beneficiaries in line with prevalent neopatrimonialist perspectives (see Zimbabwe Institute 2007). Others perceive the Minister's powers to repossess and/or cancel the lease within only 90 days of notice to reflect 'autocracy' (Vudzijena 2007). This perspective is substantiated by an alleged limited scope to appeal against such decisions because of lack of an 'independent appeal system'. Yet the lease is subject to Zimbabwe's contract law and appellate courts. These courts have heard some cases in favour of landholders, although the appeal process is cumbersome and costly to the complainants.

Formal debates on the A2 lease mainly involve the elites, including farmers, bankers, government officials, consultants, donors and cabinet ministers. The primary disagreement has concerned whether the lease should be 'tradable' and land markets reintroduced, leaving the state with the residual role of regulating land markets and the judiciary to adjudicate disputes (see Mhishi 2007; UNDP 2008). This perspective on land 'tenure security' is informed by the belief that land collateral is the only instrument that can be used or is required to borrow from private banks, since banks putatively only recognise tradable land rights as 'security'. Thus an open land market is considered necessary to manage loan defaults (Mhishi 2007). Others favour tradability because it allows farmers who want to 'exit' to sell their fixed investments (Hungwe 2006[9]), or that land markets will remove the unproductive and speculative lease holders and promote investment and productivity (e.g. Sukume 2007; Vudzijena 2007). Others have proposed a gradual movement towards land markets through providing tradable lease holdings with an option to buy the land after some years (Zimbabwe Institute 2007). The UNDP had also proposed that the permissory and customary tenures be converted to tradable lease holdings with an option to buy (UNDP 2008).

The social and economic cost-benefits of these proposals have hardly been studied in light of contrary international evidence (see Migot-Adholla 1994). The GoZ leans towards a regulated land lease market which prohibits the sales of leases to multiple landholders and foreigners, rather than an open land market (GoZ 2009). Moreover, retention of existing land laws which provide government the right of first refusal in all agricultural land sales enables the GoZ to restrict the scope of 'buyers' and prevent land concentration, if it so wishes.

But the agricultural land tenure policy remains inconsistent because some freehold agricultural land tenures continue to exist in Zimbabwe on a few large-scale farms and agro-industrial estates which were not acquired by the state. It is estimated that about 1,000 black and white landowners still held such titles as of 2010 (GoZ, Ministry of Land, Land Reform and Resettlement 2010). Conservancies have been converted into 25-year leases, now involving new black beneficiaries as shareholders. The remaining white-owned farms continue to be subjected to state acquisitions for redistribution. More recently, most of the agro-estates were acquired by the state and leased back to the existing operators, who are required to off-load shares to indigenous persons. The operators are required to cede 51 per cent of their shareholding equity to indigenous persons.

Private, but informal, land rental markets, which involve about 25 per cent of the land beneficiaries, also shape the existing agricultural landed property rights as the practise has not been vigorously policed. Meanwhile, some public estates are being leased on unclear terms to tenant graziers, including to displaced corporate farming entities, former white farmers and some elite black cattle owners (GoZ, MLRR 2010).

Thus, Zimbabwe's land reform programme, following the Fast Track Land Reform process, has led to land redistribution to twice as many beneficiary families as had been planned in the early 1980s. This increase has substantially reduced the overall scale of land concentration and expanded the numbers of those involved in farming, despite the retention of agro-industrial estates and conservancies. This outcome has reconfigured the fundamental basis of settler-colonial agrarian relations, including racial discrimination and foreign domination in the control of land and consequently of labour relations. Private property rights and markets in agricultural land were substantially extinguished in favour of state- allocated land user rights, although advocacy for the commodification of land through freehold tenure, tradable leasesand land rental markets has been growing, especially among the middle class farmers and capital. Whereas class- based struggles over land ownership and labour remain central to on-going agrarian change, racial relations of land ownership were largely redressed, as were a wider range of socio-political and cultural relations in society, as we discuss next.

Identity and nationality issues emerging from the FTLRP process

The FTLRP redistribution also restructured a wide range of social relations of agrarian production and social reproduction, which had been constructed over 120 years of colonial and post-independence rule. Restructuring was accomplished by broadening social access to land and socialising the land tenure system, through expanding the public property regime, while retaining customary land tenures. Redistribution reversed racial patterns ofland ownership, broadened the ethno-regional distribution of land and marginally altered gender relations of access to land, as we elaborate on below (Moyo 2011a). Redistribution also unravelled the unequal political and racial power relations and the related labour agrarian relations associated with inequitable control over the labour of landless people, while reversing the territorial segregation which had resulted from monopolistic control over large tracts of land and natural resources by a few landowners. This process has broadened access to various natural resources connected to land control such as water, indigenous forestsand wildlife, which, for many people, has reinforced their spiritual connections to nature and their history, as embedded in the indigenous land tenure.

These changes represent social, culturaland symbolic progress on the steep and long road to social and structural transformation. However, the land redistribution process did not reverse all the regressive social and agrarian relations, especially those evoked by patriarchal hierarchy and unequal power relations, such that land access biases against women, youthand perceived immigrants persist. There are also notable degrees of exclusion from access to land based on ethno-regional and nationality difference, although we contend that new forms of class difference in land ownership pervade most of the inequalities in land ownership and labour relations which obtain todayand struggle over the unequal distribution of agrarian surpluses continues on a qualitatively altered plane (see Chambati 2011, Chapter 5).

Race dimensions of the land redistribution

The racial and foreign nationality dimensions of land ownership animated most of the social struggles embedded in the informal and formal politics of the land redistribution process, given the settler-colonial legacy of dispossession. The mobilisation of demands for access to land were often structured on the basis of indigeneityand, within provinces, struggles for access to land were often mobilised around ethno-regional identity and 'belonging'. Consequently, the relatively limited amount of land allocated to former white farmers has raised questions concerning whether the land reform was racially discriminating against white citizens, as some former white farmers have argued in courts and at the SADC tribunal. Indeed, it appears that the state recognises 'indigeneity' to Zimbabwe in a narrow sense, with some 'coloureds' arguing that they did not benefit sufficiently from the FTLRP. Some critics of the FTLRP even suggest that this outcome has limited the citizenship rights of white farmers (Hammar and Raftopolous 2003).

By 2007, about 725 white farmers were still holding over one million hectares throughout the eight provinces (GoZ, MLRR statistics). Many of these white farmers held relatively large landholdings in the drier southern provinces (Midlands, Matabeleland and Masvingo), as well as in one high rainfall province (Mashonaland East). More than 12 per cent of these farms were over 2,000 hectares in size, with only 295 of them holding below 500 hectares each. At the end of 2007, white farmers comprised about 4 per cent of the new large 'commercial' farming sector. Following the former farmers litigation at the SADC tribunal in 2007, in a context of the election campaigns of 2008, many of these farms, including those of the litigants, were expropriated. By the end of 2011, around 300 white farmers were still on some farms.

The white population is today (2011) below 75,000 or below half a per cent of the total population. The remaining white farmers constitute about 6 per cent of the new 3,000 large-scale farms, or 0.9 per cent of the 22,000 A2 farmers. This figure excludes the mainly foreign white-owned large agro-industrial corporations which hold large areas. As such, white representation is, in crude terms, proportionate, although more white farmers could have been retained on the basis of their farming skills and experience. It can hardly be argued that land ownership is racially inequitable. The FTLRP has led, instead, to a 'de-racialisation' of 'commercial farming' in general, as over 80 per cent of the new middle-scale and large-scale capitalist farms are owned by blacks (see Table 2.2).

Many would agree that more of the truly productive former white landowners could have been retained on smaller land subdivisions like those of most A2 farmers, based on the principle of right (following the 'one person one farm, within the maximum farm size range' policy), rather than as a matter of historical course derived from racial privilege. In the event, the negotiated land transfer process failed and the number of the former white landholders who would want to remain farming on downsized farms is not publicly known. Nor is the GoZ's intention on this specified, largely because litigations by former landowners over land persist. Some in government favour accommodating former white landholders on downsized leasehold farms, while others do not.

However, a number of black large-scale or A2 farmers have hired white farm managers, who were either former landowners or farm managersand they are paid salaries and/or in shares of the farm produce. This arrangement occurs especially in high value enterprises (e.g., tobacco, dairy, export beef, horticulture, bananas, etc) that require large financial commitments, specialised imported inputsand established export markets. It has been suggested that some former white farmers control A2 farms through a subletting system in which black owners operate as 'fronts', although the evidence on this is difficult to verify. Once established, some new black agrarian capitalists are forging alliances with white farmers and agro-industry and financial capital in business partnershipsand these increasingly demand the re-introduction of private property in agricultural land and advocate for neoliberal economic and agricultural policies.

More commonly, some former white farmers have moved up or downstream of the farming 'value chain' by acting as contract financiers and marketers or supervisors of the farming operations of contracted new farmers. As such, they have retained financial interest and market influence in areas such as poultry, tobacco, export beef and horticulture. But currently, such businesses are also required to sell 51 per cent of their shares to indigenous persons under the law enacted in 2009.

Overall, the racial redistribution outcome is seen by some proponents of land reform to be a historical necessity to resolve national questions of development and democratisation (see Moyo and Yeros 2005; Nkomo 2001). But since the racial balancing of capitalist farming attained by the FTLRP involves a component of intra-class transfer of land between races, it might be argued (as Borras 2005 generally does) that this aspect does not qualify to be considered redistributive. In our view, this is true with regard to the large-scale A2 farms.

Ethno-regional outcomes of the land redistribution

Agrarian relations are still coloured by power relations derived from ethno-regional identity. Land redistribution re-linked people with their original 'homes and ancestral spirits', providing scope to re-mobilise lineage based on ethnic ties and territoriality (Mazoe focus group discussion 2005[10]; Mkodzongi 2011). Often, these affinities were used to exclude those defined as not belonging, although the evidence suggests that this exclusion operates unevenly among the provinces and peri-urban areas (Moyo 2011a). There are claims that the FTLRP generally excluded some people from accessing A2 plots on ethno-regional grounds within the provinces, especially in the A2 schemes, although the available data required confirming this assertion is limited. More research on this question is required.

There are varying degrees of either under-representation or over-representation of some ethnic groups in land access in some provinces. One alleged case of over-representation relates to people from 'Mashonaland' within the 'Matabeleland provinces', particularly in the conservancies and peri-urban farms. There were also a few high- profile cases of individual A2 farmers who were evicted from A2 or their own commercial farms in some provinces on ethno-regional grounds. At the local level, it is sometimes argued that certain clans and lineage family groups considered 'founders of those communities' and their extended families gained more access to A1 land than others did. Some multiple A2 farm holders are said to use relatives as 'front' owners. More micro-studies are required to clarify such processes.

Since the mobilisation of land occupations was not even among the provinces, some groups of families of given ethno-regional backgrounds actually gained more than others. In Matabeleland South, for instance, families of land occupiers from the Midlands province were more actively engaged in occupying large swathes of the Debshan farms owned by the Oppenheimer family, to the chagrin of the landless peoples around Shangani and Fort Rixon (interviews in Bulawayo 2004[11]). War veterans from Matabeleland blamed this disparity on the 'passivity' of some 'communities' in the Matabeleland provinces during the land occupations, allegedly because of their allegiance to the opposition party (MDC), which allegedly opposed the land occupations.

Mashonaland West was reportedly a notable ethno-regional 'hotspot' on land allocations by 2004. A radical war veterans group in an association called Mwati alleged that some of the provincial ZANU-PF leaders at the time were targeting over 50 of the A2 beneficiaries for eviction, including senior ZANU-PF and state security leaders, putatively because they did not originate from that province. In the process, about 20 farms there were being 'hoarded' to avoid their going into the 'wrong hands' (AIAS Dialogue 2004). One of the chiefs in Mazowe district was, until 2008, involved in recurrent confrontations with the provincial land officials over their demarcation of a sizeable amount of land for A2 farms, protesting that too much land was being allocated to 'outsiders' through this scheme at the expense of locals in both the A2 and A1 schemes.

It is also common to hear people say that when they had applied for A2 land and presented their National Identity Card (the serial numbers of which indicate provinces of birth), if they came from the wrong provinces they did not get land. The A2 application process became ethno-regionally structured. Some elites, during the 2000 to 2002 period, sought land near the more ethnically cosmopolitan towns where they lived (particularly Harare and Bulawayo), while others sought land near their Communal Area 'home' (*kumusha*) districts. Eventually most applicants resorted to bidding for land where they 'belong', as conflicts increasingly arose between A2 beneficiaries who 'belong' to given districts and those deemed not to. Indeed, some local elites were at the forefront in advocating for the exclusion of 'strangers' and, during the height of land bidding (2000-03), there were many 'evictions' or unfair rejections of applicants on ethno-regional grounds. Consequently, access to land, particularly in the A2 scheme, tended to be partly shaped by ethno-regional affinities throughout all the provinces, although the policy was that the A2 scheme was 'national'. This process reflects simmering ethnicised intra-class competition for land, replicating incipient tendencies found during the 1990s (Moyo 1995, 1999). Some policy elites, however, believed that this approach was necessary to avoid the kind of ethnic clashes over land that Kenya has experienced (personal communication).

Ethno-regional and national identity in Zimbabwe (as elsewhere) tend, however, to be dynamic or malleable social constructs, having been shaped by colonial displacements and regional administrative fiat, including gerrymandered ethno-regional chieftaincies. This malleability has subsequently been mobilised by the provincialisation of political party mobilisation structures and demands on the state, vis-à-vis others. It can be expected that future land struggles may evolve around the reconstruction of ethno-regional identities, as land hunger increases again due to demographic growth (if unemployment persists), especially where the FTLRP beneficiaries have large landholdings.

Nationality, citizenship and migrant labour after land distribution

Foreign land holdings in Zimbabwe were relatively large in proportion compared to the scale of land grabbing that is underway elsewhere in Africa, but still less than is the case in South Africa and Namibia (see Moyo 2011b). As discussed above, the FTLRP slightly altered the pattern of foreign land ownership in Zimbabwe, most of which comprised private agro-industrial estates largely owned by transnational firms (TNCs)and some large-scale farms owned by declared foreigners, who were protected by Bilateral Investment Promotion and Protection Agreements (BIPPAs).

Before the FTLRP, there were about 267 (on 500,000 hectares) foreign owned farms differentiated in terms of size. These 267 farms were owned by individual persons from 13 countries, 65 per cent of whom were from nine European nations (mostly German, Dutch, Swiss and Italian); Americans owned 2 and 3.9 per cent of these 267 farms in terms of count and area, respectively. Foreigners from three countries in the South (South Africa, Mauritius and Indonesia) held 32.6 and 26.8 per cent in terms of number of farms and area, respectively. About 20 per cent of the foreign-owned farms ranged in size from 2,000 to over 50,000 hectares. Foreign land ownership was even more substantial in area terms when the agro-industrial estates and conservancies with over 450,000 hectares are included.

A commonplace, related, but not legally robust, view is that, since most of the former white landowners were British citizens (despite also holding Zimbabwean citizenship), their ownership of land in Zimbabwe also represented foreign land ownership. This argument suggests that land ownership inequality based on nationality was even more widespread. This has often been referred to as the 'kith and kinship' problem, which underlies the conflict between the Zimbabwean and UK governments regarding compensation for redistributed land (see AAPPG 2009).

After the FTLRP, the persistence of extremely large and under-used foreign-owned estates has tended to contradict the redistributive objective of land reform. It was the local grievances and agitation which drove their redistribution after 2006. Over 20 per cent of the foreign BIPPA- protected farms and substantial parts of most of the agro-industrial estates were settled by 'illegal' land occupiers after 2005. The FTLRP gradually whittled down the llarge- scale 'foreign- owned' estates, mostly in the Mashonaland provinces and Matabeleland. Smaller amounts of the core estate lands, such as the highly-capitalised agro-industrial sugar and tea estates in Masvingo and Manicaland, were also expropriated. The majority of these TNCs' agro-industrial estates are now owned by transnational firms of South African origin (e.g., Tongaat Hullett Limited) involved in sugar production conglomerates. Others, involving mainly white Zimbabwean and British capital such as Tanganda Tea Company and Ariston Holdings Limited, involved in tea and coffee, were only marginally expropriated (Moyo 2011b). White family- owned estates that were expropriated were involved in tobacco, livestock, wheat and grain (e.g., the Charter Estates, Ariston Holdings, Nicolle Brothers farmsand the Oppenheimer's estate). Even smaller sections of the foreign- owned forest plantations in Manicaland were expropriated during the FTLRP (Moyo 2011b). It is notable, however, that over 20,000 'illegal settlers' were still occupying the agro-industrial estates and conservancies as of 2010.

Another dimension of the nationality question and land reform is the limited degree to which potential land reform beneficiaries among the former farm workers on the LSCF farms gained land. About 30 per cent of these are of foreign parentage, but de facto citizens of Zimbabwe who qualified to benefit from the redistribution. About 10 per cent of the land beneficiaries were former farm workers, who were allocated A1 or A2 plots, some as farm workers. A few of these joined the 'land occupations' (Sadomba 2008; Moyo et al 2009), while others benefited as members of Communal Area structures, rather than as farm workers (for details, see Chambati 2009; Chapter 5). Some of them could not resettle in Communal Areas as they have limited kinship ties there and avenues for gaining land to settle there. Thus, many former farm workers still live within the redistributed farming areas and provide casual and permanent labour to A2 and A1 farmers, or were retained by the large farms and estates. Some new farmers tended to treat farm workers as thieves, given high levels of stock theft. Some labelled them foreigners and/or 'reactionaries' who had opposed the land reform. Moreover, farm workers' residential land rights and access to small food security plots continue, as before 2000, to be informal and tied to their provision of specific labour services to landowners. Around 13 per cent of the farm workers had experienced violent confrontations with the new landholders as a result of these land and employment conflicts as of 2007 (see Zimbabwe Human Rights NGO Forum 2007).

Land access, local politics and recognition

A number of the critics of the FTLRP have been pre-occupied with the operation of political patronage (and corruption) during the process, citing anecdotal evidence or using a limited number of redistributed forms to argue that land allocation policy decisions were shaped by allegedly ZANU-

PF allegiances. Indeed, neopatrimonial relations are considered *ipso facto* to define Zimbabwean state-societal relations, particularly as directed by the ruling party, ZANU-PF (see Raftopolous 2009). Whether the distribution was predominantly related to the political affiliation or 'connection' of the beneficiaries is an issue which most of the critics have not substantiated. To do so would require that access is systematically examined in terms of the correlation of farm size distribution with immutable political party 'affiliation' and rank within theseand this would need to be tested against the influences of class status on the nature of beneficiaries across the provinces.

Nonetheless, navigating the sensitive political dynamics associated with land occupations and government land allocation processesand facing organised opposition by white land owners and the hyper-attentive 'independent' media, was a crucial aspect of bidding for land. There are different dimensions of political connectivity which operated, locally and nationally, including to party leaders and officials and the bureaucracy in charge of land reform. A variety of other social affiliations, such as professional networks, familial or clan memberships and membership in social associations (including churches, etc.) were also important dimensions of land bidding (see also Scoones et al 2010). Access to A2 land allocations was more often shaped by the brokering of connections to the bureaucracy rather than the party hierarchy, while participation in the land occupations and negotiations with local leadership structures was a more broadly based networking process during the A1 allocations.

While most the leadership of the opposition party (the MDC) distanced themselves from the land reform process, many people who were not necessarily ZANU-PF voters sought land allocations. Among the A1 beneficiaries, many belonged to provinces and local areas that voted against ZANU-PF in 2005 and 2008. Moreover, the membership of political parties has been malleable over the last decade. Furthermore, it is a highly elusive variable to measure, given its political sensitivities. Multiple and tactical political allegiances were found to operate in reality (Mkodzongi 2011), with party allegiances 'instrumentalised' (e.g., Mhondoro District). The principal-agent dilemma, which makes for uncertainty in the pay-offs expected of patronage relations, as has been noted elsewhere (de Grassi 2008), generally applied during the FTLRP process. Moreover, it is not uncommon in Zimbabwean electoral campaigns for voters to follow the advice, 'eat or drink the offerings and vote wherever you want'.

There is a belief that the liberation war veterans, who led the land occupations, gained a substantial amount of the transferred land. To the contrary, many war veterans did not get landand those who did comprised less than 8 per cent of the land beneficiaries (Moyo et al 2009). Moreover, some members of the Zimbabwe National Liberation War Veterans Association (ZNLWVA) who did get land complained that they were being dispossessed of the land that they had occupied, largely because they were opposed to some elites getting larger plots (Sadomba 2008). Many of the war veterans who got land were peasants of limited means and education (AIAS 2007).

Contrary to the media- driven assumption that only cronies of the ruling party benefited from land redistribution, empirical data demonstrate that more 'ordinary' people (poor peasants, workers and the unemployed) benefited from land redistribution (see AIAS 2007; Scoones et al 2010; Matondi 2011). Over 75 per cent of the beneficiaries in A1 farms and/or the small-scale family A2 farm units were peasants with rather limited formal connections to political parties.

Party political mobilisation and fragmentation over land has largely been a petty-bourgeois accumulation contest over A2 land allocations, more so since the leadership of the ruling party had reigned in its radical elements, particularly among the lower-echelons of the war veterans association from 2004 (Moyo and Yeros 2007). Power struggles within the ruling party shifted from the radical nationalist political unity associated with the Fast Track period towards factionalism associated with the succession contest. Currently, ideological differences across political parties are focused on the privatisation of redistributed land, with ZANU-PF being focused on maintaining the peasantry's support, through providing access to farming inputs (see Chapter 6). But political mobilisation and fragmentation over access to land between ZANU-PF and the MDC and within the former have been less visible than other divisions. Factionalism has not fully degenerated along the Shona-Ndebele ethnic line, although this partly obtains around electoral tactics (Moyo and Yeros 2007), while the rural-urban divide continues to shape ZANU-PF vs. MDC political mobilisation. Despite this divide, party politics and ethno-chauvinism are more centred on differences over the regional distribution of state support to farming and class differences over the role of the state, although the fact of having promoted land redistribution still benefits ZANU-PF electorally.

Instead, local politics are being re-shaped by the changing local administrative and political power relations that resulted from replacing white farmers' control over land, territory and labour. Local influence is now more broadly diffused, but the landless are the most vulnerable. Territorial reconfiguration has enabled freer flows of people, goodsand services, particularly in labour mobility, popular petty tradingand non-farm activities. Their regulation is beyond the reach of under-resourced local administrative structures. Local power struggles mainly involve lineage-clan leaders, chieftaincies, farmer and social associationsand local bureaucracies. The powers wielded by war veteran leaders of the land occupations have been displaced. Sparse local government authorities are ill-equipped to regulate the expanded land administration regime and ubiquitous natural resources and mineral extraction. The hereditary chiefs demand more powers to fill these regulation gaps (Charumbira 2010[12]).

The FTLRP land redistribution partly addressed outstanding national questions, which the decolonisation process evaded. Many beneficiaries say land reform helped achieve what the liberation war was meant to bring (Sadomba 2008). The scope of sovereignty and self-determination in such areas is considered to have been enlarged for some, who refer to the reforms in terms of regaining territorial autonomy. Accompanying the transfer of land as an object is also the transmission of a range of intrinsic social values, such as the symbolic and spiritual value attached to land by many people in Zimbabwe. The colonial land grab had only recently undermined their social basis. Many beneficiaries interviewed claim that the land redistribution restored their identity (e.g., in relation to ancestral graves, etc) and has re-established their 'belonging' within the given territories.

Other social benefits are realised from the more equitable political control over the rural territory, including the freer movement of people, goodsand services. Small-scale mining (especially of gold) has also proliferated, reflecting the 'liberation of mineral resources, which had been hidden under the monopolistic LSCF farms' (Kwekwe interview 2006[13]). Unfortunately the state has tried (albeit unsuccessfully) to evict gold panners at the behest of elites facing farm labour shortages. While land disputes emerged in the newly- resettled areas over the competing interests of the new miners and farmers, social reproduction can nonetheless be based on broader access to natural resources and minerals. Larger sections of the rural population now seek autonomous self-employment in farming, natural resources extraction, mining and commerce in the hitherto secluded private properties (see Moyo et al 2009).

Increased access to these varied resources and more autonomous social reproduction indicate that much more has been gained from the land reform than the pre-occupation only with the material gains of access to farming land reveals. The land reform has also altered wider social relations in society by enhancing the recognition of the socio-political aspirations of various classes and social groupings in both material and symbolic terms, through the reconfiguration of the national and local political landscape and diversifying access to a wider rural economy. However, such gains were differentiated and contested along various identity- based cleavages such as ethnicity and genderand Zimbabwe's land reform process generated much external opposition, particularly from European actors whose land was repossessed.

Gender dimensions of the land redistribution programme

The FTLRP increased women's access to land ownership. This change occurred because women's advocacy groups (such as the Women's Land Lobby Group) were among the few NGOs who openly supported the land reform by demanding access to the expropriated land. This relatively redistributive gender outcome has endured despite the open and clandestine resistance it faces from some dominant patriarchs within the state apparatus, among some customary leadersand within some lineage household leaderships.

A larger proportion of black women, between 12 and 18 per cent, now own land in their own right (Buka 2002; Utete 2003; GoZ 2007), compared to the 4 per cent of white women who owned LSCF lands (Rugube et al 2003; Moyo 1999) and the 5 per cent of black women who controlled land in previous resettlement areas and communal lands. Other studies suggest that women 'beneficiaries in their own right' range between 10 and 28 per cent of the total (WLZ 2007). Women also benefited from access to land as spouses, implying a subordinate level of control over such land. Research is yet to quantify the quality of such access in newly- redistributed areas under the prevailing patriarchal system (see Jirira and Halimana 2008).

Gendered land access inequities mostly originated at the point when women who were applying for land faced bureaucratic bottlenecks in a male-dominated beneficiary selection processand because women lacked adequate information on selection procedures (Midzi and Jowa 2007). Nonetheless, the increased access to land by women in both A1 and A2 areas suggests a new dynamic in the gender relations in land access and use. Indeed, more women have been offered land in their individual right under the Fast Track Programme than in the past. Such women landholders do not seem to predominantly come from the 'vulnerable' groups, such as widows and divorcees, as obtains in communal and older resettlement areas.

Redistributive land reform did not, however, reverse the fundamental inequities evoked by patriarchal power relations. Land access biases against women, youthand immigrantsand the exploitation of female labour through male control of products are common (see also Makura-Paradza 2010). While more women secured their own land than in previous reforms, husbands still dominate agrarian transactions (WLZ 2007; Moyo 2011a).

Gender relations of land tenure generally entail repressive customary and policy- based patriarchal relations within communal and permissory tenure areas, in relation to inheritance rights, rights on divorce, control of incomeand so forth (see Chingarande 2008). Unlike the earlier resettlement permit, the draft A1 land permit proposes to strengthen women's land tenure rights and security, although less is proposed on the wider gender front. It provides for joint 'spouse ownership' registration on the permit. This means, in theory, that men can no longer legally dispose of the land use rights or exclude women (for whatever reason: separation, divorce, widows), without the consent of their wife. The official selection system for the A2 scheme scores women higher at the starting line, although this has not adequately increased their access. Reportedly (WLZ 2007), women tended to use their husbands' name in applying for land, with the expected or implied danger that the men in this process had 'gifted' control over land by women without an 'independent' physical address.

The majority of the 'offer letters' (in A2 schemes) and A1 permits have been issued in the names of the male spouses. There are also reports that some women, who had been given these tenure documents as individuals, had gone back to reverse this by getting government officials to re-issue them in their husbands' name, contrary to the policy of joint tenure (Ministry of Land officials personal communication). GoZ officials argue that the policy does not allow them to 'force' applicants applying individually or jointly to register jointly and/or to refuse the reversal of joint land offers, as this would be regarded as an intrusion into matrimonial affairs and because their powers to insist on joint registration are not enforceable in law. Thus, while officials are expected to encourage joint registration, gender-biased officials may not do soand the practise varies among provinces. Nonetheless, some women claim that land reform liberated them from the customary tenure rules typical of the Communal Areasand they are more optimistic about waging their land struggles vis-à-vis the state (WFLA 2009).

The changing agrarian structure and class dynamics after the FTLRP

Fast Track Land Reform also undermined the underlying class logic of settler-colonial agrarian relations founded on monopoly control over land which deprived peasants of land based social reproductionand compelled cheap agrarian labour supplies. While the close association of class differentiation with race in terms of the development of capitalist farming and the labour process had been substantially altered, new less racially- defined agrarian classes have emerged.

Zimbabwe's agrarian structure now comprises four relatively distinct farm categories, constructed historically by land dispossession and the racially discriminatory state allocation of varied landholding sizes through different forms of land tenure to different farmers (Table 2.2). The FTLRP has diluted this racial criterion, but access to varied sizes of landholdings is now differentiated mostly according to the social status of landholders, conceived in terms of their declared and perceived differentiated capacities to 'invest' in farming vis-à-vis actual and the official perceived perceptions of the need for land to enhance basic social reproduction. In practise, the agrarian structure and class content of land ownership is differentiated according to various processes of agrarian class formation and struggles, being based particularly on the varied intensities of wage-labour utilisation in relation to the persistence of landlessness (mainly among permanent agricultural labourers) and the limited access to inputs by the poorer peasants.

Historically, colonial land policies had led to the demise of the peasantry (also called small producers here), but from 1980 their fortunes rose somewhat. The FTLRP beneficiaries have expanded their numbers and land base, but the increase has also created conditions for the construction of more and widely differentiated classes of capitalist farmers in terms of land size and capital intensification. These changes have come at the expense of large-scale capitalist farmers. Furthermore, the FTLRP has retained the presence and influences of large agro-industrial capital, involved directly in production on estates or plantations (and conservancies). The scale of state farming lands also decreased, but remains influential, while landlessness among agricultural workers and aspiring peasants was slightly reduced, although it persists at a relatively lower level. This suggests that state policy deliberately promoted the emergence of a tri-modal agrarian structure, comprising the peasantry, capitalist farmers and plantation capital.

The emerging agrarian classes

Agrarian relations among the peasantry continue to be defined mainly by self-employment of family labour towards producing foods for auto-consumption and selling some surpluses. They have differentiated capacities to hire limited labour and some provide labour services to others. Most of the families hold customary rights to arable and homestead plots and common grazing areas in Communal Areas, while their A1 beneficiary counterparts hold state permits for similar family and common land rights. About 30 per cent of the A1 beneficiaries are made up of urban workers and a few former farm workers (Moyo et al 2009). Eighteen per cent of the land beneficiaries retain homes and plots in Communal Areas to diversify their reproduction and production using extended family resources (ibid). Sharing land with extended family members and sub-letting to others is commonly practiced.

In general, the new peasantry has smaller farm sizes than their A2 counterparts. Since the large majority of the beneficiaries had their origin directly in the Communal Areas, the pre-existing peasantry expanded its landholdings (Moyo and Yeros 2005; Moyo et al 2009). Re-peasantisation has therefore been a significant phenomenon of agrarian change under the Fast Track Land Reform, with the entry of urban working class elements into the A1 and resettlement schemes leading to the growth of a class of new petty commodity producers, which now account for 93.7 per cent of total new farming establishments since 2005 (Moyo and Yeros 2005).

There is substantial class differentiation within the peasantry, some of which is concealed inter alia by agro-ecological variation in sizes of land entitlements, off-farm incomesand other local processes of economic and political power-building reflected in inequalities in assets and influences over access to agricultural resources. The 'better-off peasantry', which historically comprised less than 10 per cent of the peasantry in Communal Areas (Moyo 1995; Maast 1996), has expanded. Under both adverse and positive economic conditions, peasant differentiation is expected to continue, as is the operation of informal land markets, within communal and newly-redistributed areas, due to differentiated access to labour, remittancesand land (Moyo and Yeros 2005).

The range of capitalist farmers also tripled in numbers, but their landholdings were down-sized by over 60 per cent. Over 31,000 middle-scale and large-scale capitalist farmers, most of whom are blacks, now exist. Two thirds of them got land as A2 beneficiaries in all provinces on varied land sizes (Figure 2.2) and with varied farm assets. These rely on relatively larger amounts of hired labour than on family labour (see Chambati 2011). They hold land through tenures amenable to market transactions, including mainly leases, while a few retain freehold title. The majority of them originate from the middle class, including currently or formerly employed professionals, small non-farm capitalistsand rural 'elites', including chiefs and some better-off peasants, as well as some working class people (AIAS 2007). Those with larger-scale farms tend to be better educated and linked to employment and business (Moyo 2011a) and are better placed to negotiate political power and mobilise resources. A few hire farm managers, while some rent land (AIAS 2007), claiming their land sizes are too small to be 'viable'. Some hold multiple farms (Moyo 2011a).

Figure 2.1 Farm size allocations by model

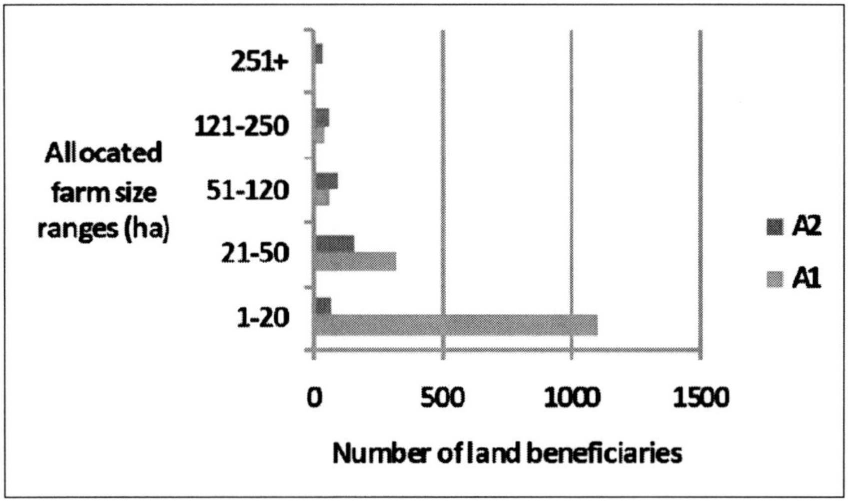

Source: AIAS Baseline Survey (2007)

A renewed 'merchant path' of agrarian social relations has emerged. This development can be seen in the increased number of urban professionals, the petty-bourgeoisie, bureaucrats and private sector managers occupying about 20 per cent of the acquired land. These small- and middle-sized capitalist farmers are, however, blurred by their differentiated levels of capital intensification and use of hired labour compared to own family labour. As will be discussed in Chapter 6, only a few of the new middle-scale farmland beneficiaries have access to the farming infrastructure and machinery necessary to intensify production, as this is partly influenced by the pace and direction of ongoing changes in the wider agrarian markets. Nonetheless, agrarian structural change has opened up diverse, 'productive'and 'non-racial' paths to rural social transformation.

Figure 2.2 Classification of all A2 farm sizes

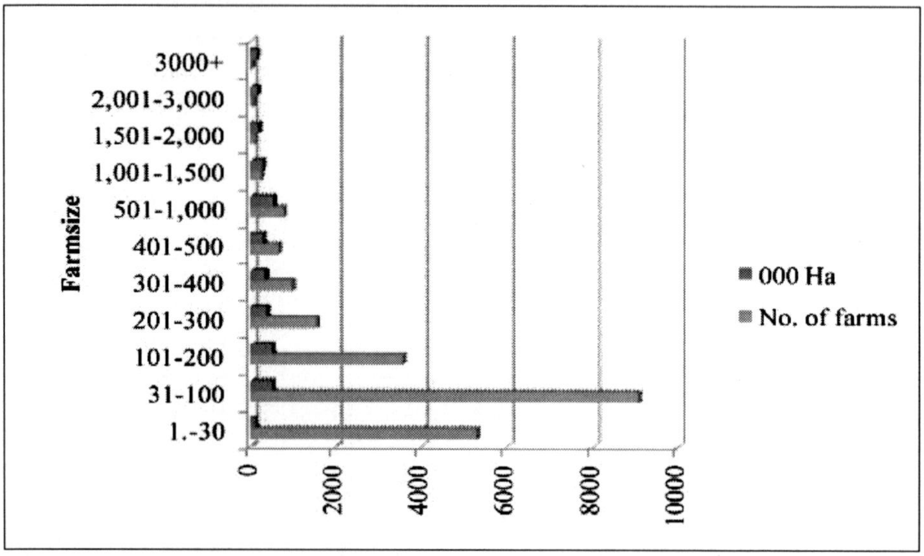

Source: Compiled by Sam Moyo from GoZ (2009)

There is an ongoing reconfiguration of the competing categories of the medium-sized and larger capitalist farmers, given that some of the middle farmers gained more access to state- subsidised means of production such as inputs, credit and machinery, largely because of their better contacts in the state and influence over the policy-making process (Moyo and Yeros 2005; Moyo 2011c). These small-scale and medium-sized farmers are broadly spread out among the provinces, with Mashonaland having created larger numbers of small farmers compared to Matabeleland, which had larger maximum farm size prescriptions associated with lower 'agro-ecological potential'.

At the same time, the land reform downsized, but retained, Large-Scale Commercial Farms by reducing their overall numbers, particularly among whites and by reducing their average landholding sizes. Prior to 2000, the large-scale capitalist farmers were highly mechanised, used agro-chemicals and fertilizers intensively and hired labour extensively. Post FTLRP, large capitalist farms now range in land size from between 300 and 500 hectares in the higher potential regions to 1,500 hectares in the drier areas, while corporate farms range in land size from 1,500 to over 5,000 hectares. Altogether, there are now a total of about 1,500 large scale capitalist farmers with average landholdings hovering around 1,000 hectares. These comprise black and white large-scale individual farmers, most of whom acquired land through the FTLRP or retained this through the Indigenous Farm Settlement Scheme before 2000. If we include those with over 300 hectares, around which point the number of farmers and size of area converge, we find that there are close to 3,000 mostly black farmers who can today be considered large-scale farmers (see Figure 2.2).

In addition to the redistribution of land, a range of on-farm infrastructures or 'immovables' such as farm houses, barns, bore holes, workshops, sheds, irrigation piping and off-farm infrastructures (dams, roads, electricity lines, etc.) left on the farms have provided additional assets to the beneficiaries. This infrastructure has broadened the horizon of physical infrastructures used by a wider range of smaller landholders, compared to the Communal Areas' conditions. One third of the A2 plots gained some of these infrastructures on an individual basis, as the rest got under-developed parts of farms, called 'plain' land. In A1 areas, most of this infrastructure is shared among the beneficiaries, including their being used as social amenities and as other public service facilities. The idea of irrigation using boreholes with motorised pumps and other mechanical handling structures has gained wider use among small landholders. But since access to these was highly skewed, this has sharpened class relations and other social differences.

The agro-industrial estates were reduced to 240 establishments, mostly owned by large-scale capital covering over one million hectares or 3 per cent of all the farming land (Moyo 2011b). They still hold freehold property in vertically integrated enclaves, including tourism conservancies and state estates. They hire large amounts of permanent and seasonal labour (Chambati 2011) and contract an expanding number of outgrowers. The latter comprise small and medium-ized land beneficiaries, relying on family and hired labour. The state has retained its plantations and expanded production through partnerships with capital. The indigenisation policy intends to redistribute the estates' shareholdings to locals. The estates owned by public trusts were largely spared, although some of them sublet their land out to 'elites'.

State farms have remained central to Zimbabwe's agrarian structure since the 1960s and about 10 of these were in place by 1980. Most of the state land was alienated by the colonial state from indigenous populations and some of the agricultural lands and forests were converted into freehold titles owned by the state, while some communal lands were converted into leasehold properties managed (and/or owned) by the state. Before the FTLRP, the state's ARDA farmed on 19 large-scale estates as a wholly state-owned private corporation. The ARDA estates were intended to promote agricultural 'development', but have tended to be run on a 'commercial' basis. Most of them were highly capitalised, especially with irrigation resources and were mandated to produce 'strategic' commodities, including those which were being imported (see Moyo 2011a, 2011b). By 2009, ARDA had increased its farms to 24, covering over 115,601 hectares (GoZ 2009) and had entered into estate investment partnerships with domestic and foreign partners. A number of Communal Area families who had occupied this land were deemed to be 'illegal' and evicted in 2010.[14]

Other state parastatals, such as the Cold Storage Commission (CSC), National Railways of Zimbabwe (NRZ), local authorities and security forces still own some of the large tracts of land, which they had before 1999, whose utilisation and investment arrangements are discussed elsewhere (Moyo 2011b). The land owned by these parastatals, however, decreased from 256,435 hectares in 1999 to 179,944 hectares in 2010, as some of the farms, particularly those owned by the NRZ (57 per cent), CSC (37 per cent) and local authorities (28 per cent) were acquired and redistributed during the FTLRP.

Landlessness and agrarian labour relations

The persistence of large-scale landholdings has meant the exclusion of potential land reform beneficiaries (Moyo 2011a) and fuels the 'illegal' occupation of lands (GoZ 2009). The policy of limiting access by former farm workers to redistributed land was partially motivated by the desire for cheap labour supplies. Landless people and poorer peasants still provide some farm labour services at low wage rates (Chambati 2011; Chapter 6). Many landless farm labourers reside precariously on new landholdings, perpetuating exploitation practises via tenancy. Since the pre-2000 relations of agrarian labour were undermined, agrarian labour shortages on capitalist farms have become common as the number of fulltime labourers has declined.

The current process of intensive labour exploitation, based on the existing manipulative labour recruitment system, is largely associated with the insecure labour tenancy among farm workers who are only allowed to live in the inherited or newly- built farm compounds, but are not provided with their own land, at least for housing. This potentially sustains the practise of labour 'bonding' and patronage, which enabled 'semi-forced' and 'unfree' labour conditions in a situation where the state provides limited rural labour protection, apparently due to capacity limitations and given that the agricultural labour unions seem to be off-compass.

Thus, despite the progressive outcomes of land redistribution, agrarian relations are imbued with salient struggles over access to land and labour. The inter-class imbalances in land redistribution (i.e., between A1 and A2 schemes), has to a significant degree, diluted the redistributive character of the FTLRP, as exploitative labour relations persist. Nonetheless, FTLRP redistribution has, in general, reversed the widespread subordination of labour which land dispossession had enabled. While the unequal control over labour power between the peasant (land-short, landless and poor) and the new farmer (landed and capitalist) remains, some of its exploitative features have been altered.

Land redistribution has allowed a large section of rural peasant society to use their own family labour for their own social reproduction in newly-gained farming lands, giving them wider ('livelihoods') options, including selling their labour within a differentiated farming set-up and in local non-farm activities. Although many of the former LSCF farm workers, half of whom were part-time peasants (semi-proletarian), did not gain access to land, even they have become relatively freer to sell their labour to many new small-to medium-sized farmers. This income is in addition to their access to small plots to cultivate 'subsistence' crops, albeit under poorly- defined or 'squatting' tenures.

Land redistribution has also opened new avenues for rural labour as workers or self-employed operators in small mining (especially that of gold), wildlife exploitation and fuel-wood and timber extraction. These activities have arisen as a result of the exposure of previously privately controlled natural resources to more people and the loosening of private property protection security systems. The dynamics of competing access to these resources is one of the main sources of land conflicts, as is their related effect of reducing farm labour supplies.

Emerging land marketisation and related class dynamics

Notwithstanding the formally declared absence of freehold lands in newly-distributed areas, a degree of land sales, sub-plot letting and plot rentals or informal land markets has been brewing (Sukume and Moyo 2003), despite the restrictions imposed by the current A1 and A2 tenures. Unequal land and labour relations are thus also being fuelled by tendencies towards land concentration through informal land rentals. About 25 per cent of the land beneficiaries sub-let or share their land (Moyo et al 2009) without official sanction. Some of these lessors lack production inputs or face social calamities such as illness or death (Mhondoro field interviews 2008). Others sublet land for speculative reasons or seek to maximise incomes from farming partnerships (ibid). Large-scale re-concentration of agricultural lands is, however, restricted by state ownership of redistributed lands and natural resources. Moreover, A2 farmers have no legal right to evict informally- settled farm workers.

Admittedly, few of the interviewed A2 farmers would openly declare that they were engaged in land rentals. Some of the A2 farmers declared themselves short of either arable or grazing land in relation to their current scale of production, claiming higher capacities to utilise rented land (Sukume and Moyo 2003, Moyo et al 2009). At times, such informal land rental arrangements were sanctioned by the local land authorities, especially where land is underused and/or on unallocated lands. About 26 per cent of the A1 land beneficiaries shared land, while only 15 per cent of the A2 households did so (Moyo et al 2009). Land sharing was extremely high in Kwekwe District (90 per cent) and Goromonzi District (27 per cent); while elsewhere, land sharing was on average below 15 per cent.

Two- thirds of such land sharing was with relatives and friends, as well as with adult family relatives. The rest of the land was shared with former and current farm workers, squatters, gold miners and millers. A few of the households even shared land with the former commercial farmers. Land sharing varied among the districts, with 8 per cent of the Kwekwe District beneficiaries reporting sharing land with former commercial farmers (Moyo et al 2009). At that time, the negotiated 'co-existence' between land beneficiaries and LSCF farmers was more common there. Furthermore, just over 20 per cent of the Kwekwe District beneficiary households reported sharing land with gold miners and millers, reflecting the 'gold rush' experienced there since the FTLRP. Further research is required to unravel the exchange relationships underlying such land 'sharing'.

Demands for the conversion of agricultural land to marketable tenures are thus a salient feature of the intra-elite and inter-class struggles over the control of land since the FTLRP. Some elites hold on to multiple and over-sized farms which they believe freehold tenure can protect, while other black and white elites rent land informally from some smaller landholders in Communal Areas, A1 areas and among A2 farmers and lease some state lands at little cost. Some even seek to evict economically and politically weaker landholders.

The privatisation of land tenure would also reinforce unequal access to natural resources such as water, woodlands and wildlife. Moreover, given continued landlessness, privatising property rights could enable new landholders to evict agricultural workers and prevent many land bidders from gaining access to the remaining Large-Scale Farms. This sequence of events would only reinforce the persistent super-exploitation of labour, which is the key motor of class formation. Thus, class relations continue to shape the politics of land, as the new capitalist farmers and the agro-industrial estates retain their advantage in the control of land and bidding for labour and lead the accumulation processes by virtue of their better access to other means of production (credit and technology) and influence over the policy-making process itself (see Chapter 6).

But demands for freehold land tenure among some A2 land holders do not only represent their desire for collaterable land tenure and a belief in the legal superiority of freehold tenure. They also reflect fears of a real threat to their relatively larger landholdings from the demands of the landless people who were emboldened by the radicalised FTLRP process to contest the social legitimacy of larger-scale landholdings. There is also a perception among the 'excluded' that many elites received more land than they can use at their expense, such that large-scale landholders are fighting on the back foot as popular ('illegal') land occupations persist.

However, the politics of land at the local level are being mobilised through experiences of struggles for land and agrarian production, despite cultural and ethno-regional differences, towards defending the new land rights, as well as access to farming inputs. As argued elsewhere (Moyo 2011c), various forms of local association, including churches, women's groups, farmer's clubs, local liberation war veterans' and collaborators' and other development associations, farm workers' associations, as well as kinship networks, shape such struggles (see also Chapters 4, 5 and 7). The formal politics of land is otherwise pre-occupied with intra-elite struggles for inclusion in the state's redistribution of land and input subsidies and over the distribution of the shareholdings of remaining conservancies and agro-industries under the indigenisation mantra, to the chagrin of landless people. Some politically influential and wealthier classes use administrative fiat, ethno-regional sentiments and sometimes force, to expand their landholdings. Many landless people continue to 'illegally' occupy land and poach resources, as local authorities, provincial politicians, chiefs and land movement leaders compete to mediate persistent land struggles. A few civil society organisations call for more land to be redistributed to farm workers, women and youths. As a result, the government is working on a land audit framework which, among other things, seeks to broaden the inclusion of these groups and of other politically excluded persons.

Concluding remarks

The cumulative outcome of three decades of land reform in Zimbabwe has been redistributive in scale and breadth, as popular access to agricultural land expanded despite the inequalities which remain. This outcome contradicts dominant narratives which allege that the FTLRP mainly benefited black 'elites' and cronies of the ruling ZANU-PF. State- derived land user rights are now dominant and (for now) they have contained the growth of inequitable land markets, while limiting the powers of the new capitalist farmers to re-establish exploitative farm-labour-tenancy, despite the continued exploitation of landless labourers. The unequal political power relations shaped by the erstwhile racially-monopolistic landholding structure have also been undermined, reversing the loss of local territorial sovereignty and spatial segregation. Broader access to natural resources, such as woodlands, wildlife and water, has enhanced the 'recognition' of wider societal rights and values.

However, substantial areas of large-scale foreign and state- owned agricultural estates were retained, ostensibly on the grounds of promoting agro-industrial and wider development, although their shareholdings are gradually being 'indigenised'. This anomaly circumscribes the scope for even more extensive land redistribution. Not surprisingly, such lands continue to be 'illegally' occupied by peasants. Second generation land questions include struggles for the redistribution of multiple and oversized landholdings held by the new capitalist farmers, to redress the exclusion of some former farm workers, landless peasants and various classes of women. Although land concentration exists on the margins of land ownership and the current redistribution promises more equitable agrarian change, it highlights the potential polarisation of agrarian reform policy. The tri-modal landholding structure has obtained substantial social legitimacy within the cross-class alliance that defends the land redistribution, although the failure of the state to ameliorate the gains realised by large-scale farmers is opposed at the popular level and among sections of the petty-bourgeoisie.

The Zimbabwe experience illustrates that, despite the hegemony of neoliberalism, radical land reform can be mobilised nationally and involve various classes, while transcending other divides such as rural-urban, worker-peasant and ethno-regional differences. Implementing radical land reform required decentralised structures and coherent leadership, which the liberation war veterans stimulated (see Moyo and Yeros 2007; Sadomba 2008). Both direct popular action through land occupations and state expropriations, led by the petty-bourgeoisie within and outside the state, shaped the actual redistribution process by balancing the demands of popular and other classes. This process arose from the long- drawn mobilisation of various socio-political forces around the historically specific national questions raised by settler colonialism and the post-independence constraints to social transformation imposed by neoliberalism. This experience differs from the standard formulations regarding the class basis of revolutionary transformations since the 1950s (see Borras 2005). In Zimbabwe's case, radicalism also provoked intensive external sanctions and support for a political democracy movement, aimed at effecting regime change, while mobilising against the radicalisation of neighbouring former settler-colonial-countries. This in turn provoked greater authoritarian rule and repression of the opposition, leading to a deeper democratic deficit.

The class struggles encountered also query over-generalisations about the neopatrimonial nature of African political and agricultural policy regimes (see de Grassi 2008), including claims that patronage dominated Zimbabwe's land reform. Instead, there was a deliberate balancing of class-based and ethno-regional demands by the decentralised land movements and bureaucracies involved in land allocations, which the 'central command' structures of the state and land movements monitored. Even the opposition's critique fomented such balancing. Ethno-regional pressures from some central and local elites to exclude 'outsiders' obtained, but they were not universally accepted. While urbanites influenced land allocations, this influence operated within limits, debunking the alleged 'urban bias' of the FTLRP process. Moreover, since capital was not totally ousted from Zimbabwe's agrarian political economy, internal class contradictions have enabled international capital to influence agrarian change as discussed in Chapter 6. In historical perspective and despite its many contradictions, the FTLRP placed brakes on foreign land grabbing and offered scope for progressive agrarian struggles in the former settler-colony.

Notes

1. Some reworked sections of this chapter have been published in S. Moyo, JPS 1, S. Moyo, JPS 2 and ROAPE.
2. This is called 'de-congestion' in Zimbabwe's policy, although the concept and targets are vague.
3. Hunzvi, 2000, Public statement on RBZ/ZTC, May 2000.
4. A1 targeted landless and poor famili es, providing land use permits on small plots for residence, cropping and common grazing. A2 targeted new 'commercial' farmers, providing larger individual plots on long lease to beneficiaries with skills and/or resources.
5. In 2002, the Government of Zimbabwe (GoZ) enacted a law to expropriate 'movable' farm properties (tractors, irrigation pumps, etc.) which were being warehoused by former farmers or exported to neighbouring countries, to prevent this and their "grabbing" by some new farmers (and criminals).
6. For instance, when 77 farmers took the GoZ to the SADC tribunal in 2007 (before the 2008 election), more farms were expropriated. Some land occupations around 2003 were considered to involve opportunists taking advantage of the political conflict and labelled a third force by some ZANU-PF leaders.
7. S. Moyo, 2005, Interviews, Centenary.
8. S.Moyo, 2008, field interviews, Mhondoro.
9. Hungwe, 2006, Personal Communication, Harare.
10. S. Moyo, 2005, Mazoe focus group discussion.
11. S. Moyo, 2004, Interviews in Bulawayo.
12. Chief, Charumbira, 2010, Statement at a COPAC meeting, Harare (4 January 2010).
13. S. Moyo, 2006, interview, Kwekwe.
14. It is reported that Garahwa (Chipinge) Communal Area residents had occupied this land and were essentially reclaiming it from ARDA, but that ARDA and some local chiefs and leaders had agreed for ARDA to develop land.

References

AAPPG, 2009, 'Land in Zimbabwe: past mistakes, future prospects', A report by the Africa All Party Parliamentary Group (AAPPG). (http://www.royalafricansociety.org/images/stories/pdf_files/aappg_report_land_in_zimbabwe.pdf), December.

AIAS, 2004, 'National Stakeholders Dialogue on Land and Agrarian Reform Report', Organized by the African Institute for Agrarian Studies in conjunction with the Ministry of Special Affairs in the Office of the President Responsible for Lands, Land Reform and Resettlement, 15–16 July, Sheraton Hotel, Harare, Zimbabwe.

AIAS Baseline Survey, 2007, 'Inter-district household and whole farm survey data-base', African Institute for Agrarian Studies (AIAS).

Bond, P., 2005, 'Zimbabwe's Hide and Seek with the IMF', *Review of African Political Economy*, Vol. 106, pp. 609-19.

Borras, S.M. Jr., 2005, 'Can redistributive reform be achieved via market-based land transfer schemes? Lessons and evidence from the Philippines', *Journal of Development Studies,* Vol. 41, No. 1, pp. 90–134.

Buka, F., 2002, 'A preliminary audit report of Land Reform Programme', Unpublished report, Government of Zimbabwe.

Chambati, W., 2009, 'Land reform and changing agrarian labour processes in Zimbabwe', MA Thesis, South Africa: University of Witwatersrand.

Chambati ,W., 2011, 'From land dispossession to land repossession: restructuring of agrarian labour relations in Zimbabwe', *Journal of Peasant Studies,* Vol. 38, No. 15, pp.1048-1068.

Chambati, W. and Magaramombe, G., 2008, 'The abandoned question: farm workers', in S.Moyo, K. Helliker, T. Murisa, eds., Contested terrain: land reform and civil society in contemporary Zimbabwe, Pietermaritzburg: S&S Publishers, pp. 207–239.

Moyo, S., Helliker, K., Murisa, T. eds, Contested terrain: land reform and civil society in contemporary Zimbabwe, Pietermaritzburg: S&S Publishers, pp. 207–239.

Chingarande, S., 2008, 'Gender and the struggle for land equity', in S. Moyo, K. Helliker and T. Murisa, eds., Contested terrain: land reform and civil society in contemporary Zimbabwe, Pietermaritzburg, South Africa: S&S publishers. pp. 275–304.

Cliffe, L., J. Alexander, B. Cousins and R. Gaidzanwa, 2011, 'An Overview of Fast Track Land Reform in Zimbabwe: Editorial introduction', Journal of Peasant Studies, Vol. 38, No.5, pp.907-938.

Cusworth, J. and J. Walker, 1988, *Land Resettlement in Zimbabwe: A Preliminary Evaluation*, London: ODA, Evaluation report EV434.

De Grassi, A., 2008, '"Neopatrimonialism" and Agricultural Development in Africa: Contributions and Limitations of a Contested Concept', *African Studies Review,* Vol. 51, No. 3, pp. 107–33.

DFID, [1997] 2000, 'Making Markets Work Better for the Poor: A Framework Paper', London: Department for International Development (DFID).

GoZ, 1986, *The First Five Development Plan, 1986-1990*, Harare: Government Printers.

GoZ, 1992, *Land Acquisition Act* (Chapter 20:10), Harare: Government Printers.

GoZ, 1998, *Land Reform and Resettlement Programme Phase II: A Policy Framework and Project Document*, Harare: Government Printers. September 1998.

GoZ, 2001, *Fast Track Land Reform Programme*, Harare: Government Printers (Draft 2000).

GoZ, 2004, *Wildlife Based Land Reform Policy*, Revised Draft, 2 April 2004.

GoZ, MLRR, 2009, 'Memorandum to Cabinet by the Minister of Lands and Rural Resettlement, Hon. H.M. Murerwa (M.P.) on the Update on Land Reform and Resettlement Programme', Ministry of Lands and Rural Resettlement, September 2009.

GoZ, MLRR, 2010, Various unpublished Government of Zimbabwe landownership data.

GoZ, MoF, 2009, Finance Bill (No. 3), Ministry of Finance, Government of Zimbabwe.

Hammar, H., Raftopoulos, B. and Jensen, S. eds, 2003, *Zimbabwe's unfinished business: rethinking land, state and nation in the context of crisis*, Harare: Weaver Press.

Herbst, J., 1990, *State Politics in Zimbabwe*, Harare: University of Zimbabwe Press.

JAG (Justice for Agriculture), 2006, 'Justice for Agriculture viewpoint on "99-year leases"'. (http://www.kubatana.net/html/archive/agric/061201jag.asp?sector=LANDR&year=2006&range_start=1). 1 December.

JAG/RAU, 2008, *Land, retribution and elections: post-election violence on Zimbabwe's remaining farms 2008*, A report prepared by the Justice for Agriculture (JAG) Trust and the Research and Advocacy Unit. May.

Jirira, K.M. and Halimana, M.C., 2008, *A gender audit of women and land rights in Zimbabwe,* Desk review final report, Prepared for The Zimbabwe Women's Resource Centre and Network (ZWRCN)'s Project for the European Commission and Action Aid UK on women, land and economic rights.

Johnson, D., 2009, 'Mamdani, Moyo and Deep Thinkers of Zimbabwe', (http://www.pambazuka.org/ en/category/comment/54039), 30 October.

Maast, M., 1996, 'The harvest of independence: commodity boom and socio-economic differentiation among peasants in Zimbabwe', PhD. Thesis, Roskilde University.

Makura-Paradza, G.G., 2010, *Single women, land and livelihood vulnerability in a communal area in Zimbabwe*, AWLAE Series No. 9, Wageningen: Wageningen Academic Publishers.

Matondi, P., ed., (2011), *Inside the political economy of redistributive land and agrarian reforms in Mazoe, Shamva and Mangwe districts in Zimbabwe,* Harare: Ruzivo Trust.

Mhishi, M.L., 2007, 'A general overview of the legal aspects of leasehold tenure, conveyancing and land transfer', Unpublished paper prepared for the AIAS Land Tenure Policy Review Dialogue, Harare.

Midzi, P. and Jowa, E., 2007, 'Land Administration in Zimbabwe', Unpublished paper prepared for the AIAS Land Tenure Policy Review Dialogue, Harare.

Migot–Adholla, S.E., 1994, 'Land, Security of Tenure and Productivity in Ghana', in J.W. Bruce and S.E. Migot-Adholla, eds, *Searching for Land Tenure Security in Africa*, Dubuque: Kendall/Hunt Publishing Company, pp. 169-198.

Mkodzongi, G., 2011, '*Land Occupations and the Quest for Livelihoods in Mhondoro Ngezi Area*, Zimbabwe'. Mimeo.

Moyo, S., 1986, 'The Land Question', in I. Mandaza, ed., *Zimbabwe: The Political Economy of Transition 1980-1986*, Dakar: CODESRIA.

Moyo, S., 1995, *The land question in Zimbabwe*, Harare: SAPES Books.

Moyo, S., 1999, 'The political economy of land acquisition in Zimbabwe, 1990–1999', *Journal of Southern African Studies*, Vol. 26, No. 1, pp. 5–28.

Moyo, S., 2000, *Land reform under structural adjustment in Zimbabwe: land use change in Mashonaland provinces*, Uppsala: Nordic Africa Institute.

Moyo, S., 2001, 'The Land Occupation Movement and Democratization in Zimbabwe: Contradictions of neoliberalism', *Millennium Journal of International Studies*, Vol. 30, No. 2, pp. 311-330.

Moyo, S., 2005a, 'Land policy, poverty reduction and public action in Zimbabwe', in A. Haroon Akram-Lodhi, S.M. Borras Jr, C. Kay, eds, *Land, Poverty and Livelihoods in an Era of Globalisation: Perspective from developing and transition countries*, London and New York: Routledge ISS Studies in Rural Livelihoods, pp. 344-382.

Moyo, S., 2007, 'Emerging land tenure issues in Zimbabwe', AIAS Monograph, Harare: African Institute for Agrarian Studies.

Moyo, S., 2011a, 'Three decades of agrarian reform in Zimbabwe', *Journal of Peasant Studies*, Vol. 38, No. 3, pp. 493-531.

Moyo, S., 2011b, 'Land concentration and accumulation after redistributive reform in post-settler Zimbabwe', *Review of African Political Economy*, Vol. 38, No. 128, pp. 257-276.

Moyo, S., 2011c, 'Changing agrarian relations after redistributive land reform in Zimbabwe', *Journal of Peasant Studies*, Vol. 38, No. 5, pp. 907–934.

Moyo, S., 2011d, 'Changing agrarian relations after redistributive land reform in Zimbabwe', *Journal of Peasant Studies*, Vol. 38, No. 5, pp. 939-966.

Moyo, S. and Yeros, P., 2005, 'Land occupations and land reform in Zimbabwe: towards the National Democratic Revolution', in S. Moyo, P. Yeros, eds., R*eclaiming the land: the resurgence of rural movements in Africa, Asia and Latin America*, London: Zed Books, pp. 165–208.

Moyo, S. and Yeros, P., 2007, 'The radicalised state: Zimbabwe's interrupted revolution', *Review of African Political Economy*, No. 111, pp. 103-121.

Moyo, S. and Yeros, P., 2009, Zimbabwe ten years on: results and prospects, 25 February.

Moyo, S., Chambati, W., Murisa, T., Siziba, D., Dangwa, C., Mujeyi, K.and Nyoni, N. 2009, *Fast Track Land Reform Baseline Survey in Zimbabwe: Trends and Tendencies, 2005/06*, Harare: African Institute for Agrarian Studies (AIAS).

Mtizwa-Mangiza, N.D and Helmsing, A.H.J., eds, 1991, *Rural Development and Planning in Zimbabwe*, Hants: Avebury.

Nkomo, M.J., 2001, *The story of my life,* Harare: Sapes Book.

Raftopoulos, B., 2003, 'The State in Crisis: Authoritarian Nationalism, Selective Citizenship and Distortions of Democracy in Zimbabwe', in A. Hammar, Raftopoulos, B. Jensen, S., eds, *Zimbabwe's Unfinished Business: Rethinking Land, State and Nation in the Context of Crisis,* Harare: Weaver Press.

Richardson, C., 2005, 'The loss of property rights and the collapse of Zimbabwe', *CATO Journal,* Vol. 25, No. 2, pp. 541-565.

Robertson, J., 2011, 'A Macroeconomic Policy Framework for Economic Stabilisation in Zimbabwe', in H. Besada, ed., *Zimbabwe Picking up the Pieces,* New York: Palgrave MacMillan. pp. 83-106.

Rugube, L., *et al,* 2003, 'Land transactions monitoring and evaluation of public and private land markets in redistributing land in Zimbabwe', Paper prepared for Land Tenure Center, University of Wisconsin-Madison and Centre for Applied Social Science, University of Zimbabwe.

Sadomba, W., 2008, 'War veterans in Zimbabwe: complexities of a liberation movement in an African post-colonial settler society', PhD. Thesis, Wageningin University.

Scoones, I., N. Marongwe, S., Mavedzenge, B., Murimbarimba, F., Mahenehene, J. and Sukume, C. 2010, *Zimbabwe's Land Reform: Myths and Realities,* Harare: Weaver Press.

Shivji, I.G., S. Moyo, W. Ncube and D. Gunby, 1998, 'National Land Policy Draft', a draft discussion paper prepared for the Government of Zimbabwe, Harare: FAO and Ministry of Lands and Agriculture.

Short, Clare, 1997, 'One bad letter with long-lasting consequences: A letter from Britain's then secretary of state for international development, Clare Short to Zimbabwe's then minister of agriculture and land, Kumbirai Kangai, repudiating Britain's colonial responsibility for land reform in Zimbabwe', *New African,* May 2007.

Sukume, C., 2007, 'Land Tenure Policy-Overall Conceptual Framework and Issues', unpublished paper prepared for the AIAS Land Tenure Policy Review Dialogue, Harare.

Sukume, C. and Moyo, S., 2003, *Farm sizes, decongestion and land use: implications of the Fast Track Land Redistribution Programme in Zimbabwe,* Harare: AIAS Mimeo.

Tshuma, L., 1997, *A matter of (in)justice: Law, state and the agrarian question in Zimbabwe,* Harare: SAPES Books.

UNDP, 2008, 'Comprehensive Economic Recovery in Zimbabwe: A Discussion Document', Harare: UNDP.

Utete, 2003, *Report of the Presidential Land Review Committee,* under the chairmanship of Dr Charles M.B. Utete, Vols. 1 and 2: Main report to his Excellency, the President of the Republic of Zimbabwe, Presidential Land Review Committee (PLRC). Harare: Government Printers.

Vudzijena, V., 2007, 'Investment and Productivity Prospects in the A2 Resettlement Areas', Unpublished paper prepared for the AIAS Land Tenure Policy Review Dialogue, Harare.

Weiner, D., Moyo, S. Manslow B. and Okeefe, P. 1985, 'Land Use and Agricultural Productivity in Zimbabwe', Journal of Modern African Studies, Vol. 23, No. 2.

WFLA, 2009, 'The socio-economic and political impact of the Fast Track Land Reform Programme on Zimbabwean women: research extract', Mimeo.

Whitesun Foundation, 1983, Land Reform in Zimbabwe, Harare: Whitesun Foundation.

Willems, W., 2005, 'Peaceful demonstrators, violent invaders: Representations of the land question in the Zimbabwean press', *World Development*, Vol. 32, No. 10, pp. 1767-1783.

WLZ, 2007, 'A rapid appraisal in select provinces to determine the extent of women's allocation to land during the land reform and resettlement programme, 2002-2005', Unpublished paper, Women and Land in Zimbabwe (WLZ), Harare.

World Bank, 1982, 'Zimbabwe: small farm credit project', Staff appraisal report No. 3888, Zimbabwe: Southern Agriculture Division, Eastern Africa Projects Department; Washington, DC: World Bank.

World Bank, 1999, 'Project Appraisal Document on Zimbabwe Land Reform Support Project', Report No. 19618-ZW, July, Washington, DC: World Bank.

World Bank, forthcoming, Zimbabwe Agriculture Policy Paper, Washington, DC: World Bank.

Yeros, P., 2002, 'The political economy of civilisation: peasant-workers in Zimbabwe and the neo-colonial world', PhD. Thesis, University of London.

Zimbabwe Human Rights NGO Forum, 2007, 'Adding insult to injury: a preliminary report on human rights violations on commercial farms, 2000 to 2005', A report prepared by the Zimbabwe Human Rights NGO Forum and the Justice for Agriculture Trust [JAG] in Zimbabwe. June 2007.

Zimbabwe Institute, 2007, *Understanding the impact of the Fast Track Land Reform Resettlement Programme*, Cape Town: Zimbabwe Institute.

ZIJRI, 2001, Zimbabwe Joint Resettlement Initiative (ZIJRI) Proposal, Harare: Commercial Farmers Union.

3

A Decade of Zimbabwe's Land Revolution: The Politics of the War Veteran Vanguard

Zvakanyorwa Wilbert Sadomba

Introduction

The Zimbabwe state, governed since 1980 by a nationalist elite with origins in the liberation movement, has experienced complex dynamics and changes regarding class relations and power in a post-colonial settler economy. The state reached a climax of political polarisation during this last decade, from 2000 to 2010. In the first two decades of independence, the ruling nationalist class had enjoyed an alliance with settler capital forged during peace negotiations in 1979 at Lancaster House (see Horne 2001[1] and Selby 2006). The alliance antagonised and negated the aspirations of the liberation struggle expressed symbolically and concretely in terms of reversing a century old grievance over unequal colonial land ownership structures. War veterans were an 'embodiment' of this anti-colonial demand (Kriger 1995), although a scattered peasant movement had dominated land struggles until 1996 (see Moyo 2001). These war veterans, as a social category, were constituted by a movement of former military youth and so-called former refugees, whose nucleus were fighters of the Zimbabwe's liberation war[2]. The conflict between the neocolonial state on the one hand and peasants and war veterans, on the other, intensified during the 1990s. The state had successfully managed to suppress the organisation of war veterans during the 1980s. However, in 1997, it conceded to provide for their welfare and financial demands and, as part of the conditions of a truce entered between war veterans and President Robert Mugabe, promised to redistribute land. The state did not honour this promise. Under war veteran leadership, the land movement then became more militant, challenging settler capital, the state, ZANU-PF elites and President Mugabe from 1998, generating an unfolding drama of sharp class conflicts in the polarisation of land politics and state/society relations.

The post-independence era has largely exhibited the inherent contradictions of the state/society relations found in neocolonial and settler dominated capitalist settings (Sadomba 2008). Many scholars of Zimbabwe's crisis have not identified this contradiction due to the failure to understand the class position of the state itself, a critical point which Borras (2001) observes in the case of the Philippines. This omission leads to an erroneous assumption that 'the state is autonomous in making policy choices...even when these run counter to the interests of the dominant classes or groups in society' (Borras 2001: 545). The Zimbabwean state, being essentially a bourgeois neocolonial establishment, promoted interests and values that were opposed to those of the peasants, rural and urban workers and marginalised war veterans who comprised the land movement. War veterans led the land movement, culminating in intense political and social conflicts based on divergent class interests, challenging settler capital, the emerging black bourgeoisie and the ruling elite, transforming it into a powerful revolution. Analogously in Latin America, the state mediated land conflicts with a bias towards the elites and capitalists, against the poor and marginalised. The 'state apparatus' became a 'source of accumulation' for both 'state actors who are also businesspersons/ landowners and for capitalists who are not formally part of the state', but are nevertheless favoured by the state through 'subsidies and protective trade policies' (Das 2007:5). The Zimbabwean state is no different and the land revolutionaries continued to challenge it at different levels and with varied intensity. Whether 'a revolutionary rupture with the capitalist system is on the cards' and under whose leadership remains, an intriguing question which this chapter partly explores (Moyo and Yeros 2007: 105).

Historical background

The 1970s guerrilla war that ended with the Lancaster House negotiations in 1979 was led by veteran nationalists, except for a brief two year period from 1975-77 when guerrillas disowned nationalist parties and fought as a united movement of the Zimbabwe People's Revolutionary Army (ZIPRA) and the Zimbabwe National Liberation Army (ZANLA), under the Zimbabwe People's

Army (ZIPA). Owing to partisan cleavages of its leaders, ZIPA fractured (Sadomba 2011), giving way to nationalist control of the guerrilla movement again. A decisive military intervention by the Front for the Liberation of Mozambique (FRELIMO), led to the incarceration of ZIPA leaders in Mozambique and a simultaneous rise by Robert Mugabe to the helm in 1977[3]. The political myopia of ZIPA leaders and their failure to cope with the internal power dynamics of the liberation movement is a weakness that has pervaded and sustained itself in the war veterans' movement, with disastrous consequences to the general liberation movement. ZIPA's demise lay in its failure to transcend partisan cleavages and the self-centred nature of some of its leadership like Mhanda himself, which led to continued division between the foundational liberation armies of ZANLA and ZIPRA (Sadomba 2011). Worse than ZIPA's lack of vision, war veterans have not put their whole weight into internally transforming the ZANU-PF movement into a revolutionary party and state power has remained, therefore, in the hands of an elite bourgeois leadership.

Peasant occupation movements, which had gained momentum during the liberation war, intensified after independence, with the state reaction continuously changing. Between 1980 and 1984, the state appeased the peasants through 'accelerated' regularisation of land occupations. The short-lived resettlement programme (mainly confined to marginal agricultural land) was later followed by the brutal suppression of land movement actors, now labelled squatters (Moyo 2001). From 1985, government resettlement policy tilted towards allocating land to what it considered capable small farmers and by 1990 towards the black elite and state functionaries who were allocated large commercial farms at the expense of the land movement and marginalised war veterans (see Moyo 1995).

During the 1980s, war veterans' (particularly the nucleus of former guerrillas) attempts at forming an organisation to protect their interests and those of the liberation war were systematically repressed by the state and, in particular, the dissenting former ZIPRA guerrillas, who unfortunately were backed by Apartheid South Africa[4], were thwarted. After the Unity Accord was signed between ZANU-PF and PF-ZAPU in 1987, war veterans reorganised themselves and formed the Zimbabwe National Liberation War Veterans Association (ZNLWVA) in 1989, leading to various episodes of rebellion against the state, President Mugabe and ZANU-PF during the 1990s. They put forward demands to President Mugabe at an inaugural meeting held at Chinhoyi, which can be summarised as a return to the liberation agenda and pressure for their recognition and benefits. The President did not honour any of those demands and war veterans started to stage street demonstrations demanding to meet him, but President Mugabe turned his back on them. Towards the end of 1997, a truce between President Mugabe – representing ZANU-PF and the state – and war veterans was negotiated, signalling a new political era of war veteran dominance in politics. War veterans were each awarded Z$50,000.00 as disbursement for unpaid demobilisation backdated to 1980. This agreement for payment has been seen wrongly by scholars and analysts such as Raftopoulos and Mlambo (2009), Amanda Hammar (2009) and Geoffrey Feltoe (2004) as a process of cooptation of war veterans.

It is this victory by the war veterans over ZANU-PF, the state and President Mugabe that undermined the 1979 Lancaster House compromise and 'accelerated the deteriorating relationship between [white] farmers and the state', eventually leading to the demise of the 'alliance' between settler farmers and the ZANU-PF elites (Selby 2006: 257). White commercial farmers 'resolved to "internationalise" the issue' 'in the hope that external awareness would arbitrate the process' (Selby 2006: 257), but the Zimbabwe Congress of Trade Unions (ZCTU) was much ahead of them in this strategy, for:

> A month before the New Labour Party was voted into power in Britain … European Trade Unions ... [through] the Danish Trade Union Council posted Georg Limke in late 1996 to … turn the trade union movement in Zimbabwe into a political party. Therein lay the evolutionary roots of the Movement for Democratic Change (MDC) (Mudenge 2004a: 10).

The ZCTU had been fighting for autonomy from the state from the beginning of the 1990s and, by aligning with international donors, they plunged Zimbabwe's politics into the neoliberal regime change agendas designed by western powers. This tendency of civil society is described by Masunungure as:

> ... shackled[,] ... characterised by a debilitating irony: it agitated for autonomy vis-à-vis the state but did not enjoy such autonomy vis-à-vis international donors and partners. Because of the financial and material umbilical cord between the two unequal partners and the asymmetrical relationships attendant thereof, the Zimbabwean civil society community absorbed the international donor agenda hook, line and sinker (2008: 64).

In this context, I have argued that the MDC emerged to replace ZANU-PF elites as surrogates to the Lancaster House 'alliance' (Sadomba 2008: 165; Sadomba 2008a: 165; Sadomba 2008: 8; Sadomba 2011: 279). Rather than ending, the Lancaster House alliance went through a metamorphosis. This was the beginning of a new alliance and new mix of players to serve the same purpose, though that of rescuing the interests of settler and international capital that was now threatened by rebellious veterans and, later, by a tsunami of the land hungry and a reserve army of the unemployed. The nationalist elites had outlived their usefullness as they could not manage to hold back these revolutionaries. This reconfiguration of political forces involved the MDC, white farmers and settler capital. These players chose to internationalise the issues and drifted away from domestic mediation, thereby undermining state autonomy and compromising on national sovereignty, resulting in adversarial relations that led to complex political diplomacy. The MDC was appealing to former colonial powers, rather than regional powers such as the Southern African Development Community (SADC) or the African Union (AU) and it sharpened conflicts between African states and the white world of Europe, the USA and Australia. The mobilisation of European states, international donors and financial institutions and the western media in defence of international and settler capital, against the Zimbabwe state, the ZANU-PF ruling party and President Robert Mugabe, was another turning point in the country's politics. Conflict between the two groups has led to a theatrical diplomatic antagonism attracting worldwide attention and polarisation.[5]

Two decades of President Mugabe's leadership had therefore reversed the ideological gains of the liberation war and effectively protected the interests of white capital. Simultaneously, it suppressed all voices of dissent with appeasement of the peasants through a cosmetic resettlement programme. Power became more concentrated and centralised, crushing PF-ZAPU, which was a potential alternative to the leadership of the liberation movement. With this, a de facto one-party state under President Robert Mugabe reigned and the Lancaster House alliance consolidated. Major internal opposition to the ruling elite developed in the 1990s. Coupled with an economic downturn and social strife, the stage was set for a war veteran-led land occupations revolution, which took shape in 1998 and spearheaded occupations up to and beyond the eruption of 2000. From this period, the position of the state in relation to the land movement shifted many times, as did the position of President Mugabe and ZANU-PF, with the MDC alliance[6] acting as a catalytic agent (Andrew and Sadomba 2006; Sadomba 2011; McCandless 2011).

The land occupations revolution of 1998-2002

Most scholars of Zimbabwe's land conflicts (e.g., Hammar 2009) do not distinguish the various phases and salient points concerning the unfolding land struggle from 1998, except Moyo (2001) who identified four phases of land occupations from 1980, including a variety of policy shifts from 2001 and 2003 (Moyo 2005). This paper argues that there were two distinct land occupation periods, which were 1998 to 2002[7] and thereafter.

When war veterans forced the state and President Mugabe to the negotiation table in 1997, they agreed that white commercial farms would be 'seized' and distributed to the land hungry, with 20 per cent of the land for the war veterans. Government reacted by immediately designating 1,471 commercial farms for compulsory acquisition (Moyo 2001). This was contested legally by white commercial farmers and no land redistribution materialised (Moyo 1999). War veterans reacted in isolated group initiatives by mobilising traditional leaders across the country (Interview K 2004[8]), leading to more than 30 war veteran-led occupations (Sadomba 2004; Marongwe 2003; Moyo 2001), including the most outspoken Svosve occupations in mid-1998.

The veterans-led land occupations were qualitatively different from previous peasant led land occupations in a number of ways, but mainly because this new leadership intensified the land struggles to a level of deep class antagonism and strategized its organisation. First, the occupations were militant – being confrontational where white farmers resisted – clearly borrowing 'aggressive' and surprise attack tactics from the guerrilla experiences of the armed struggle. Second, the land occupation movement became potentially more socially inclusive, by destroying the rural/urban divide that characterised previous land occupations and by incorporating state organs where war veterans were concentrated, such as the uniformed forces. In this sense, the 'local orientation' of the peasant land movement was 'transcended and peasants entered national politics', developing an 'alliance with the workers' (Das 2007:10). Thirdly, war veteran leadership introduced new ideologies, liberation war metaphors and symbolism and guerrilla tactics. These tactics included operating in small independent units that were autonomous, politicising the masses and establishing bases as command centres. Fourthly, war veteran leadership was vital in challenging the monopolisation of the cultural capital of the liberation war and history by nationalist politicians and ZANU-PF. All in all, the war veteran leadership of the land movement radically shifted 'grassroots agency' from being merely 'confined to and aimed at a power structure within its own immediate vicinity' to challenging the 'state at the national level' where class 'power is concentrated' (Das 2007:8).

The land occupations of 1998 were targeted at farms which the government of Zimbabwe had designated for acquisition, but could not acquire due to both litigation commenced by white farmers and to lack of will, according to war veterans (Interviews P 2000, DTM 2000-2012, DM 2000-2008, Muchaneta 2004[9]). War veterans mobilised peasants and occupied the farms, challenging the state laxity regarding solving inherited racial imbalances in land ownership. It is important to remember that these land occupations constituted a building up of successful confrontation with the state, having pinned President Robert Mugabe[10] to negotiate and agree a truce. As such, land occupations were a continuation of the war veterans' challenge to the state, which had started at the end of the previous decade and sharpened at Chinhoyi in 1992 (Sadomba 2008). Few scholars have noticed this linkage (e.g., Moyo 2001; Sadomba 2008, 2011). The land occupations were, therefore, a manifestation of class struggle and war veterans were quite conscious of this aspect.[11]

The state reaction to the 1998 land occupations was draconian. The Zimbabwe Republic Police (ZRP), the Central Intelligence Organisation (CIO) and the white- dominated judiciary attacked the land revolutionary actors by torching their shelters, scattering them in nearby mountains and bushes and finally arresting and slamming them with all sorts of judicial punishments (Sadomba 2008). White farmers naturally aligned with the state to suppress the occupations. Some ZANU-PF elites, the state and President Robert Mugabe, were at this stage in the middle of the road and therefore were protecting their class interests and not necessarily serving the Lancaster House alliance.

The activism of war veterans up to the 1997 truce had created scepticism within ZANU-PF 'elites' and President Mugabe and their relationship remained strained throughout the early occupation period of 1998-2000 (Sadomba 2008). However, the entry of war veterans into the farms revealed political strategies by the white commercial farmers, particularly their 'internationalisation of the issue', which was seen as an affront by war veterans.[12]

The labour movement represented by the ZCTU evolved into a constitutional movement, the National Constitutional Assembly (NCA) led by ZCTU President, Morgan Tsvangirai and some University of Zimbabwe activists such as Lovemore Madhuku. The NCA was vocal about the need to have a new Zimbabwean drafted constitution, citing weaknesses in the Lancaster Constitution of 1979. It forced government to institute a Constitutional Commission in April 1999, chaired by Justice Chidyausiku. The greater part of 1999 was therefore spent working on the new constitution and land occupations were carried out with limited media publicity. However, war veterans believed that the new constitution would have clauses that would allow land to be expropriated from the white farmers to resettle the land hungry. As such, war veterans had particular interest in the constitution for this purpose.

On September 11, the MDC was finally formed, evolving from the labour and later, constitutional movements, led by Morgan Tsvangirai.[13] The formation of the MDC had catalytic effects on the unfolding revolution led by war veterans. It seems that throughout 1998 and the first three quarters of 1999, neither the opposition nor the constitution preoccupied the state or President Mugabe. Considering that the NCA and other civil society movements had remained just pressure groups without a political party formation, it could be safely concluded that the main focus of the elite and the state was the land revolution that was spreading. On the one hand, the white farmers took government to court and resisted compulsory acquisition; on the other, war veterans started mobilising peasants to occupy land nationwide (Marongwe 2003). In 1998, a donor conference on land was held in Harare at a time when the occupations were raging and the state was fire- fighting them. It is also important to note that, during this period, war veterans were antagonistic to President Mugabe, threatening to disown him as the patron of their association (Sadomba 2008). Dr Hunzvi, then Chairman of the ZNLWVA, was arrested on charges of embezzlement, but also implicated on forming a hit squad to assassinate some senior ZANU-PF members and government ministers. Relations between them had continued to sour even after the truce. ZANU-PF elites had become more and more isolated, with forces from within (e.g., war veterans and peasants) and from without (e.g., opposition civil society movements) converging to attack the ruling oligarchy.

The situation was, however, more complex in that the new alliance of civil society, settler and international capital also competed with war veterans because the ideologies of the two sides were in direct conflict (i.e., with that of capital, domestic or international). In fact, ZANU-PF elites and the new alliance had common class interests which differed from those of marginalised war veterans and the land- hungry. It is, therefore, important to note that the difference between ZANU-PF elites and the MDC was a power and not an ideological struggle. Both of them were desperate for an opinion poll for the pending elections.[14]

The government hurriedly prepared for a referendum, despite the absence of statutory instruments for it (Madhuku 2000: 55). Section 57 (Sub-section 2 (1) of the draft contained a clause on the land issue that effectively maintained the spirit and content of the Lancaster House constitution, proposing compensation for agricultural land acquired for resettlement by the government. It said:

> where agricultural land is acquired compulsorily for the resettlement of people in accordance with a programme of land reform, any compensation payable must reflect an equitable balance between the public interest and the interest of those from whom the land is acquired.

This section angered the war veterans, 300 of whom organised a demonstration against the draft constitution, threatening to mobilise the electorate to vote against it (Participant observation 2000[15]). They petitioned the British High Commissioner and Emmerson Munangagwa, the Minister of Justice:

> ... demanding amendments in the draft constitution section dealing with land redistribution. The section [said] government [would] compensate farmers whose lands [would] have been acquired but the war vets demanded that no compensation be paid. ZNLWVA Harare branch Chairman, Douglas Mahiya [said], 'We are saying the price of the land has been paid by the blood of the people who died during the war' (*Daily News,* 12 January 2000).

President Mugabe then changed the clause of the draft constitution, using powers conferred by the Act.[16] This clause triggered various counteractions from the white community, both inside and outside Zimbabwe, now campaigning to vote against the draft constitution. The mobilisation intensified and the heightened participation of the white constituency was unparalleled since independence, illustrating the gravity of issues at stake.

Climax of the revolution 2000-2002

When the referendum was finally held on 11 and 12 February 2000, the 'no' vote prevailed and the land issue exploded, with nationwide occupations spreading at an unprecedented speed. This new level of occupations was signified by the occupation of a derelict farm in Masvingo Province by war veterans, followed by nationwide occupations activity. What was the state reaction to these nationwide occupations? What was the position of civil society and the opposition movement? And, finally, what was the position of ZANU-PF and President Mugabe?

The defeat of the draft constitution, which was largely about the land, but also about preparing for President Mugabe's exit, was widely interpreted as indicating ZANU-PF's impending defeat in the next elections. In this sense, the referendum had served its purpose for both ZANU-PF and the MDC.

Owing to isolation from the liberation movement, settler and international capital and weakened by war veterans' attacks and the opposition coalition now led by the MDC, the ZANU-PF ruling class was desperate. President Mugabe realised that war veterans and the surging land revolution were an asset in manoeuvring this new development. Tactically, he decided to 'hijack' the land movement in a bid to use its cultural capital against the MDC and particularly against white commercial farmers. He started to work towards what many thought was a genuine alliance with the land movement, particularly the war veterans who led it, from around February 2000 (Sadomba 2008).

Government enacted and amended various pieces of legislation protecting occupiers and breaking the resistance of white commercial farmers.[17] These enactments should be viewed against the backdrop of heightened MDC activities against land occupations, such as the increased demonization of President Mugabe, the intensified diplomatic onslaught and sanctions, all of which catalysed the situation. It prompted both war veterans and the ruling elite to increase mobilisation against settler farmland and capital. President Mugabe hardened his stance against the white farmers progressively as Masunungure (2004: 176-7) and Feltoe (2004: 200) observe:

> In early April 2000, the president said no white commercial farmer would be chased away from Zimbabwe ... But as time went on the anti-white rhetoric intensified and ... in December 2001, the president was really on a *warpath.*

The organisation of the land movement did not have a conventional hierarchical formation, having been rooted in the structures, ethos and practices of guerrilla strategies and tactics and in the local traditional agro-religious formations. These two forms of movement and social organisation resulted in horizontally-structured, locally-organised units of occupation with no centralised command. As such, the attempt to hijack the movement by co-opting the leaders of the Zimbabwe National Liberation War Veterans Association (ZNLWVA) was futile (Sadomba 2008). This is because the structures of the ZNLWVA did not initiate the nationwide occupations, nor did they control them. MT, a war veteran who coordinated occupation in Mazowe and Matepatepa, was clear that, 'The whole thing was spontaneous [with] no central organising platform that gave any direction ... nobody told us to do anything' (Interview MT 2001)[18].

The horizontal organisation of the land occupations revolution also explains the nationwide spontaneity in relation to land grievances and the antagonistic level it had reached. The land revolution assumed its organisational structure from the guerrilla war which operates in small units and isolated activities, the mosaic of which aggregates into a complete pattern of struggle. The structure made cooptation by the state difficult, as well as impossible to confront by the Commercial Farmers Union. It was elusive to potential enemies, flexible and efficacious as it suited local conditions and atmosphere. In short, the horizontal and dispersed nature of the land movement was ideal for land occupations, but, later, this proved a limitation for state attacks during the Murambatsvina period when the ZNLWVA was also weak and partially co-opted.

The land movement also engulfed the urban landless, who occupied land for urban housing and agriculture. According to Masuko,

> The fast track housing cooperatives were born out of structures of the War Veterans Association that led the land occupations in both urban and rural areas. From the year 2000 they moved into open space belonging to councils within the urban areas and onto privately-owned White capitalist farms around urban areas. Housing cooperatives were formed on all occupied land. Members were drawn from the community and were [the] homeless poor ... Thirty nine co-operatives [were] formed between 2000 and 2003 (Masuko 2008: 191).

These housing cooperatives were also characteristically different from the preceding ones that started in mid 1980s. In Harare, war veterans formed a union in 2001 called the Greater Harare Housing Cooperative. Before they were smashed by Operation Murambatsvina in 2005, the union in Harare alone had achieved various stages of housing construction and development on allocated stands: 1,712 completed houses, 734 houses at roof level, 473 at window level, 1,332 at slab level and 2,026 foundations. In Harare alone and for projects registered with the Greater Harare Housing Union, there was a total of 10,097 houses in construction progress, on land distributed through the war veteran land movement. In comparison, nationally, actual houses constructed annually between 1985 and 2000 ranged from 15,000 to 20,000 (Masuko 2008: 186). The construction of about 10,097 houses in two years without state assistance was phenomenal. This figure can be juxtaposed with progressively dwindling numbers from 1,500 stands officially allocated in Harare in 1999 to 220 stands in 2004. The progress in housing provision in Harare alone was at least two-thirds of the national annual output and more than 1,000 times of official land allocations in Harare.

With this, the war veteran-led occupation movement had done away with the political rural and urban divide along ZANU-PF/MDC partisan lines. It became a movement that united the poor classes of the peasantry, farm workers, urban workers and the reserve army of the unemployed, thereby raising the land question and struggles to their true national character. It is this factor that is fundamental in explaining the class conflicts of Murambatsvina which scholars have not debated. They have put their whole weight on humanitarian statistics, but have failed to use the data to explain the class antagonisms at the core of the clean-up. In cases such as the UN report by Tibaijuka (2005), figures are exaggerated (Moyo and Yeros 2005), as figures seemed to be an end in themselves rather than a means to unravelling a more complex interplay of forces. The picture is even better illuminated when Murambatsvina is viewed from its attack on indigenous industrialisation, which was threatening international capital and rising national capitalists (Sadomba 2011).

This class alliance, forged through the land revolution (Sadomba 2008; Masuko 2008), has escaped the analysis of many scholars and researchers. The majority of this poor urban working class comprised workers retrenched in the 1990s as formal industry shrank[19] under ESAP and later under European and American sanctions. These retrenchments, coupled with a severe housing shortage, had caused congestion, owing to illegal construction and extension of outbuildings on high and low density properties. By the time Murambatsvina struck in 2005, more than 80 per cent of housing stands in Harare had such illegal constructions (Masuko 2008; Toriro 2006), illustrating the extent of the social base for mobilisation at the disposal of the land movement.

This was the situation at the height of urbanisation. Lynch (2000) pointed out the characteristic paradox of the inverse relation of urban influx and shrinkage of resources or means of livelihood in Third World cities. The situation of Zimbabwe during the ESAP and especially during the occupation and Fast Track periods seems quite different and this might explain how the country has managed to sustain its economy even under sanctions and mismanagement at the national level, especially before Operation Murambatsvina in 2005. The Zimbabwean working class is highly skilled, disciplined and educated. As a result, the shrinkage of industry gave birth to a vibrant informal sector of small-scale manufacturers in different fields with trained artisans and technicians producing high quality products (Sadomba and Mujeyi forthcoming). More than 60 per cent of the urban working class in Harare were in the informal sector, absorbing labour from the rural areas (Masuko 2003: 186). Almost all the different sectors of industry were duplicated, albeit at a lower scale, in the informal sector, competing now with the established capitalist large-scale industries (Spencer was a qualified artisan, manufacturing agricultural equipment and supplying the SADC region, Interview 2009[20]).

This economic condition swiftly changed at the commencement of the occupations in 2000. There was sudden capital flight resulting from the imposition of sanctions and as a backlash to expropriation of settler capital. The working class structure also, as suddenly, transformed from formal to informal self-employment. The power of trade unions on the working class withered with this changing environment. Land occupations in the urban and rural areas were a major turning point since the formation of the MDC in 1999 and they eroded its base, which heavily relied on trade union politics. Regimentation of the workers was no longer possible for both the state and for the MDC. Structural changes of and within the working class explain the failures of stay-aways called for by the MDC after 2000 that had succeeded earlier.

The land movement had developed far-reaching objectives beyond redistribution and these included 'restoration of dignity of Zimbabweans', 'equitable distribution of land', 'restoration of cultural values', 'urban health' and demonstrating the potential of 'self-help' (Masuko 2008: 200). The formation of housing cooperatives and unions by war veterans in urban areas, the reconfiguration of industry that ensued and the resuscitation of war time base committees, show that the war veteran leadership and the land movement broadly was not 'based on a single issue' as Moyo and Yeros (2005: 190) have argued, but, in fact, a broader ideological struggle with evolving 'democratic peasant worker organisational structures'.

Differences in ideology between the state, ZANU-PF elites and the land movement caused antagonistic clashes at the district level, proving that co-optation of ZNLWVA leadership at the national level was not automatic; in fact, it was resisted at the base of the pyramid owing to divergent objectives and the conflicting ideology of the FTLRP. The objective of war veterans was for land to be given to the land-hungry people as outlined by DTM (a former ZIPA commander and leader of land occupations in Mazowe):

> Some people have been saying the land issue has been on the agenda because ZANU-PF wants to use it to gain some political mileage. But I, as well as my colleagues, War Veterans, we have a genuine desire to have the land issue resolved once and for all now. Political mileage would be a downstream benefit rather than the main objective. What we want to do is actually to give land to the people. If ZANU-PF as a party is going to benefit by that, well, there is nothing wrong with that. But it will be wrong for anybody to assume that we are doing this so as to bolster the position of ZANU-PF. That is not the case. We genuinely want to resettle people. The poverty that is quite abundant among our people can only be ameliorated… rectified… corrected, if people get land. There are no jobs in town, there are no jobs in industry and most of our people are on the land (Interview DTM 2000).

However, war veterans were also aware that ZANU-PF elites were opposed to them and the final objectives of the land movement. According to a war veteran leader of occupations holding a political science degree, they knew that ZANU-PF elites and government officials 'dislike', 'fear' and feel 'unease' about war veterans, preferring to 'keep' them at a 'distance' as a result (Interview DTM 2000). At Shiro Farm in Goromonzi in 1998, war veterans threatened the ZANU-PF senior ministers and politburo members that after repossessing land they would return to overhaul the party leadership and allow the marginalised peasants and war veterans to take over (Interviews DM 2000; K 2004; Muchaneta 2004). However, the intriguing question is, if war veterans saw the need to change the leadership of the party and government since the Chinhoyi meeting of 1992, why did they not take it as a priority? Why did war veterans not take over the party machinery first as a prelude to taking over state power? What exactly did they mean about their return? This could be attributed to limited theoretical understanding of revolution, especially from a Marxist viewpoint. Failure to prioritise seizure of state power with a view to establishing one that protected the interests of the working masses reflected considerable myopia and theoretical immaturity on the part of the veterans.

This was the leadership dilemma that the war veterans were grappling with during the revolution. As Mudenge later confirmed, President Mugabe had decided to unleash the state machinery – specifically the army – to silence war veterans (2004a). So war veterans 'knew that they risked becoming targets of state violence', which they had to avoid. At the same time, they sought to win the ZANU-PF elites away from the Lancaster alliance, resulting in the 'interaction between the needs of politicians for a constituency, [and] of people for land' (Alexander 2003: 97). War veterans faced the threat of ZANU-PF elites going back into the Lancaster House alliance. Strategic compromise seemed inevitable and ZANU-PF's gain of 'downstream benefits' was certainly one of them.

The MDC, the Commercial Farmers Union (CFU), private and international media and anti-revolution scholars of neoliberal traditions painted the occupations as chaotic. They claimed that occupiers were murdering, raping and torturing people on the farms. They depicted war veterans as rogue elements sent into the farms by a beleaguered ZANU-PF, an aged president and a desperate state to suppress the opposition party. Evidence on the ground suggests otherwise. Land occupations were far from being chaotic; they were orderly, principled, with a few violent clashes of mainly minor assaults. Interviews with many white commercial farmers, a white lawyer who represented farmers and farm workers, not to mention the occupiers themselves, show this. During the occupation period, war veterans had specific rules of operation that they followed, which included not to take any farm property, not to destroy produce or kill animals, to request from the owner anything they needed, etc (Interviews, P 2000; Bota 2004; Personal communication with white farmer H 2000-2003[21]; Interview, white farmer BT 2004).[22] War veterans were given more rules by the spirit mediums before they went to occupy land. These included prohibition of any type of sex during occupations and in the occupied areas (Participant observation at Nyabira, Mazowe and Matepatepa, 2000-2004).[23]

However criminal elements took advantage of the situation as a white farmer clarified:

> War veterans would not fall in the criminal element [group] who I know took advantage of the whole situation to… gain from what was going on at the time … those people can't be war veterans. So they were abusing the name of war veterans … Genuine war veterans were after the land. This was pretty much as straight forward as that (Interview, B 2004).

For politically strategic reasons, the new alliance of the MDC, settlers and international capital on the one hand and, ZANU-PF elites on the other, portrayed war veterans as barbaric, unintelligent and incapable of ruling the country or leading the people. White commercial farmers themselves confirmed that war veterans were not violent and did not loot the farms; they were clearly after the land.

During the occupation period, white commercial farmers failed to respond positively to the pragmatic proposals of war veterans to share excess land. War veterans actually approached the farmers with the idea of not driving them off the land, but to share land that was in excess and was underutilised or applied the one household-one-farm principle.

> We first negotiated with the farmers [to share land] and entered into written agreements before even occupying. However when we now made a follow-up to implement the agreements at the time of elections, the farmers started to change their mind and they were now saying they signed under duress. We then realized that these people were dishonest and they were not serious; we were just wasting our time (Interview, P 2000).

Another reason why commercial farmers kept vacillating was that farmers expected the MDC and Morgan Tsvangirai to win the 2002 elections, hoping that he would then reverse the land reform. This expectation destroyed chances of negotiation and dialogue, giving the ZANU-PF elites the opportunity to start fresh occupations that were not based on the criteria that the war veterans had used. The ruling party elites were punishing the white settler farmers for aligning with the MDC and therefore sharing land with them was out of question. In some cases, war veterans went ahead to mobilise the land movement actors in defence of some white commercial farmers against the state and ZANU-PF wave of occupations on occupied farms. In one situation, armed police squads in riot gear came from Harare to disperse demonstrators who had locked up the settled war veteran who had driven the farmer out against the wishes of the land revolution actors. According to revolutionaries, this war veteran, a medical practitioner and herbalist, was being sponsored by the highest offices of the land as a gesture of appreciation for treating HIV in one of the relatives, whereas the white commercial farmers (husband and wife) were understood to have been supportive of the marginalised through sponsoring health programmes and the district hospital (personal communication with the demonstrators, personal observation G Farm Concession 2004).[24]

The state, as an actor, intensified occupations and targeted critical white commercial farmers to break the backbone of the MDC. Both the MDC and ZANU-PF engaged in violent political clashes during the electoral period and this clouded the land occupations. The MDC, in alliance with white commercial farmers, actively organised gangs to attack occupiers in their bases and even at their homes. The first deployment of occupiers was attacked such that in many areas occupations were postponed until reinforcement was mobilised from elsewhere (Interview, DTM 2000). However, some MDC members supported land occupations.[25] For example, Munyaradzi Gwisai openly and publicly supported them and he was removed from the party for his utterances. Another person was Learnmore Jongwe, then spokesman for the MDC, who went with a group of MDC supporters to join occupiers in Nyabira (Interview, DM 2001).

Although the FTLRP was started in mid-2000, it took time for it to be felt on the ground and even more time for the state to ultimately control the land movement. By the end of the year, the state had failed to penetrate the movement and bring it under control. Instead, it met with stiff resistance from the land movement, prompting the state to convene a high-powered meeting between some war veterans and the state on 18 December 2000. The meeting was aimed at disempowering war veterans and reinforcing the authority of the task force structures driving the Fast Track. Nevertheless, the land movement actors continued to resist state manoeuvres. With imminent presidential elections (in 2002) the state and ZANU-PF proceeded cautiously in relation to the FTLRP, fearing that war veterans would mobilise voters against President Mugabe.

After the presidential elections of 2002, which were won by President Mugabe, the state now implemented its Fast Track Programme in full throttle, drastically changing the situation in the farms in order to control the movement. ZANU-PF leadership at the local level, chiefs and civil servants were mobilised by the state against the war veteran leadership. Civil servants and local leaders started to target war veterans for removal from the farms. The Presidential Land Review Committee observed that although war veterans were happy that the government at last 'had heeded the call for land redistribution … [however] their members had not benefited as promised … [and] land was allocated on regional lines…' (Utete 2003: 35). From 2002/3, the state managed to usurp the leadership of the land movement from the war veterans, marking the end of the occupation period.[26]

The Fast Track Land Reform Programme

The FTLRP was an immediate strategy formulated by government and the ruling ZANU-PF to deal with the revolution. At this time, the state shifted its role from one of the actors of the land movement to that of a power above the movement, exercising the authority of 'legalizing and regulating the occupations' (van der Haar 2005: 5). However, latently its objective was to usurp control of the land revolution from the war veteran leadership and sway it from its original objective of land redistribution to the land hungry. Through the FTLRP, the state regained legitimacy and assumed authority to take charge of and structure the land occupations. Implementation focused on attacking and weakening the land revolution leadership. Organisationally, the FTLRP had a national task force to study the movement, create structures and re-establish state control. A National Task Force, led by the controversial Minister of Local Government Ignatius Chombo,[27] decentralised its operations forming Provincial and District Lands Committees. District committees and lower- tier structures (called Committees of Seven) were frontline structures of the FTLRP, created specifically to negate war veteran leadership of the movement at the grassroots level of the revolution.

The first manifest clash between the state and the land revolutionaries was based on the new structures imposed by the state. At the national level, there was neither representation of war veterans nor any actors of the

CARDIFF UNIVERSITY ★ PRIFYSGOL CAERDYDD ★

revolution. At provincial levels, war veterans employed by government were used to represent land occupiers, despite the fact that these, in many cases, had nothing to do with the occupations. At district level, the most senior civil servant – District Administrator (DA) – and the District Lands Committee sidelined the actual war veterans who were leaders of the land movement and replaced them with hand-picked individuals who they preferred to represent the land movement constituency. Even these hand-picked individuals became a minority in a committee of about fifteen people. Many such tactics were used, but the land revolutionaries resisted, resulting in serious clashes, which, at times, degenerated into physical assaults (Sadomba 2008).

Overtly, the FTLRP involved land assessment to determine carrying capacity, demarcation into plots, settler selection and finally land allocation. Three tier tenure systems resulted where A1 plots were based on a communitarian policy and A2 was for commercial farming; communal lands remained unchanged. The objective of the land revolution was in line with the A1 model where as many peasant farmers as possible would be resettled through the scheme. The A2 model became very controversial as it was distributed for patrimonial reasons, handled directly by the Minister of Lands and Agriculture, then Joseph Made.

Scholars have failed to analyse the particular development of the war veteran- led revolution from 1997. This is erroneous in that failure to distinguish the various phases conceals many factors that help us understand the dynamic metamorphosis from a 'single issue' movement to a revolution. The land occupations differed markedly between the nature, approach, objective and motive of war veteran-led occupations and state and ZANU-PF- led occupations during the Fast Track period. The ZANU-PF and state-orchestrated occupations were mainly invasions of land already occupied by revolutionaries, aimed at dispossessing occupiers in order to give it to ZANU-PF elites, senior civil servants, or relatives of those in the system. War veterans dubbed this wave of occupations *jambanja on jambanja*[28], meaning that they were occupations of already occupied land by revolutionaries. The Fast Track was not about bringing order to a disorderly operation, as claimed, but on the contrary, it started to introduce disorder and new waves of occupations.

The process of the FTLRP was summarised in a document presented to the Provincial Stakeholder Dialogue held from 23 to 24 August 2004, organised by the African Institute of Agrarian Studies. War veterans wrote:

> Arrests of land occupiers has been orchestrated and well planned so much that strategies are made to create crimes where war veterans [are] fast-tracked to cells,

> court and jail. It's a well organized syndicate of officials from the mass that is used ... police details who arrest, magistrates and his public prosecutor who make sure you [go] to jail. When others [occupiers] realize this humiliation, they ... go back to [their] towns of origin and the so-called politicians become happy and celebrate. But can we say they will have solved the problem? No! ... Already there is political discontent and distortion in the Agrarian Revolution (Mashonaland West War Veterans Association 2004).

As soon as the Fast Track Programme took hold, its implementers started to weed out war veterans and other revolutionaries, opening commercial farms for elite settlement. These elites were mainly senior government officials, senior members of the uniformed forces, party loyalists, relatives and the ruling oligarchy, who were given whole farms to themselves measuring hundreds or even thousands of hectares. These 'chefs', as the elite are commonly called, chose prime land with good infrastructure and farm houses, chasing away the revolutionaries. In contrast, revolutionaries were allocated A1 plots that were several subdivisions of a farm (as small as six hectares per household) according to official government policy. Moreover, the government input scheme favoured these large- scale commercial A2 farmers more than the small A1 farmers. For example, A1 farmers, occupying 98 per cent of the resettled land, got less of the funds, with the balance going to A2 farms. In 2006, government budgeted a paltry Z$1 trillion for '2005-6 season crop input finance to support A1 and communal farmers', comprising more than a million household farmers, to be conservative. In comparison, A2 farmers got, through the Central Bank programme called the Agricultural Sector Productivity Enhancement Fund (ASPEF), 'ZW $7 trillion and other private financing schemes' (Gono 2008).

Many no longer had the energy to fight and they simply returned to their houses in towns or to their rural homes. The Fast Track was marked by many violent clashes between the state and the land revolutionaries. Moreover, many of those who were given the land for large- scale A2 commercial farming, where land movement actors had been removed, were not capitalist farmers and were accused of asset stripping[29] (Mashonaland West War Veterans Association 2004).

Murambatsvina period

This chapter argues that the decade of Zimbabwe's revolution spearheaded by war veterans is a tale of class conflict within the liberation movement and without. Class antagonisms reached their climax during the Murambatsvina[30] phase.

In this section, we examine the revolution during the Murambatsvina period from mid 2004 to 2008. Many scholars and analysts have looked at Murambatsvina in partisan terms, arguing that it was retribution against MDC supporters. This chapter disagrees with this analysis and argues that Murambatsvina was an attack on the revolution. It further argues that Murambatsvina occurred both in rural as well as urban areas and it started before 2005. Murambatsvina was imbedded in an overall strategy to deal systematically with the revolution – in which war veterans were the vanguard – that had been fully developed by the state from the rupture in 2000 (Sadomba 2011).

The state designed a strategy comprising three options for formal intervention into the land revolution. The first option was simple cooptation of the movement through ZNLWVA leadership structures aimed to diffuse the movement's autonomy and to subdue it. This strategy was embarked on soon after the February referendum in 2000, when President Mugabe invited Hunzvi to spearhead the electoral campaign at the ZANU-PF Politburo post-mortem meeting that was held a week after the referendum.[31] The second option was to create parallel state structures that would antagonise those of the land revolution in the hope that the later would succumb. The third and last resort was to smash the revolution violently and dissipate it. The Fast Track was a process of executing these three options in that order and Operation Murambatsvina was a culmination of that long-term strategy. During execution, the options overlapped and backdated, although they remain distinguishable and severable.

The attempt by the ruling clique to co-opt the revolution for its purposes failed for a number of reasons, including, but not limited to, the complex horizontality of the revolution's organisation, localisation, divisions within state organs,[32] and effective negation of land revolution structures that continuously made war veterans suspicious about the actual agenda of the state and ZANU-PF elites. Although parallel structures were created in the form of Task Force Committees, land committees, village Committees of Seven and rejuvenation of traditional leadership, the land revolution under war veteran leadership did not succumb. This left the state with no option but to implement the third alternative, then code-named Murabatsvina.

Operation Murambatsvina/Restore Order began much earlier than 2005, as a continuation of clashes between state organs and the anti-settler revolutionaries, particularly war veteran leaders. In Mashonaland West and Central, for example, there were continuous brutal evictions of occupiers by the state in Zvimba, Mazowe and Shamva districts, implicating provincial governors such as Elliot Manyika and ministers such as Ignatius Chombo. The onslaught on Roger Boka by the ruling oligarchy is also seen as part of the Murambatsvina strategy (Sadomba 2011). Tactically, government postponed the widespread violent onslaught on the land revolution until the general elections of 2005.[33] It is notable that ZANU-PF won the elections overwhelmingly, reflecting the 'downstream benefits' of the revolution, both in rural and urban areas (Masuko 2008; Sadomba 2008). It is illustrative that ZANU-PF regained the only seat in the traditional stronghold city of Harare, both in the 2005 and 2008 parliamentary elections and that the seat was won by a war veteran candidate, Nyanhongo, who clearly rose from being councillor because of his articulation of the land revolution objectives.

As a long-term strategy to consolidate the land revolution, war veterans decided to take over the political leadership of ZANU-PF by mobilising support through the land revolution. Their first step was to strengthen the ZNLWVA. To do so, they had to identify a courageous leader for the association, after the death of Dr Hunzvi. Jabulani Sibanda, then chair of Matabeleland Province, had emerged as a fearless leader when he publicly denounced the 'old guard' politicians of Matabeleland, including stalwarts like Joseph Msika, then member of both the state and ZANU-PF Presidium. The state and the ZANU-PF ruling elite therefore backed Joseph Chinotimba, an outspoken land occupier during *Jambanja*, who received extensive television coverage by the state station, the Zimbabwe Television (ZTV). The Joint Operations Command tried to influence[34] the choice of war veterans, but to no avail and Sibanda became the new chair, unopposed (personal observation ZNLWVA national congress, 2004, Mutare). However, he also was later co-opted and he organised an attempt of a million- (wo)man march in Harare in 2007 to support President Mugabe's candidature for presidential elections in 2008. For this new role, he was rewarded with a brand new twin cab vehicle and a house in an up-market suburb of the capital, Harare. In addition, his farm was heavily equipped, according to information given at a ZNLWVA meeting held at the ZANU-PF provincial offices along Harare's Fourth Street (Personal observation, January 2008[35]). This co-option of the ZNLWVA's national executive dashed the plans of the association and the revolutionaries.

The war veterans' second step was to get into Parliament in massive numbers. Many registered for ZANU-PF primary elections, but were removed from the list by the party elite and were replaced by counter revolutionaries. Ironically, war veterans campaigned for these imposed candidates: for example, they campaigned for a relative of President Mugabe's, young Patrick Zhuwao, in the general elections who was preferred instead of the war veteran who had led occupations in Nyabira. He was given the post of Minister of Science and Technology. The vision of war veterans and their political tactics are in this sense, baffling. However, this reflects the complexity of the situation where a revolution confronts catalysing neocolonial tendencies and the imminent threat of international capital and military forces (displayed by the MDC as surrogates of European powers), nationalist bourgeois tendencies (among ZANU-PF elites), settler and international capital. Determining the priority enemy at any given time might be tricky and debatable. Why did war veterans not insist on getting into Parliament when ZANU-PF was at its lowest point and they (war veterans) were powerful, instead of accepting to back imposed and little-known candidates? Indeed, this weakened this revolution substantially. The criticism of Moyo and Yeros (2005) is relevant in this regard. War veterans, despite ideological clarity and long term-strategies, were tactically sterile. A retreat at this point was tantamount to bolstering the position of ZANU-PF elites, giving them the tactical advantage which they, led by the ingenious President Mugabe, were quick to exploit and they swiftly smashed the movement through Murambatsvina. This tactical error grossly and dearly cost Zimbabwe's revolution.

Soon after the general elections in 2005, the postponed 'violent retribution by the state', to borrow Jun Borras' words (2001: 548), was commenced on the land revolution.[36] The operation started by demolishing houses of cooperatives in the urban areas.[37] The demolition was done by local authority operatives using earth-moving equipment accompanied by the police and army. 'Illegal' structures in high-density properties were also razed to the ground as were the established informal sector production sites and workshops. As there was not enough warning, property was lost and, worse still, the means of urban livelihood were destroyed as means of production for the small--scale manufacturers were crushed in the process. Above all, the operation was life-threatening as it left many families without housing, exposed in cold winter winds. The effect on the urban land movement was clear as Masuko writes:

> ... in doing so (government) dashed the hopes of the low income urban homeless and of one of the most radical housing developments ever initiated in Zimbabwe. However ... the occupiers remained on the occupied farms minus all the structures that they had built ... (Masuko 2008: 204).

The most intriguing question that scholars have glossed over or totally ignored is, what was the motive of the regime in carrying out the operation? ZANU-PF had clearly started to regain popularity through the land movement, winning more than two-thirds majority in the 2005 general elections[38] (Masuko 2008). Why did President Robert Mugabe and ZANU-PF not forge genuine unity with war veterans and the land revolutionaries as opposed to a marriage of convenience? The answer seems to lie in the intrinsic class contradictions and class struggle. War veterans feel that Operation Murambatsvina was targeted at them specifically as the vanguard of the revolution. Operation Murambatsvina was not the only operation of this period. Another one was Operation Chikorokoza Chapera that was carried out in 2006. This was a rural operation that focused on specific occupied farms and mineral exploitation sites that had become the new source of livelihood for dispersed Murambatsvina victims and the rural masses under economic siege. Although Chikorokoza Chapera was countrywide, the most brutal attacks were in the Chimanimani gold and the Chiadzwa diamond mines, both in Manicaland, the war veteran strong-hold. These were seen as ethnic attacks on the constituency of war veterans that contributed between 75 and 85 per cent of people who joined the liberation war. War veterans led these mining operations (participant observation, Chiadzwa 2006[39], Sadomba 2011).

Structural reconfigurations also occurred during the Murambatsvina period. The state, ZANU-PF and President Mugabe, aware of the cruel attacks they had made on the revolutionaries, particularly war veterans, decided to forge a new alliance. This time they chose the traditional leaders – who had been sidelined since independence owing to co-optation of most of them by the colonial state during the liberation struggle – to replace the mobilisation role of the revolutionary actors in ZANU-PF. The countryside was not being democratised by going back to traditional authority. Rather, this structure was, as in the case of the Philippines, being elevated and entrenched into an elite 'to dominate the rural polity'. With state resources and delegated powers, this elite could 'use extensive patronage networks that combine (partial) provision of daily subsistence needs of rural poor households with the threat and/or actual use of violence' (Borras 2001: 550).

First, chiefs were allocated prime land with beautiful farm houses and infrastructure. In addition, they were given grants of seed and chemical fertilisers. They were also given double cab vehicles for personal transport and administrative personnel, including secretaries and messengers. The powers of traditional leaders were also increased and they were given more functions as commissioners of oaths. In 2006, new agricultural programmes were initiated by the Reserve Bank of Zimbabwe (RBZ). These were the Productive Sector Finance Facility (PSF) in 2003 and the Agricultural Sector

Productivity Enhancement Facility (ASPEF) in June 2005 (Gono 2008: 148-150). The farm mechanisation programme under the ASPEF included many schemes for drought relief: food, seed, fuel, livestock, liquid money and farm equipment such as tractors and combine harvesters. These traditional leaders and elites were a bourgeois class under formation. Chiefs were not only direct beneficiaries of this project, but also distributors, giving them the extra advantage of consolidating their social networks. The Basic Commodity Supply Side Intervention Facility (BACOSSI), 'under which primary, secondary and tertiary producers and suppliers in targeted key sectors of the economy were afforded concessional production-linked financial support for working capital requirements' (Gono 2008: 151), was also implemented. However, this exerted high inflationary pressures on the economy as it distorted prices and as some of the inputs were abused and were not channelled into production.[40]

Much has been written and debated[41] about the evil nature with which Operation Murambatsvina was carried out by the state (Tibaijuka 2005; Toriro 2005; Masuko 2008; Mhiripiri 2008; Mlambo 2008; Moore 2008; Vambe 2008), but little or no analysis has been offered on the class nature of the state operation. As a result, the analysis is, at best, shallow and, at worst, confused. For example, simple empirical facts are contested, like who was targeted by Murambatsvina. Vambe argues that 'both rural and urban areas; ZANU-PF supporters and MDC supporters and non-aligned, were targeted' (2008: 3). However, others see the operation as partisan, attacking MDC city strongholds as 'punishment' for 'voting for MDC' and the desire of the ruling party to reverse modernity - symbolised by the urban working class – attempting to unwind the time of the urbanites to 'year zero' rural homelands (Moore 2008: 28). In desperate defence of state action, Mahoso tried to separate Murambatsvina from the land revolution itself, saying 'the African land reclamation movement [was] rural and [had] little to do with urban slum clearance' (2008: 160). Vambe's view is correct and supported by empirical evidence. Moore's argument can be challenged on grounds of lack of empirical data based on fieldwork demanded by such a controversial study. It sounds more reasonable that urbanites who voted ZANU-PF in 2005 were influenced mainly by the land revolution and this signified a shift from the MDC, considering that the party's land policy and alliance with white commercial farmers were seen as negating the land occupation revolution. Moreover, Morgan Tsvangirai said that land occupiers were spreading in a slovenly fashion like sprouting mushrooms and warned against starvation under indigenous farmers, thereby enhancing this perception about the MDC.

This chapter argues that the land revolution of Zimbabwe for the decade 2000 - 2010 was the climax of class conflict. The politics of power at this juncture transcended partisan interests as the real bone of contention was protection of class interests and class domination by ZANU-PF elites and the petty-bourgeoisie against peasants, rural and urban working classes. Political power was under formidable threat from the land revolution that had now mobilised both urban workers and peasants. At no point in Zimbabwe's liberation struggle had such a powerful alliance of the urban working class and rural peasants been forged; moreover, never had such an alliance been seen at such a national scale. The ZANU-PF ruling elite, petty and rising national bourgeoisie were worried by the imminent power shifts threatening to take place in favour of the lower classes, comprising the land revolutionaries led by war veterans.

The myth that war veterans were incapable of leading the Zimbabwean society had been utterly dispelled and a revolutionary climate had developed. According to a war veteran leader of occupations in Mazowe District, 'The situation had presented itself' (Interview, DTM 2000). The leadership capabilities of war veterans had been demonstrated when they organised the land- hungry, homeless, informal-sector producers and farm workers, sending unequivocal signals that it was only a matter of time before the movement took over state power. This possibility, of course, sent shivers through the ruling elite, who immediately took the third option – the real 'hidden dimension of operation Murambatsvina' – a violent retributive class attack on the urban and rural poor revolutionaries. The impact of the housing cooperatives and unions are illustrative of this new and rising power of the peasants and workers with marginalised war veterans as the vanguard, against both capital and elitism.

A question that has been debated is whether or not the land revolution dissipated and disintegrated after Murambatsvina. What became of the land movement and what is its status today? This question can be answered by viewing the agency of the revolutionary actors from 2005. Many war veterans, who were interviewed in connection with Operation Murambatsvina were bitter. More than 10,000 properties at different stages of development were destroyed, including and especially those of the war veterans.[42] Members of the land occupation revolution were scattered across the country as a result of Operation Murambatsvina and Operation Chikorokoza Chapera.[43] Counter strategies by war veterans included ousting ZANU-PF elites in the 2008 elections[44] (Interview, Muchaneta 2006). Dispersal of revolutionaries and Murambatsvina victims effectively also spread mobilising agents against the ruling ZANU-PF elite, President Mugabe and their bourgeois counterparts.

Strategically, war veterans mobilised the ZANU-PF electorate for the Parliamentary elections of 2008. However, the politburo sought ways of weeding out war veterans by applying unconstitutional qualifications, including that a ZANU-PF Parliamentary candidate had to have been in the provincial executive for at least five years. In 1980, at Zimbabwe's independence, ZANU-PF had issued a directive barring war veterans from participating in the leadership of the party at any level, which condition was only lifted during the occupation period, by default. This directive had been communicated by Witness Mangwende at 88 Manica Road, then ZANU-PF Headquarters (participant observation, 1980). It was therefore impossible that under normal circumstances one would have risen through the ranks to occupy a provincial level post, so this was clearly done to exclude war veterans.[45] Many war veterans lost their meagre income campaigning to be parliamentarians, only to be weeded out.[46] Das is correct in his observation that the sheer numbers of the land movement 'constitute[d] a political threat to [the] regime overlooking their interests, either through elections or through non-electoral agency' (Das 2007). Jabulani Sibanda, who tried to silence war veterans sidelined in the primary elections, was viciously snapped at in a meeting of war veterans (Personal observation, January 2008, Fourth Street Offices). The angry crowd threatened the doom of the party in the 2008 elections. Some war veteran candidates, for example in a Marondera constituency and in Mutasa, refused to step down with disastrous consequences to the ruling party. Others took the primary election irregularities to the High Court, but many others simply withdrew, including war veterans in Goromonzi, Zvimba, Domboshawa and Harare (Personal observation 2008).

The ballot became the new 'weapon of the weak', now mobilised by the scattered Murambatsvina victims, comprising urban informal sector producers, urban homeless and some dispossessed A1 settlers and marginalised war veterans. War veterans and revolutionaries were disgruntled by the process and the sidelining of their candidates. This anger, disillusionment and mobilisation by Murabatsvina victims changed the traditional voting behaviour in the rural areas, leading to ZANU-PF general defeat, losing parliamentary majority to the MDC. Numerically, President Mugabe was also defeated by Morgan Tsvangirai in the March 29th elections. With panic, the state reacted by unleashing retributive violence on both the rural and urban electorate in a military operation codenamed Operation Mavhotera Papi? (Operation whom did you vote for?). The army was engaged to 'mobilise' or is it to 'coerce' voters. A shift in the use of the army instead of war veterans clearly explains that the state had terminated its alliance with the latter. A re-run of the presidential elections was marred by organised state violence, resulting in Morgan Tsvangirai's withdrawal from the race and seeking refuge in the Dutch Embassy in Zimbabwe. Murambatsvina had pushed partisan, but especially class, contradictions to their zenith.

War veterans heavily criticised regimentation, threats and violence against the electorate in the run up to the presidential run-off.[47] The elections that put President Mugabe back into power were widely condemned regionally and internationally. This situation forced ZANU-PF to concede to a power-sharing deal, forming a Government of National Unity with the two MDC parties.[48] Omission of the land movement and particularly of war veterans, in this GNU is conspicuous, raising questions regarding the future of both the GNU and the land revolution.

Rupture with the capitalist system?

Zimbabwe's decade-long experience raises a number of controversial questions regarding whether it represents a revolution and whether war veterans were a vanguard force or merely a ZANU-PF political instrument (see Moyo 2001; Moyo and Yeros 2007; Mamdani 2009; Sadomba 2008; Sadomba 2011). Moyo and Yeros argue that while a revolutionary situation had emerged, it was 'interrupted', ... '[leading] not to a revolution but a *radicalised* state [-] peripheral state which has rebelled against neo-colonialism' (2005, 2007). Interruption occurred for various reasons, largely because the process was co-opted by the state through the official FTLRP and its structures and its consistent attempts to normalise relations with capital (ibid). They also argue that it was possible to co-opt the process partly because of the war veteran leadership weaknesses, given their limited focus as a 'single issue' movement. I have differed with elements of their argument on the basis of their poor class characterisation of the political forces and weak analysis of state control (Sadomba 2011). I concluded that a revolution far beyond agrarian transformation did actually occur with ramifications on the totality of the economic, political and social life of Zimbabwe. It was an anti-settler revolution first and foremost, focusing on reversing racial resource distribution with its main leverage being a rupture of the elite-settler alliance. A century-old settler economy was reversed, democratised and de-racialised between 1997/98 and 2002/03. Nonetheless, while Moyo and Yeros concur with the idea that a wider social transformation occurred (Moyo and Yeros 2005, 2007, 2009, 2011), they consider this change to have fallen short of a revolution. They ask whether:

> a revolutionary rupture with the capitalist system is on the cards. The latter would require disciplined and durable working class organisation, revolutionary leadership and not least the implosion of the state apparatus, either through war and/or through the dissent/disintegration of the armed forces – none of which occurred in Zimbabwe (Moyo and Yeros 2007: 105).

This raises a fundamental question regarding whether a 'rupture with the capitalist system' is the *sine qua non* of a revolution? Some war veteran ideologues have different theoretical interpretations of the outcome from that of Moyo and Yeros, as explained by DTM:

> What many people would want to see, that's the democratic stage of our revolution. This is where we are; it was left uncompleted at independence. All we achieved at independence was political independence, but the actual revolution was not completed. We are now completing this one. We took the political power, but without being backed by the economic base. This is now what we are trying to do – the involvement of the indigenous people [i.e.] the majority of the people in economic activities of the country and one of them is land redistribution, making funds available to indigenous people to open up business enterprises, in industry, commerce, mining and so forth. All this was not done; the national democratic revolution [NDR] was not completed. We are now in the process of realizing this stage. But in my case, I have ideals which go far beyond that. I would want a society that is free of exploitation of man by man. I would love to have a situation where classes within society are narrowed. The [gap between the] haves and the have nots should be narrowed as much as possible, if not completely done away with. The state should own the basic means of production on behalf of the majority of the people. These ideals go far beyond ideals of a national democratic revolution because it ends here. People share the land [access], the means to go into industry and commerce and this revolution is complete. But there would still be differences in terms of wealth with stratification of society. I would want to go a step further (Interview, DTM, 2000).

This perspective suggests that a socialist revolution is attained when class differences are addressed; that is going beyond de-racialisation of ownership of the means of production. Moyo and Yeros (2005, 2007) and Moyo (2011a, 2011b) provide empirical details of the new class relations that have emerged since the FTLRP and have argued that an NDR was on the cards, but they negate this by identifying an 'interrupted' revolution and 'radicalised' state. For war veterans, the revolution they spearheaded between 1998 and 2002 was a continuum of a phased revolution characterised by different class configurations, objectives and outcomes - a 'completion' of a 'national democratic revolution', entailing backing 'political power' with an 'economic base' through redistributive objectives.

Thus a national democratic revolution is just a prelude to a socialist transformation aimed at 'narrowing' the gap between the 'haves and the have nots' or 'completely done away with', by freeing society from 'exploitation of [wo]man by [wo]man' and the state owning the 'basic means of production on behalf of the majority of the people'. DTM recognises that there was no rupture with the 'capitalist system' between 1998 and 2002, but does not consider the outcome an aborted revolution, since a revolution does not only occur when the capitalist system is destroyed. Given the history of settler colonialism and imperialist domination, how practical was it to wage a socialist revolution without undertaking an NDR? Moyo and Yeros do not specifically debate this issue that the war veterans examine and clearly articulate, arguing that a socialist revolution can be phased. There is no doubt that the land revolution managed to successfully transform the settler property regimes, although scholars have mainly focused on rural land occupations, ignoring the urban process (Moyo 2001; Moyo and Yeros 2007; Moyo et al 2009; Scoones et al 2010; Sadomba 2011). These studies have shown that this was a successful agrarian revolution by debunking the myths about the FTLRP and how it changed livelihoods. I think there is need to be more open in the application of theory than Moyo and Yeros have been, since the idiosyncratic nature of society determines the unique character of a revolution, including the appropriateness of strategies and tactics adopted and the friends and foes of that revolution. DTM is correct in arguing that revolutions are not only socialist in character and that the theory of interruption would not stand sustained interrogation. Was the French Revolution not a revolution because it did not lead to socialism? Nonetheless, Moyo and Yeros' observation that a fundamental rupture with capital did not occur has to be understood within the theory of a socialist transformation, rather than as foreclosing the definition and phasing of a 'revolution'. I now turn to an attempt at theorising Zimbabwe's experiences up 2010, based on a Marxist interpretation of what the two sides are grappling with, to offer an explanation of the current situation and a prognosis of the future of this revolution.

Conclusion: towards a theory of Zimbabwe's revolution

Applying Marxist dialectical philosophy to Zimbabwe's socio-economic development (historical materialism) that culminated in the past decade's revolution, I begin from the colonial land grabs that transformed agrarian production relations into a race and class conflict. The new and received capitalist mode of production was characterised by rapid development of scientific technology and increased labour productivity – the productive forces. With freehold tenure systems and private property protection, the country experienced unprecedented development of the agrarian economy as illustrated by large commercial farming.[49] State of the art technology and cheap African labour were in unison with the capitalist agrarian relations.

However, even under conditions of rapid technological advancement and high labour productivity, relations of production remained settler, colonial, capitalist and racist, leading to intensification of exploitation of indigenous Africans. This exploitation escalated racial tensions and conflict, eventually igniting a social revolution, the bloody liberation war of the 1960s and 1970s. The forces within this revolution, however, did not have common interests and objectives. Elite nationalists were content with removing settler, colonial and racial rule, but not capitalist relations, whereas the guerrilla combatants were mobilising peasants and workers for complete removal of colonialism and capitalism. Nationalists are class allies of capital, but war veterans and marginalised working classes are enemies of capital. Therein lay the ideological contradictions of nationalists and war veterans. The Lancaster House betrayal, the elite-settler alliance and the war veterans' economic and social disempowerment, relegation and systematic elimination[50] reflect class conflict. To the marginalised working classes and war veterans, racial dominance and oppression was replaced by elite dominance and oppression, thereby setting a perfect condition for neocolonialism. 'The actual revolution was not completed' but 'interrupted'; and the economy was not democratised, but remained governed by elite-settler capitalist relations.

These class relations did not improve in the post- independence period, but deteriorated. Adoption of ESAP in the early 1990s marked the apex of these class conflicts as international and settler capital was bolstered, strengthening the bond of the elite-settler alliance and increasing exploitation to unprecedented levels. To the marginalised war veterans and the struggling working classes, these neocolonial production relations became untenable and a social revolution was in the cards. The immutability of class and race production relations for over a century, juxtaposed with rapid technological advancement and productivity, reached a breaking point in the 1990s when war veterans started to challenge the state, ZANU-PF, President Mugabe and settler dominance in the economy. They mobilised the marginalised working classes towards the completion of Zimbabwe's 'unfinished business'.

When war veterans took over the leadership of the land occupations movement after the truce with President Mugabe, ZANU-PF and the state in

1997, this was no longer about a 'single issue' of their grievances, but a social grievance of a national nature. They then articulated, not only the land and agrarian question, but also the urban and industrial question, simultaneously attacking the landed settler bourgeoisie and organising urban housing co-operatives and informal sector establishments. Completion of this NDR raised questions far beyond the limits of removing colonial and settler systems of production. This revolution questioned elitism and nationalist garb, noting that this was a wolf in sheep's clothing, whereupon it became as much a target as settler and international capital, then breaking the tripartite alliance. International capital panicked and sought what it thought would be a better class ally than the nationalists and backed the opposition led by the MDC, isolating their former nationalist allies.

By 2000, the matrix of contradictions had reached its climax and an uncontrollable anti-state, anti-elite and anti-capital revolutionary tsunami raged. Race and class elitism were under challenge from below. However, the rupture of the Lancaster alliance called for a change in the strategy and tactics of the revolution as the elitism and nationalism on the one hand drifted from settler and international capital on the other. Nationalists sought an alliance with the revolutionaries to face their class allies of settler and international capital, the later of which had now groomed a formidable opposition MDC. This was a complex reconfiguration of political forces.

However, the alliance of nationalist elites with the revolutionaries was just a marriage of convenience since the two had antagonistic class interests. So when nationalists retained political power in the crucial 2005 elections, they unilaterally broke their alliance with the revolutionaries and mounted military style attacks on them, Operations Murambatsvina and Chikorokoza Chapera. Revolutionaries further mobilised and by 2008 the nationalist movement was isolated, suffering near defeat and they mounted the retributive violent Operation Mavhotera Papi. However, both sides of capital saw how dangerous was the game they were playing and decided to unite in the best interests of their class. They formed the GNU, an alliance against the working classes and marginalised war veterans.

Finally, what is the position and function of the state amidst this struggle? Is it a neutral arbiter of class forces and contradictions? To what extent can the state act against its class and under what specific conditions? What does radicalisation of the state denote? Moyo and Yeros do not specifically address these critical questions. How does the state radicalise and, if it does, what is the class position of the radicalised state? Is radicalisation a transformation of the state or just a tactic of survival? It is difficult to understand how a neocolonial state can choose to act against its class and serve the revolutionaries. Das argues elsewhere:

> Just as the neoliberal society is a class society, so the neoliberal state is a class state … neoliberalism had made no difference to the fact that the state must protect capitalist property relations. Indeed, government policy is much rather about the restoration of class power and increasing capitalist control over society's material resources (Das 2007: 4).

It is this class contest and arbitration by the neoliberal and neocolonial state to which scholars need to pay particular attention in this unfolding drama. The future of this revolution beyond this stalemate seems to lie in how the ideologues of the vanguard will tackle it tactically and strategically. Only then could we hope for the total emancipation of an African state from capitalism and neocolonialism to express true liberation.

Notes

1. Although some authors reject the alliance argument, Gerald Horne, studying the role of the United States in delaying Zimbabwe's independence, exposes the foundational formula for the alliance as designed by Andrew Young being 'a settlement [that] would leave Africans in charge of government and the European minority in control of the economy – an un steady alliance that continues to hold in Zimbabwe. This dispensation has been challenged ever more noisily in Zimbabwe since independence and, intermittently, in the United States' (2001: 22).
2. The term war veterans, in the context of Zimbabwe's revolution, has been defined in narrow militarist terms to mean only those cadres who wielded the gun (see Hammar and Raftopoulos 2003; Muzondidya; Vambe 2008). However, as a category of a social or political movement, they comprise a much broader group within the liberation movement, including former armed youth or militia in the former liberated and semi liberated zones at the height of the war (personal observation, 1978-79), recruits awaiting military training in so-called refugee camps when the war ended in 1979 and a growing body of youth continuously mobilized into the liberation movement then and after the war (for this characterization, see Sadomba 2008, 2011).

3. In his recent publication, Wilfred Mhanda (2011) gives details of the leadership conflicts during the ZIPA period between guerrilla leaders and nationalists up to the imprisonment of ZIPA in 1977. I have also held interviews with ZIPA commanders who were incarcerated. Mhanda's book confirms information from interviews. Mhanda adopted a self-centred and personalized approach to the leadership of ZIPA. There were no meetings held to design strategies and he sidelined senior cadres with both high levels of ideological consciousness and battlefield experience (interview, DTM 2000-2012). All interviews in this chapter were done by the author unless where specifically indicated otherwise.
4. Heidi Holland, in a documentary film ('Zimbabwe – Past the post … on a dark horse', Kevin Harris Productions (directed by Kevin Harris, 2010), emphasises that the dissident conflict was fanned by Apartheid South Africa against the ZANU-PF Government.
5. Sanctions were imposed on the country disguised as 'smart', but affecting the whole economy.
6. This alliance was composed of employers, civil society organisations, white commercial farmers, student activists and workers and was considered an 'unholy' alliance by Masunungure (2004: 171).
7. This period is quite distinct in that it was a time of marked war veteran leadership of occupying groups, with weak or no state or ZANU-PF involvement in the movement. Although the state took part, it was specifically as one of the actors, but with war veterans controlling the movement. All actors, viz war veterans, commercial farmers, peasants and farm workers have clearly distinguished this period in the interviews.
8. Z.W. Sadomba, Interview, K., 2004, at African Institute of Agrarian Studies.
9. Z.W. Sadomba, Interview, P., 2000, at his farm in Mazowe; Interview, DTM, 2000-2012 (these are longitudinal interviews which are still continuing at the time of writing); Interview, DM, 2000-2006 (these are longitudinal interviews held each year at various locations in Harare and at the occupied farms); Interview, Muchaneta, 2004-2006, longitudinal interviews held at Nyabira/Mazowe War Veterans Offices and the occupied farms.
10. From the 1992 Chinhoyi inaugural meeting of ZNLWVA, where war veterans confronted Mugabe on his leadership style and bad governance, demanding the dismissal of his cabinet and a return to the objectives of the liberation struggle, President Mugabe refused to meet the war veterans again. Their street demonstrations during the 1990s - climaxing in 'besieging'

the State House - were aimed at seeking audience with President Mugabe, who adamantly refused.

11. Many interviews held with war veterans revealed this. Interviews included O, a woman war veteran who led occupations in Domboshawa area (2004, at her farm in Hatcliff; this interview was also attended by Nancy Andrew); DM, a self-employed war veteran leader of occupations from 1998 (2000-2006); DT, a war veteran leader in Mazowe, University of Zimbabwe graduate in Political Science and former senior ZIPA commander (2000-2001, held in Concession).
12. Through their intelligence services, war veterans intercepted various communications like e-mails and secret documents circulating within the white commercial farming community, some of which are in the author's possession.
13. The formation of the MDC coincided with the period during which the Constitutional Assembly was consulting the electorate provincially. The project of the new party was to prepare for the next elections, to be held the following June. As such, the main objective of the new party was to display itself by actively working against the Constitutional Assembly programme, disrupting constitutional assembly meetings and mobilising the electorate.
14. This was clearly expressed by H (2000), a founder member of MDC, in an e-mail dated 11 February 2000.
15. Z.W. Sadomba, Participant observation, 2000, Corner House, Samora Machel Avenue.
16. Participant observation (Sadomba, January 2000): 'I participated in the demonstrations at Corner House at the intersection of Samora Machel Avenue and Leopold Takawira Street in Harare. The offices of both the Ministry of Justice (handling the constitutional process) and the British High Commission were located in that building. Munangagwa announced to the demonstrating war veterans that he had phoned Mugabe about our demands and Mugabe had promised to change the clause of the draft constitution by about two o'clock that afternoon which he did'.
17. The Land Occupiers and Protection Act 2001 and the Land Settlement Act 2000 were amendments that included Section 8, giving a maximum period of 90 days to wind off operations and vacate a designated farm.
18. Z.W. Sadomba, Interview, MT, 2001, Harare. Dr Hunzvi's statement is very clear about the role played by the ZNLWVA (*The Standard,* 15-19 March 2000), *'I must categorically state that I am not and was not responsible for the occupation of farms.'* Many researchers have erroneously dressed the

ZNLWVA in borrowed robes, giving them credit for organising land occupations. For example, Moyo and Yeros (2005: 189) claim that, '*The land occupation movement was organised by the War Veterans' Association.*' This needs correction, as the association, in its official capacity did not organise occupations, but they were organised at the local level outside the structure of the association.

19. Although scholars have documented the shrinkage of industry and growing unemployment, no studies have been carried out about the corresponding expansion of growth of informal industry and its employment capacity. A study by Sadomba and Mujeyi (2009) reveals that industry did not actually shrink, nor did employment, but rather reconfigured, absorbing highly qualified managerial and technical labour and expanding employment, explaining Zimbabwe's economic resilience in the first years of the crisis. However, this initiative was crippled by Murambatsvina.
20. Z.W. Sadomba, Interview, Spencer, 2009, at Magaba informal industry complex.
21. Z.W. Sadomba, Interview, Bota, X, 2004 at Chinhoyi Institute of Technology during the AIAS Stakeholder Workshop; Personal communication with H, a founder member of MDC, in an e-mail dated 11 February 2000; Interview, white farmer BT, 2004, Lowdale Farm.
22. The doctoral research done in the same research area by Angus Selby (2006), a son of a white commercial farmer, is also quite illustrative. It is notable that Selby fails to cite any assault from the research area, let alone any death, but he relies on newspaper reports on cases that took place outside the research area, at a national level. This article is not suggesting that there was absolutely no act of violence in the research area, but this was quite minimal. For example, one farmer was not allowed to get out of his house when he refused to share his land and he was rescued by the police. At another farm, war veterans admitted clapping and forcing the white farmer to take off his shoes and sit on the ground while they addressed farm workers. However, these were a few isolated cases and, in most cases, based on the farmer's reaction. At Duncombe, for example, the farmer, after agreeing to subdivide land and occupiers had planted their maize crop, came and ploughed them under and this caused retaliation by occupiers. They stripped tobacco leaves with whips and a new agreement for compensation was signed, this time at the police station (Personal observation 2000).
23. See Sadomba 2008. In this doctoral thesis, I include correspondence (Appendix 3) between war veterans and white commercial farmers which

reveals the non-violent and organised nature of the occupations. For example, the Townsends open their letter dated 16 August 2000 to war veterans, saying, 'Mr B and ourselves do appreciate that up to this stage your requests have been peaceful and non confrontational'. See also Z.W. Sadomba, 2006, Personal observation, Nyabira; 2000-2004, Participant observation, Nyabira, Mazowe and Matepatepa areas.

24. See also Selby (2006: 299). However, Selby did not have accurate information. Removal of Ngwenya was organised by war veterans from Concession who mobilised youths and land occupiers for this purpose. The Gaisfords were particularly defended from the beginning of occupations because of their role in supporting community health and education. They sought donations for clinics, hospitals and schools and were sitting on the board of Concession Hospital (Interview DTM 2003, war veteran leader and former civil servant). Failure to acknowledge the role of war veterans does not show the dynamics of the movement. A similar defence, by the war veterans as well, was of Hawks, next to Collingwood Farm.
25. Some supporters of the MDC from the rural areas also participated in the land occupations, but they had to disguise themselves partly because of their party position and partly because they feared the reaction of ZANU-PF supporters. This shows that land demand and the propensity to occupy cut across the political divide.
26. To give more details, 2001 was the transition from occupation to full Fast Track operation.
27. Various press reports and court cases of his divorce have revealed that Minister of Local Government, Ignatius Chombo, is a large-scale landlord owning prime urban land and multiple rural farms. Another such landlord is business tycoon and relative of President Mugabe, Phillip Chiyangwa. This has been seen as primitive accumulation backed by powerful alliance of ethnic forces.
28. 'Jambanja' is a Shona word which connotes simultaneous use of force (in this case, expropriation) and disregard or suspension of the 'received' rule of law.
29. For example, these new A2 farmers started to remove parts like plumbing materials, fancy lamp shades etc. from the farm houses to sell or to use in their houses in town.
30. 'Murambatsvina' is a Shona word which literally means 'one who rejects dirt or garbage'. It was coined by environmental health technicians because of their message of refraining from dirt. *Mu* is class one noun prefix; *ramb* is a verb root meaning refuse or refrain from; *tsvina* is noun

meaning dirt or garbage. However, *tsvina* is also euphemistically used to mean human excrement, but, in this context, excrement, as Judith Todd interprets (2007: 102), is a misnomer.

31. Z.W. Sadomba, Personal communication with the late Washington Chipfunde, a consultant to ZEXCOM (a company of war veterans) and a personal advisor of Dr Chenjerai Hunzvi, 1999-2002.
32. Dumiso Dabengwa, former Zimbabwe People's Revolutionary Army (ZIPRA) intelligence supremo, then Minister of Home Affairs, sent police to evict occupiers in March and April 2000. Joseph Musika, acting as President while Mugabe was out of the country, did the same later. In August 2000, Minister of Lands, John Nkomo, announced that occupations had to stop. War veterans actually clarified their position, telling prospective members of parliament that '... if government was saying "land to the people" as a political gimmick, we were on our part, serious.' (Interview DM 2000). In March 2000, war veterans locked ZANU-PF provincial offices and demanded an audience with President Mugabe, complaining that the ruling party and government were not pushing government and ZANU-PF to unequivocally support their land occupation initiative. President Mugabe sent Didymus Mutasa and Joseph Musika for negotiations (Interview DM 2000).
33. Government is always tactical the closer to elections the timing is: e.g. they only intensely executed the Fast Track after the 2002 presidential elections.
34. A day before the congress, the outgoing executive and provincial leaders were addressed by the Joint Operations Command (commanders of the uniformed forces and the Central Intelligence Organisation) at King George VI (KG VI) Barracks. According to a report back by one of the attendants, C the meeting had two objectives (Z.W. Sadomba, Report back by C, 2007, re meeting with Joint Command, at Magaba informal industry complex). One was to advise the leadership not to wash their dirty linen in public, meaning that their contradictions had to be shelved in the light of the focus by the international community on the events taking place in the country. The second was trying to impress upon the organisation to elect members who would be acceptable to the political leaders. It is possible that C wanted to use the report back to gain mileage as he was clearly a candidate sponsored by the politicians. He was heavily de-campaigned in Mutare and he could not become the Chair of ZNLWVA, a post that went to Jabulani Sibanda. Sibanda was preferred by war veterans for demonstrating courage against the ZANU-PF old guard in Matabeleland,

but was, however, later co-opted and became an ally of President Mugabe in the campaign for ZANU-PF congress in 2007.

35. Z.W. Sadomba, January 2008, Personal observation, Proceedings of war veterans' meeting convened by Jabulani Sibanda held at Fourth Street Offices, Harare. I attended this meeting and recorded it. There, Sibanda argued that war veterans had to campaign for politburo imposed candidates who had blocked them (war veterans) in the primaries. War veterans protested that they would not back these candidates. In addition, Sibanda campaigned for Mugabe's presidential candidature, emphasising the need to follow blindly.
36. The main characteristic feature that distinguishes the Murambatsvina period is retributive violence, epitomised by state coined operations, namely: Operation Murambatsvina, Operation Chikorokoza Chapera (mainly rural) and Operation Mavhotera Papi?
37. One of the most widely publicized cases of Murambatsvina was the destruction by a bulldozer of Chinx Chingaira's house. Chingaira, a prominent singer, was a war veteran and had acquired a stand through the housing cooperatives. He tried to stop destruction of his house by standing on top of it, but was pulled down and severely beaten by the police, warning the rest that the state meant business.
38. Credit for this victory certainly goes to the propaganda and media policies of Professor Jonathan Moyo, who was Mugabe's Information Minister and member of the ZANU-PF Politburo. Moyo used perennial tactics, including jingles and counter attacks on international and private local media focusing on the land revolution.
39. Z.W. Sadomba, Interview, war veterans and traditional leaders of Marange and Chiadzwa, 2007. I went to Chiadzwa in 2006, carrying out anthropological research in the area and at the diamond fields. I visited local traditional leaders, bush camps of the informal miners and war veterans and actually went into the mined areas which were heavily guarded by the Zimbabwe Republic Police. War veterans and local chiefs and headmen were bitter about government draconian measures. This research was prompted by a war veteran from this area, who had been dispossessed of his farm which was given to a University of Zimbabwe lecturer in Mazowe on ethnic lines, during *jambanja on jambanja* elite operations. He decided to venture into illegal diamond mining in his home area and I followed him there. One war veteran rhetorically questioned, 'Is this the *socialism* that we preached about during the war?' ('*Ndiyo here gutsaruzhinji yacho iyi iyo taitaurirana [kuhondo]*?') referring to possession of diamond

fields dubbed *mutaka waGrace newaMujuru* (the heaps of Grace and Mujuru the late general).

40. In many cases, fuel was resold on the parallel market and production vehicles like tractors were converted into taxis for desperate commuters (Personal observation 2005-2008).
41. Research in Zimbabwe has largely mimicked the political polarity, thereby clouded with non-academic pursuits by scholars.
42. Destruction of war veterans' houses, like that of Chinx Chingaira the popular musician of Chimurenga war songs and former member of ZANLA, was very conspicuous. Another war veteran leader of a housing cooperative in Malborough collapsed and died at the news of destruction of the houses.
43. The whereabouts of Murambatsvina victims and their impact wherever they went is yet to be studied. I carried out some research in 2006 in Zvimba, in 2007 in Marange (Chiadzwa diamond mines) and in 2008 in Uzumba, assessing Murambatsvina outcomes. It showed that Murambatsvina victims are spread in all social groups of the country and in all areas. Insome cases, the victims were allocated land by local leaders, establishing whole communities (Personal observation, Nyabira, 2006; Personal communication with Murambatsvina victims, Uzumba, 2008; Interview, war veterans and traditional leaders of Marange and Chiadzwa, 2007).
44. In some interviews, war veterans revealed that they had started to prepare for military defence in the occupied rural lands, anticipating state attacks after the urban clean-up. War veterans had started to organise units according to their wartime specialisations such as military engineering and intelligence, to prepare for military counter-attacks. It is not clear whether the Central Intelligence Organisation did get this information and advised the state to change tactics by embarking on Operation Chikorokoza Chapera in the rural setting.
45. However, noting that there were other elites who had to be included but did not satisfy the condition, an exemption clause was put for such members as those who had been on diplomatic missions.
46. Personal observation, Goromonzi, Chinhamora, Mutasa and Harare South constituency, 2008.
47. Personal observation, Zanadu, Concession, April-May, 2008: At a rally organised by the Zimbabwe National Army, which I attended, as war veterans we challenged the members of the army, the local MP and CIO operatives in attendance. War Veterans challenged their counterparts in ZNA as to why they (ZNA) were now coercing the masses – assaulting

them – to vote, yet the same masses would come to bases where battles were raging, to feed the combatants during the war. 'You have to ask yourself what has gone wrong with ZANU-PF instead of terming the people sell-outs' demanded war veterans. This diffused violence in the area in 2008.

48. The MDC split into two, with one group led by Arthur Mutambara and the other by Morgan Tsvangirai. The Mutambara party is known as MDC while the Tsvangirai party is referred to as MDC-T.
49. Note that when settlers decided to opt for agriculture after becoming frustrated that there were not as abundant mineral resources as had been speculated, they established a land bank as early as 1894 to finance capitalist farming. From about the 1950s, when European capitalist agriculture had finally managed to dominate African agriculture, settler farmers were at the cutting edge of global capitalist agriculture pioneering and leading in hybrid maize production seed (following only the US) and scooping several international awards, dominating tobacco farming, producing high-quality cotton and beef for the international market using state-of-the-art machinery and infrastructure.
50. NB nationalists agreed with the Smith regime and the British that the Rhodesian Security Forces (RSF) would be the sole and official defence forces during the ceasefire period, despite the RSF defeat marked by the battle of Mavhonde dubbed the 'Man-to-man Battle' by Peter Walls in August 1979. The RSF killed and maimed many guerrilla combatants during the ceasefire period, dumping them in mass graves that are being exhumed even at the time of writing this chapter. Their injuries were not treated by the new state and many died of the war wounds and health conditions (Sadomba 2011).

References

Alexander, J., ed., 2003, '*Squatters', Veterans and the State in Zimbabwe*, Harare: Weaver Press.

Andrew, N. and W. Sadomba, 2006, 'Zimbabwe: la "soif de terres" aux origines du mouvement des anciens combattants (Zimbabwe: Land hunger and the War Veteran-led Occupations Movement in 2000), *Critique Internationale*, No. 31, pp. 126-144.

Borras, S.M., 2001, 'State Society Relations in Land Reform Implementation in the Phillipines', *Development and Change*, Vol. 32, No. 30.

Daily News, 12 January 2000, 'War vets demo about land,' Staff Reporter, p.2..

Das, R.J., 2007, 'Introduction: Peasant, state and class', *Journal of Peasant Studies*, Vol. 34, No. 3 and 4, pp. 19.

Feltoe G., 2004, 'The onslaught against democracy and the rule of law in Zimbabwe in 2000', in D. Harrold-Berry, *Zimbabwe*: the Past is the Future, Harare: Weaver Press.

Gono, G., 2008, *Zimbabwe's Casino Economy: Extraordinary measures for extraordinary challenges*, Harare: Zimbabwe Publishing House.

GoZ, Constitutional Commission, 2000, T*he Draft Constitution: with amendments and clarifications*, Harare: Government Printers (dated 19 February 2000).

GoZ, 2000, 'Land Settlement Act 2000', Harare: Government Printers.

GoZ, 2001, 'Land Occupiers and Protection Act 2001', Harare: Government Printers.

H, 2000, a founder member of MDC, in an e-mail dated 11 February 2000

Haar van der, G., 2005, 'Land Reform, the State and the Zapatista uprising in Chiapas', *Journal of Peasant Studies*, Vol. 32, No. 3 and 4, pp. 17.

Hammar, A., 2009, 'A Measure of Just Demand? A response to Mamdani', Association of Concerned African Scholars.

Hammar, A. and B. Raftopolous, eds, 2003, *Zimbabwe's Unfinished Business: Rethinking land, state and nation*, Harare: Weaver Press.

Holland, H., 2008, *Dinner with Mugabe: The untold story of a freedom fighter who became a tyrant*, Johannesburg: Penguin Books.

Horne, G, 2001, *From the Barrel of the Gun: The United States and the war against Zimbabwe, 1965 - 1980*, Chapel Hill: University of Carolina Press.

Kriger, N., 1995, 'The Politics of Creating National Heroes: The search for political legitimacy and national identity', in N. Bhebhe and T.O. Ranger, eds, *Soldiers in Zimbabwe's liberation war*, Harare: University of Zimbabwe Publications.

Lynch, K., 2000, *Rural-Urban Interaction in the Developing World*, London: Routlege.

Madhuku, L., 2000, 'The Constitution-Making Process in Zimbabwe: The Way Forward', *Legal Forum*, Vol. 12, No. 1, pp. 50-60.

Mahoso, T., 2008, 'Reading the 2005 Tibaijuka Report on Zimbabwe in a Global Context', in M. Vambe, ed., *The Hidden Dimensions of Operation Murambatsvina in Zimbabwe*, Harare: Weaver Press: pp. 159-168.

Mamdani, M., 2008, 'Lessons of Zimbabwe', *London Review of Books*, Vol. 30, No. 23, Decembre 2008.

Marongwe, N., 2003, 'Farm Occupations and Occupiers in the new Politics of Land in Zimbabwe', in A. Hammar, B. Raftopolous, eds, *Zimbabwe's Unfinished Business: Rethinking land, state and nation in the context of crisis*, Harare: Weaver Press.

Mashonaland West War Veterans Association, 2004, 'Corruption document', Mashonaland West Provincial Stake-holder Dialogue, Chinhoyi Insitute of Technology, Author's archives.

Masuko, L., 2008, 'War Veterans and the Re-emergence of Housing Co-operatives', in S. Moyo, K. Helliker, T. Murisa, eds, *Contested Terrain: Land Reform and Civil Society in Contemporary Zimbabwe*, Pietermaritzburg: S&S Publishing.

Masunungure, E., 2004, 'Travails of Opposition Politics in Zimbabwe since Independence', in D. Harold-Berry, *Zimbabwe: the Past is the Future,* Harare: Weaver Press.

Masunungure, E., 2008, 'Civil Society and Land Reforms in Zimbabwe: Conceptual Considerations', in S. Moyo, K. Helliker and T. Murisa, eds, *Contested Terrain: Land Reform and Civil Society in Contemporary Zimbabwe*, Pietermaritzburg: S & S Publishing.

McCandless, E., 2011, *Polarisation and Transformation in Zimbabwe: Social Movements, Strategy Dilemmas and Change,* Lenham: Lexington Books.

Mhanda, W., 2011, *Dzino: Memories of a freedom fighter*, Harare: Weaver Press.

Mhiripiri, N., 2008, 'The Zimbabwe Government's Response to Criticism of Operation Murambatsvina', in M. Vambe, ed., T*he Hidden Dimensions of Operation Murambatsvina,* Harare: Weaver Press: pp. 149-158.

Mlambo, A., 2008, 'Historical Antecedent to Operation Murambatsvina', in M. Vambe, ed., *The Hidden Dimensions of Operation Murambatsvin*a, Harare: Weaver Press: pp. 8-23.

Moore, D., 2008, *Coercion, Consent, Context: Operation Murambatsvina and ZANU-PF's Quest for Hegemony,* Harare: Weaver Press.

Moyo, S., 1995, *The Land Question in Zimbabwe*, Harare: SAPES Trust.

Moyo, S., 1999, 'The Political Economy of land acquisition in Zimbabwe', *Journal of Southern African Studies*, Vol. 26, No. 1, pp. 5 - 28.

Moyo, S., 2001, 'The Land Movement and democratisation in Zimbabwe': contradictions of neo-liberalism', *Millennium Journal of International Studies*, Vol. 30, No. 2.

Moyo, S., 2005, 'Land policy, poverty reduction and public action in Zimbabwe', in A. Haroon Akram-Lodhi, S.M. Borras Jr. and C. Kay, eds, Land, policy and livelihoods in an era of globalisation: perspectives from developing and transition countries, Routledge and Institute of Social Studies (ISS), the Hague, Studies in Rural Livelihoods, London and New York: Routledge, pp. 344-382.

Moyo, S., 2011a, 'Three decades of agrarian reform in Zimbabwe', *Journal of Peasant Studie*s, Vol. 38, No. 3, pp. 493 - 531.

Moyo, S., 2011b, 'Land concentration and accumulation after redistributive reform in post-settler Zimbabwe', *Review of African Political Economy*, Vol. 38, No. 128, pp. 257 - 276.

Moyo, S. and Yeros, P., 2005, 'Land Occupations and Land Reform in Zimbabwe: Towards the National Democratic Revolution', in *Reclaiming the land*, London: Zed Books.

Moyo, S. and Yeros, P., 2007, 'The Radicalised State: Zimbabwe's Interrupted Revolution', *Review of Political Economy*, Vol. 34, No. 111.

Moyo, S. and Yeros, P., 2009, 'Zimbabwe ten years on: results and prospects'. (http://www.oambazuka.org/en/category/features/54037 accessed 25 February 2009).

Moyo, S. and Yeros, P., 2011, *Reclaiming the Nation: The Return of the National Question in Africa, Asia and Latin America,* London: Pluto Press.

Mudenge, I., 2004, 'Zimbabwe's land reform program (the reversal of colonial land occupation and domination): Its impact on the country's regional and international relations', *The Struggle Continues,* Harare Sheraton, ZANU-PF.

Sadomba, Z.W., 2008, 'War Veterans in Zimbabwe's Land Occupations: Complexities of a liberation movement in an African post-colonial settler society', PhD. Thesis, Wageningen University.

Sadomba, Z.W., 2008a, 'Whither Zimbabwe? Revolution, reform or reaction?', *Public debate Wilfred Mhanda,* University of Johannesburg Faculty of Humanities, University of Johannesburg Dept. of Anthropology and Development Studies.

Sadomba, Z.W., 2011, *War Veterans in Zimbabwe's Revolution: Challenging neo-colonialism & settler & international capital,* Suffolk: James Currey and Harare: Weaver Press.

Sadomba, Z.W. and Mujeyi, K., 2009, Sadomba, Z.W. and K. Mujeyi, unpublished manuscript, 'The Folly of Nipping Zimbabwe's Indigenous Industrialisation in the Bud: Murambatsvina, class, state and power'.

Selby, A., 2006, 'Commercial Farmers and the State: Interest Group Politics and Land Reform in Zimbabwe', PhD. Thesis, Oxford University.

The Standard, 15-19 March 2000.

Tibaijuka, A.K., 2005, *Report on UNDP Mission on Zimbabwe,* New York: UNDP.

Todd, J., 2007, *Through the Darkness: A Life in Zimbabwe,* Cape Town: Zebra Press.

Toriro, P., 2005, 'Zimbabwe's urban land and housing question', *Policy, planning and implementation issues,* Harare: African Institute of Agrarian Studies.

Toriro, P., 2006, 'The Urban Housing Question in the context of Operation Murambatsvina and operation Garikai/Hlalani Kuhle', Harare: African Institute of Agrarian Studies monograph series.

Utete, C.M.B., 2003, Report of the Presidential Land Review Committee, Government of Zimbabwe, Harare: Government Printers.

Vambe, M., ed. 2008, *The Hidden Dimensions of Operation Murambatsvina in Zimbabwe,* Harare: Weaver Press.

Vambe, M., 2008a, 'Introduction: Rethinking Citizen and Subject in Zimbabwe', in M. Vambe, eds, *The Hidden Dimensions of Operation Murambatsvina in Zimbabwe,* Harare: Weaver Press.

4

Nyabira-Mazowe War Veterans' Association: A Microcosm of the National Land Occupation Movement

Louis Masuko

Introduction

Land reforms have taken shape in many countries of the world, across all continents and at different stages of their respective development. In Zimbabwe, the Fast Track Land Reform Programme (FTLRP), an outcome of invasions and subsequent occupations of Large Scale Commercial Farms (LSCF), shook the Zimbabwean and Western aristocrats' establishment in 2000 and has been characterised as the 'first radical shift in agrarian property rights in the post Cold War world' (Moyo and Yeros 2005). The FTLRP radically changed, not only the unequal and inequitable land distribution in Zimbabwe, but insecurity of land tenure and unsustainable and suboptimal land use as well. It ended the hegemony of the minority whites on land and in the agriculture sector (Masuko 2004), empowered the landless black majority and set a solid pathway for solving the long standing land question in Zimbabwe.

Opinion surrounding the causes of Zimbabwe's land reform, the forces behind it, its timing, its outcome and its legitimacy differs, largely along lines of the diverse interests of the different contenders and/or their ideological inclinations. The overall controversy on the route taken by Zimbabwe's land reform is whether it was indeed a part of a broader development strategy to propel the country to a sustainable social and economic growth path (see Chambati and Moyo 2007), or meant to serve immediate political ends. The first perspective views land occupation and the FTLRP as a strategy to address the land question (Masuko 2004, 2008; Moyo and Yeros 2005; Moyo, Murisa and Helliker 2008). Central to this opinion is the understanding that the occupations were a culmination of a process that had been unfolding under the surface, to address a national question, since 1980 (Masuko 2004; Moyo and Yeros 2005; Sadomba 2008). The occupations, according to this view, were a protest by the landless against the ZANU-PF government (Masuko 2004; Moyo and Yeros 2005) for not utilising the two-thirds majority that it enjoyed in the 1990s to speed up the land acquisition and redistribution process. Indeed, this was a direct challenge to the neoliberal state, capital and the laws that protected the system (Masuko 2008).

Conversely, an opinion is advanced that the occupations and the FTLRP were a political gimmick meant to bolster the waning support for the ruling ZANU-PF party (Raftopoulos 2003; Hammar 2003; Sachikonye 2003). They highlight the strong hand of the state in the process, including the violation of the rule of law, human rights and good governance. Indeed the 'beleaguered' state had to create a crisis in order to sustain itself in power. Land occupations are viewed as an event that was only triggered by the rejection of the draft constitution of 2000 (Masiiwa and Chipungu 2004; Raftopoulos and Phimister 2004; Hammar 2003). Sustainability of the programme was brought into question as the land occupation movement was reduced to a single- issue partisan strategy focusing exclusively on the immediate question of land repossession, its distribution to Mugabe's cronies and not meant to address long-term political, social and economic questions.

Without losing focus on the substance at the core of these two contending opinions, this chapter explores the sustainability of Zimbabwe's land reform from a social movement perspective. As a case study, it uses the Nyabira-Mazowe War Veterans Association (NMWVA), one of the many micro-organisations that led the invasions and occupations of Large Scale Commercial Farms (LSCF) in Zimbabwe. Two issues come to the fore. First, this chapter considers whether what some observers have seen as one movement or a top down state intervention was indeed a national movement made up of small groups of people controlled and led at the micro level, employing different strategies and tactics to achieve the same objective. Second, the chapter illuminates how new social and political structures, networks and institutions, that never existed before the land occupations, developed in the occupied areas. Such networks and institutions, it is argued, provide the necessary social capital and social structure conducive for future movement activity to take land reform beyond the single issue of land repossession.

Three questions are raised here to provide the theoretical framework within which the issues raised above are examined and analyzed. The central questions asked include: (i) what is the process that results in individuals engaging in occupations?; (ii) how and why did people come together to challenge the existing land distribution regime and the legal system that supported it?; and (iii) were there structures and networks that existed before, during and after the occupations? Any theory that adequately addresses these questions must, first, account for an individual's willingness to expose himself or herself to exceptional physical danger, explain how individuals come together to form rebellious groups and reveal why groups of citizens are able to militantly challenge a state and its coercive apparatus and the force and strength of capital.

A framework that embraces the 'micro-mobilization process' (Snow et al 1986) that links Tilly's (1978) and Freeman's (1983) discussion of categories, networks and group mobilisation and Gurr's (1970) discussion of relative deprivation is used for this study. To provide a comprehensive context to the discussion, Gurr's (1970) notion of the 'coercive balance' between rebels and states and Tilly's (1978) discussion of state-challenger conflict is informed by Skocpol's state-centred analysis of the structural conditions which weaken a state's ability to employ its coercive capacity.

Various instruments were used to collect both primary and secondary data and information. Participatory observations were undertaken during the period between 2000 and 2003 and provided most of the baseline information used alongside a structured questionnaire administered to 24 former members of the NMWVA, focusing on those who had been allocated land, rather than on those who dropped out of the occupation and those who did not acquire land through occupation. Interviews were also carried out with the members of the executive of the association. Minutes of the NMWVA association recorded between the year 2000 and 2003 and some from the District Land Committee meetings were analyzed. Lastly, data from the baseline study conducted particularly in Zvimba District by the African Institute of Agrarian Studies (AIAS) in 2005 was also used.

Seven sections make up this chapter, including this introduction. A brief background of the Nyabira-Mazowe War Veterans Association, the origin and class structure of its members, are explained in section two. Section three describes the organisational and leadership structure of the NMWVA. In section four, an analysis is developed of the importance of groups and networks in organising people into social movements with the capacity to demand reforms that change the status of their members; and section five shows that collective action is a function of the existence of relative individual deprivation. Section six shows that by developing new institutions, networks and relations, social movements mutate, reproduce and define new demands and ways of achieving them.

This chapter highlights how land occupation and reform in Zimbabwe is a contested terrain. This is so regardless of the observations that from the second half of the 1990s into the 2000s, the balance of class forces within ZANU-PF and society in general was tipped in favour of radical nationalist solutions (see Moyo and Yeros 2005; Masuko forthcoming). The journey to explore the contestations commences with section two below. In this section, the social fabric of NMWVA, one of the micro-organisms that actively executed the occupations that led to the FTLRP, is examined.

Origin and class composition of the NMWVA membership

A social movement is defined by Garreton as a collective action with some stability over time and some degree of organisation and oriented towards change or conservation of society or some sphere of it (1997). The land occupation social movement in Zimbabwe was an outcome of the integration of small and manageable units at the local level (see also Sadomba 2008). The Nyabira-Mazowe War Veterans Association (NMWVA) was one of these small micro-organisations whose origins, membership and class composition was diverse.

The NMWVA derived its name from the districts that it occupied during the land occupations and the group of people who led the process, the veterans of the war of liberation struggle. Nyabira is an area within Zvimba District in the Mashonaland West Province and shares boundaries with Mazowe District in the Mashonaland Central Province. The NMWVA as an organisation occupied 10 farms in Mashonaland West Province and 12 farms in Mashonaland Central Province, covering an area of plus or minus 44 thousand hectares. The organisation was formed in February 2000, immediately after the first occupations by the war veterans from the City of Harare that saw most farms surrounding the city being occupied.

With a robust membership base of more than 200 paid- up members (excluding spouses and siblings) by mid 2000 (NMWVA Membership Register 2000), the NMWVA took the occupations further, deeper into the districts adjacent to the City of Harare and as far as the Great Dyke to the north- west. In Mashonaland West, the NMWVA collaborated and networked with the Hunyani War Veterans Association and individual spiritual leaders such as Sekuru Mushowe of Hunyani Hills, to the west and south, respectively, while in Mashonaland Central, they collaborated with two more small movements led by war veterans Cde. Urimbo and Cde. Zimbabwe, respectively, who occupied southern parts of Mazowe District.

The Nyabira-Mazowe War Veteran Association was formed by war veterans of the Second Chimurenga, at the onset of the land occupations, to organise the occupation process and was integrated into the national land occupation movement, not necessarily along vertical structures, but horizontally through networking with other occupying groups. In other areas, however, there were no such organisations as the occupiers rallied behind a particular war veteran or a group of war veterans (e.g., South-west Mazowe and South- east Mazowe), the area Member of Parliament and traditional leaders (e.g., Hunyani Hills area occupations led by Sekuru Mushowe), who in turn sought to formalise their occupations by appealing to war veterans (see Moyo and Yeros 2005).

Zvimba District (in which the Nyabira area falls) and Mazowe District are much closer to Harare (they share boundaries) than to communal areas (which were the centre of many occupation groups and initiatives) (see Sadomba 2008) and the need for such organised occupation was critical to minimise vandalism and conflict between occupiers themselves, who tended to concentrate on fewer farms closer to town. These areas had no formal political party or Zimbabwe National Liberation War Veteran Association (ZNLWVA) structures, except for the Nyawo dance groups and pretty little or no initiative to occupy farms was likely to come from within the LSCFs themselves (i.e. from farm workers). Such Large Scale Commercial Farming areas could be described as 'liberated zones' for the white farmers and the sanctity of private property ruled supreme. These were areas where only the laws of the white farmers were observed.

Although the name suggests that the movement comprised war veterans only, membership included members of ZANU-PF, the landless residents and the youth, who resided mainly, but not exclusively, in the high-density suburbs of Warren Park, Kuwadzana and Dzivarasekwa and the low-density suburbs of Mabelreign, Marlborough and Mount Hampden. The six suburbs are on the western and north-western side of the City of Harare and adjacent to the two districts of Zvimba and Mazowe. NMWVA membership cut across class lines to include industrial and farm workers, managers and directors, academics, self employed-indigenous people and the unemployed (see Table 4.1). Two groups were dominant: the retired professionals in the private sector and those who were self-employed. Data from an AIAS baseline study show that in Zvimba District, in general, the dominant groups were the private sector unskilled employees and uniformed personnel (Table 4.1). The difference basically reflects the urban-rural settings of the survey areas. The number of farm workers in both cases is not very significant.

Table 4.1: Class structure of the NMWVA farm occupiers and land beneficiaries

NMWVA																
Category	**Private Sector**		**Private Sector Unskilled**		**Self-employed**		**Civil Servants Managers /Skilled**		**Civil Servants Unskilled**		**Uniformed**		**Farm Workers**		**Total**	
	No	**%**	**No**	**%**	**No**	**%**	**No**	**%**	**No**	**%**	**No**	**%**	**No**	**%**	**No**	**%**
Employed	1	8			8	62			2	15	1	8	1	8	13	100
Unemployed (former professional)	7	63	1	9			2	18			1	9	1	8	13	100
Total	8	33	1	4	8	33	2	8	2	8	2	8	1	4	24	100
Land Beneficiaries															**N/A**	
Employed	86	10.8	200	25	53	6.5	65	8.1	41	2			4	5	235	29.4
Unemployed (former professional)	84	6.9	413	34.5	101	8.4	41	3.4	85	7.1	221	18.5	20	1.7	221	18.5
Total	170	9	613	30	154	7.5	106	6.5	126	5	221	18.5	24	2.2	456	24

Source: Masuko 2008 and AIAS Baseline Study (2005): Zvimba District

The urban character of the membership of the NMWVA was strong with 54 per cent of the respondents originating from both the high and low density suburbs of Harare (Table 4.2). About 4 per cent came from the large-scale farming areas within the occupied farms. Forty-two per cent originated from communal areas within the respective provinces of Mashonaland West and Central. Even those that came from communal areas knew of the occupation from their Harare connections. In contrast, 62.5 per cent of the occupiers in the AIAS Baseline Survey originated from the communal areas and only 22 per cent from urban areas.[1] Those from the Diaspora and other farming schemes (*e.g.* LSCF areas) accounted for 8 per cent of the total sample. What stood out from this data is that urban industrial workers and the urban unemployed, contrary to the belief held by the opposition political parties, saw an opportunity to extricate themselves from urban poverty and a chance to own a piece of land. The strength of the NMWVA resided in their diversity in terms of class, social background and skills possessed.[2]

Table 4.2: Origin of land occupiers: NMWVA and national

	NMWVA		**National (AIAS sample)**	
Area of Origin	**No of people**	%	**No of people**	%
Urban Areas	13	54	452	22.9
LSCF Areas	1	4	159	8.1
Communal Areas	10	42	1227	62.1
Other			131	6.9
Total	24	100	1969	100

Source: Masuko 2008 and AIAS Baseline Study 2005

Membership at the inception of the association stretched across all the individual categories outlined by Snow (1986), that is: adherents, those individuals who identified with the occupying organisation and were willing to act on that identification; free riders, a pool of potentials who identified but did not contribute; and those who did not identify with the organisation and by default were unwilling to act. However, the majority of the latter two groups of individuals who could be referred to as the Potential Organisation Members (POMS) did not, in the case of the NMWVA, last the distance as the land occupation process intensified.

Starting in June 2000, new criteria to qualify for membership[3] in the association were outlined to screen the POMS from those who identified with and participated in the activities of the association. These included consistent physical participation and contributions in cash and in kind to support the day-to-day operations of the occupying movement. The former involved attending meetings held every Saturday, carrying out the actual invasions according to the strategy mapped out for the particular farm and putting up temporary shelter and tilling at least a hectare of land on the occupied farm, themselves a sign of the transformation from land invasion to land occupation (see Moyo 2001). Cash contributions took the form of a joining fee and a monthly subscription, as determined by the leadership and approved by the Saturday general meetings. Any other commitments in the form of vehicles for transporting the members during invasions and occupations, fuel, stationery, etc, were voluntary.

With these changes in the recruitment criteria, the NMWVA transformed itself between February and June 2000, from what Truman (1951) identified as a latent organisation or the un-mobilised sentiment pool (McCarthy 1987) to an organisation with potential for collective action and later to a full Social Movement Organisation (SMOs) (Garreton 1997). The former two stages of development of an organisation have only the potential to act, because they lack the requisite number and/or types of resources (i.e. ideological cohesion, material resources, human participation and communal structures/organisation) necessary to act.

To signal its transformation from a land invasion movement to an occupation movement, the NMWVA's strategy changed from simply stationing youths and war veterans on compounds and holding bases to establishing permanent bases (command structures) on each of the farms they occupied. Members committed themselves to constructing temporary shelters on the occupied farms; a register of paid up members was compiled and the association mobilised enough human power, financial means and transport to implement its programme (NMWVA Executive Committee Meeting Minutes 2000). In other words, it had already acquired both the numbers and types of resources necessary to act. The numbers were made up of around two hundred members, their spouses and the youths. It is important to mention that, before the changes in recruitment criteria, the participants under the NMWVA's command were plus or minus one thousand.

What emerges is that indeed there were many micro-organisations that executed the invasions and occupations and operated independently of the state. The urban element and the class diversity of the members of the NMWVA confirm that even the urbanites felt deprived of a resource that was by birth theirs and that land was not an issue only for the rural peasants. The capacity to mobilise people and resources and to execute the occupation without assistance from the state or any other external influences, was clearly demonstrated in the manner the NMWVA transformed from a latent organisation to an organisation with potential for collective action and later to a full Social Movement Organisation (SMOs). Section three below describes how the association managed the people and resources and the occupation process itself.

Organisational and leadership structure of the NMWVA

This section briefly discusses the hierarchical structure of the NMWVA. It then outlines how the different posts within the hierarchy were occupied and who was eligible to occupy which posts, discussing the responsibilities of the different functionaries. The section also seeks to show that while the NMWVA was a micro-organisation, it had far-reaching links that made it part of the much larger national invasion and occupation movement.

The NMWVA, like many rebellious organisations had both administrative and operational structures. Administratively, a chairperson occupied the top post and was supported by a deputy chairperson and functionary departments which were headed by experts; these included the secretariat, treasury, security and planning. Being the overall leader of the association, the chairperson was the undisputed custodian of the association and was appointed by a caucus of war veterans. She or he was the link between the association and the state, politicians, donors, public enterprises and other occupation groups. Over and above, the chairperson was responsible for maintaining the members' interest in the activities and their coherence as an association

The operations of the association were the direct responsibility of the deputy chairperson, popularly known as the Zone Commander, who reported directly to the chairperson. The title of Zone Commander owes its origin to the fact that Mazowe District was divided into six zones for the purpose of coordination (see Sadomba 2008) and each zone was to be led by a war veteran. The seventh zone which was grudgingly 'recognized' by the District Land Committee in 2000 fell in the area occupied by the NMWVA and was part of the Mazowe West constituency. Operations involved farm invasions and occupations, evictions, demonstrations, security and surveillance, organising Saturday general meetings, etc. While deputising the chairperson came with the title, the Deputy Chairperson also acted in the absence of the chairperson. But more importantly, each of the two top most leaders of the association would not act unilaterally without consulting the other or, if need be, a caucus of war veterans within the operation zone.

Each occupied farm was referred to as a base and had its own structure that resembled that at the zone level, headed by a 'base commander'. However, during the invasion stage (first three months from February 2000), only one farm would be used as a holding base. Such a farm, generally strategically located, was used as the launch pad from which more farms would be invaded. At the end of each day, the invaders would retreat to the holding base where tactics for the next assignments would be discussed. 'Base commanders', who were appointees of the executive of the association and were, in all cases, war veterans, were the link between the executive committee and the membership between meetings. Although base commanders were not part of the executive of the association, they attended weekly executive meetings held once every week in Harare. Agenda items at these meetings included problems encountered at the occupied farms, concerns of the membership, new farms to be occupied, planned meetings with different stakeholders and resource and membership mobilisation. Strategies and tactics were also reviewed during the same meetings (see NMWVA Executive Committee Meeting Minutes 2000-2003).

General meetings were held every Saturday at a central place in the occupied districts or at an identified farm, particularly where problems that required the attention of the executive would have been reported. Grievances from the membership were discussed and solutions reached unanimously or by majority vote by show of hands. These were also report- back meetings, where the executive was expected to provide answers to questions and problems raised during the previous general meetings. These meetings, in a very big way, kept the interest of the membership in the business of the association very strong. They were basically designed to consolidate the ideological position by galvanising individual deprivation into collective deprivation and continuously reminding the membership of the source of their deprivation. Although tactics to occupy new farms or to consolidate the position on the occupied farms were discussed, this was done only if the tactics were to be implemented on that day, straight from the meeting, for security reasons. Such strategies and/or tactics proved to be the most effective and popular tactics, as the strength that lay in numbers was easily put to effective use.

Although democratic practices were evident in the way the association conducted its business, its leaders were not elected by the majority of its members. Leaders were chosen from among war veterans by war veterans. This approach to leadership applied to both the structures at the association level and the farm or base level. Non- war veteran members of the association were also handpicked by the war veterans on the basis of trust. This approach was primarily to keep strategies and tactics closely guarded, as it was difficult to trust non-war veterans with issues involving tactics and strategies.[4]

As trust was built over time, non- war veterans were co-opted into the executive, but such co-options were restricted to areas that needed expertise. This was necessitated by the paucity of war veterans in the structures or war veterans with expertise in a particular area. The planning department, for example, was staffed by retired AREX personnel who were familiar with the occupied areas and, in the majority of cases, with the farmers too and possessed skills in land surveying; secretarial work within the secretariat department was done by experienced personnel, some with shorthand skills to record the minutes of the meetings; and, while the treasury department was led by a war veteran, trained bookkeepers recorded the transactions. The same approach to leadership applied at the farm level. Up to today, these non- war veteran office bearers are recognised by the generality of the membership and farm workers in the areas of resettlement as war veterans.

From the description above, the leadership of the association was not democratically elected, but appointed. This approach to leadership survived the test of time because members of the association held the belief that war veterans had the keys to many doors and had the stamina to stand up to politicians and farmers and indeed had direct links to Officers-in-Charge at the different police stations who were themselves, in the majority of cases, also war veterans. However, evidence from Saturday meetings suggest that democratic practices determined the outcome at these meetings, either through consensus or voting by show of hands. The latter tenet of democracy has its critics, but it served the association well under the circumstances. This is shown by the successes recorded by the association in occupying farms and in resettling its members on these farms (and on some farms as far as Glendale) between 2000 and 2004. The association's leadership and organisational structure coordinated the mobilisation, management and deployment of members and resources, while simultaneously networking with the world outside of the association. Section four below articulates some of the strategies and tactics used during the lifespan of the association.

Micro-mobilisation of land movements and relations with the state

Few studies provide empirical evidence on how and why people came together to form the land occupation movement to challenge the existing land distribution regime and the legal system that supported it. Few studies account for the individuals' willingness to expose themselves to exceptional physical danger, or for how groups of citizens are able to militantly challenge a state and its coercive apparatus and the force and strength of capital. A critical explanatory variable in any theory of rebellion is the mobilisation of individuals into organised groups of people (SMOs), including the interaction between individuals and groups in terms of recruitment and maintaining commitment of resources to build capacity for collective action. This section seeks to show that salient shared traits and strong networking (frequent interaction) must both be present or forged for an organisation to act collectively.

Both the critical shared traits and frequent interaction among members of the NMWVA existed before the formation of the NMWVA. Documented evidence shows that Zimbabweans, from time immemorial, have formed groups and associations through which they share, not only their problems and achievements, but express their anger and dislike of the system. These include burial societies, boxing clubs, trade unions, savings clubs, cooking and sewing clubs, housing cooperatives, formal and informal business associations, *etc.* In the case of the NMWVA, two organisations that provided the platform for frequent interaction were the Zimbabwe National Liberation War Veterans Association (ZNLWVA) and ZANU-PF. Most of the occupiers were either members of ZANU-PF or ZNLWVA. They were also members of indigenous business groups, localised savings clubs, housing cooperatives, or burial societies. Moreover, thc majority were also related to each other, either because they came from the same rural areas (through homeboy drinking clubs) or were blood relatives.

Of the NMWVA members and would-be land occupiers interviewed, 96 per cent got their information on the land reform either through their political party structure or from war veterans, or both (Table 4.3). Yet the AIAS baseline survey shows that over 57 per cent of the wider land beneficiaries got to know of the land reform through their political party structures. Otherwise such information got to settlers (read as occupiers) through the media (accounting for 19.6 per cent), other villagers (7.3 per cent) and Agriculture Extension Officers AREX (5.1 per cent). Chiefs accounted for less than 5 per cent of the sources of information and, ironically, the war veterans, who were the organisers and leaders of the occupation, were cited by only 0.4 per cent of the respondents as their source.

Table 4.3: WVA and national sample

	NMWVA		National Sample	
Source of Information	**No of respondents**	**%**	**No of respondents**	**%**
Chief			93	4.8
Media	1	4	377	19.6
Other Villagers			141	7.3
RDC			78	4.1
AREX			98	5.1
Political Party	12	50	1111	57.9
DA			13	0.7
Political Party/War Vets	5	21		
War Veterans	6	25	8	0.4
Total	24	100	1919	100

Source: Masuko 2008 and AIAS Baseline Study 2005

The ruling political party, ZANU-PF, had widespread structures that started from the cell level, to ward, district and province, feeding into the national level with members at all these levels. These consistently met and interacted well before 2000. However, from the onset of the land reform, the information on the opportunity to own a piece of land was shared between ZANU-PF activists and their relatives or friends who were not active party members. Most could not afford to miss the unique opportunity to own a piece of land and joining the NMWVA was simply based on their willingness to commit themselves to the cause of the organisation. There was no vetting to flush out non- ZANU-PF occupiers.

War veterans generated and processed information on occupations and opportunities to own a piece of farming land without formally applying for it through the official channels. They disseminated information on their achievements, how they were 'taming' the white farmers and even what rich pickings were to be found within the occupied farms (such as scrap metal, cheap beef from farm butcheries, etc). This information would be transmitted to the ZANU-PF members through the party structures (see also Sadomba 2008). While the war veteran structures also filtered down to ward level, they were basically at the service of war veterans. But the symbiotic[5] relations that existed between the ZNLWVA and the ruling party made it easy for the war veterans to use and channel information through the structures of the ZANU-PF.

The majority of those who got information through this channel participated in the initial occupations and were founder members of the NMWVA. Most of those that got information through the party structures were individuals who identified with the occupying organisation and were willing to become adherents of the cause. The same applies to the other 20 per cent that got the information through other villagers, chiefs and agricultural extension officers. These sources of information are within the sphere of influence of ZANU-PF structures to the extent that even information on agriculture field days is announced at ZANU-PF meetings.

The late- comers to the occupations were basically those who reacted to media information. This group comprises those who joined the land occupations three to four months after the initial occupations. Although late occupiers could be among those associated with the NMWVA, the majority of them fell into the group of persons with the potential to identify with the land reform cause, but did not contribute resources or were free riders. Many did not identify with the NMWVA organisation and by default were unwilling to act on the ground, although some were willing to provide resources and let the movement work for them. The initial goal was to get their share of the land and then reduce their role in collective action. Expectations were that the process was short and swift and not protracted and not a transformative struggle of the opposites that required physical participation and contribution of resources to sustain it. Some did not want to be identified with the ruling party, the war veterans and the perceived lawlessness, in spite of their desire to own a piece of land. In this group, urban upper-middle and high- income class, were some members of the opposition party (MDC) who feared reprisals from their leadership, those in the NGO sector and, surprisingly, some of the senior members of the ruling party itself who equated occupations to lawlessness and therefore a liability to their political ambitions.

The large scale-land invasions that commenced in February 2000 took both the state and agrarian capital by surprise. Even the very few, such as senior ZANU-PF officials who had some information about the impending land occupations, did not believe that any such nationwide protests would actually happen. This scepticism derived from the coercive force at the disposal of the state, the laws protecting private property and the strong networks, arms and ammunition registered in the names of the Large Scale Commercial Farmers. They believed that the strong ties that had been developed over the years between the white farmers and the senior politicians within ZANU-PF and the state managers were insurmountable! The latter two had acquired farms through purchases on the open market, through different land reform programmes such as the Tenant Farmer Scheme of the mid 1990s. Being the new landlords, the two were reading from the same script as the Large Scale Commercial Farmers and had developed personal and working relations with the latter. Senior politicians and state managers would protect their interests and by design those of the Large Scale Commercial Farmers. Calls by responsible ministers and technocrats to send the police and army to flush out the invaders, besides the true intensions to safeguard the rule of law, showed the class instincts of the propertied class of politicians. Potentially embarrassing clashes between classes were avoided following the call by the President not to pit 'war veterans against war veterans' (Masuko 2004).

However, the land invasions and later the land occupations demonstrated the relative autonomy of the state in Zimbabwe, as conceptualised elsewhere (see Bonano 1988) and that the class contradictions that are inscribed in the very nature of the state (Poulantzas 1975) can enable radical reform. The NMWVA therefore took advantage of the technical paralysis caused by the class contradictions within the state after the invasions to compete for legitimacy with the state (see 'multiple sovereignty' by Tilly 1978). While the state could not be seen to be protecting the white farmer (by enforcing the laws and flexing its coercive power), it equally and at the same time could not lose legitimacy by going against equitable land redistribution, itself the ultimate objective of the occupations.

Taking advantage of this technical paralysis, the NMWVA, through the District Administrator's office, demanded to be part of the District Land Committee, a government institution for land identification and resettlement representing Mazowe West Ward 23 which fell within Zone 7 of the movement's spatial structure. This denial was in spite of the presence of a war veteran district representative, as prescribed within the official documentation, on the composition of the committee. As most of the members of the NMWVA were not from the communal areas of Chiweshe or Mashonaland Central Province, there was strong suspicion that their demands could not be well represented by those whose interests were influenced by the power balance in Chiweshe communal lands. This mistrust did not go down well with the local politicians, the war veteran representative and the Zone 6 commander under whose command Zone 7 fell and remained a source of friction throughout the land occupations.

However, a compromise was reached and the NMWVA was only grudgingly accepted into the District Land Committee. This acceptance was in acknowledgement of the capacity of the association to influence political outcomes within the area they had occupied and the networks that had evolved between the NMWVA and other occupying groups in Mazowe District. More importantly, occupations were still at their peak, the political situation was volatile and the area was under the control of the NMWVA, which had for the first time established ZANU-PF and war veteran structures within the area. Mazowe District War Veterans Association structures only served war veterans that were concentrated in the Chiweshe communal farming area. The compromise showed the state's concern for stability, hence its desire to create 'proper channels' within which to bargain. Indeed, the state and politicians from ZANU-PF sought to engage the leaders of the NMWVA through their representation in the land committee in order to extract political mileage from the structures and the vote of the masses of farm workers that the association had established and mobilised, respectively, in Ward 23 of Mazowe West constituency.

Because of the temporary technical paralysis in local governance in 2000, the state opted to rule primarily through consensus, rather than coercion as highlighted above. It tended to 'care' about representation of the major constituencies in Mazowe District. Major constituencies ultimately meant those groups with some capacity to disrupt the established order, be it through a capital or labour strike, electoral abstention or challenge, or more overt forms of mass defiance. The NMWVA had its strength in numbers and capacity to influence the politics of Mazowe West Ward 23 and Zvimba constituency, particularly between 2000 and 2001 as it, besides establishing structures, had developed strong linkages with other occupying groups in the two districts and with spirit mediums such as Sekuru Mushowe of the Hunyani Hills.

Relations changed after July 2001 with the intervention of the state through legislative mechanisms, particularly the launch of the FTLRP in 2000. The Land Occupiers Act 2001 and the Abuja Agreement of 2001 signalled the official commencement of the Fast Track Land Reform. These developments allowed the politicians and the bureaucrats to develop hegemony over land occupations and ultimately to own the land revolution. The state started to encourage alternative representatives to compete for the allegiance of the land movement's mass base. The offices of the Provincial Governor and District Administrator became very visible in Mazowe West, communicating the message that they were the only authority responsible for land allocation and therefore all those who wanted land should register with them. To drive the point home, the farm occupied by the Chairperson of the NMWVA and his deputy was officially allocated to the Chair and Secretary General of the ZNLWVA. It only came to the attention of the latter that actually the Provincial and District leadership had used divide and rule tactics when the NMWVA had moved in to evict the national leadership. They had been informed by the officials that the farm had not yet been allocated, which indeed was the official position. What they were not told was that it had been occupied by some of their own. The situation was only diffused after ZANU-PF intervened and assured the NMWVA that their leaders would be settled on any other farm of their choice.

The displacement of the leadership of the NMWVA was the culmination of a concerted strategy by senior provincial politicians that included the unwarranted arrests of the war veterans and protection of the white farmers in the area. The NMWVA was perceived as unwilling to negotiate on the local government's terms, because it exposed the corrupt and anti-reform relationships that existed between senior politicians and state bureaucrats on the one hand and the white farmers on the other. Those relationships entailed the delisting on technical grounds of farms owned by 'friendly whites' that had been designated for distribution, providing security to white farmers against land occupation by the police on instruction from politicians in the District (see District Land Committee Minutes 2003). Although there were no party structures in the LSCF areas, politicians and state managers had long established relationships with white farmers based on mutual interests. The white farmers made huge donations during functions such as political rallies, birthday parties, ZANU-PF victory celebrations and occasionally directly to individual senior politicians and state managers. District and provincial officials involved in farm designation and allocation took advantage of their positions to also extract rents from the white farmers in exchange for protection of their farms from occupiers and from designation.

Networking and establishing good working relations with the local police saved the NMWVA members, in particular war veterans, from imminent arrest. For example, instructions from some senior ZANU-PF officials to arrest all war veterans in order to paralyze the association were given to the Officer in Charge at Nyabira Police Station in 2002. Instead of implementing the instructions, police alerted the association of the instruction to forcefully evict them. And, true to their anti-corruption position, the association tracked down the official through ZANU-PF structures and confronted him at his home in Concession. It emerged that the culprit, a central committee member, had actually bypassed ZANU-PF structures and acted unilaterally.

This action compromised the structures and procedures of ZANU-PF and the party official was made to acknowledge this and offered apologies both to the war veterans and his subordinates in the District Coordination Committee (DCC) of ZANU-PF.

The strength and legitimacy of the NMWVA was under scrutiny and threatened by both the technocrats and politicians at the district level, although it enjoyed support from uniformed forces (police and the army). Eventually in 2001, the NMWVA was pushed out of the land committee, leaving it only one representative. The state argued that the law recognised only one representative of war veterans in the land committee at any level (see Masiyiwa and Chapungu 2004). This was at the crucial time when land allocation decisions were being taken in the context in which the need to alleviate overcrowding in Chiwese Communal areas was pressing. Members of the NMWVA were seen as aliens to the district. They were referred to in the land committee meetings as '*Vanhu veku Harare*' (people from Harare). On the other hand, the 'geese' that laid the 'golden eggs' for politicians and district lands officials were concentrated in the farms that the NMWVA occupied, as indicated above.

The official Fast Track Land Reform Programme land allocation began in 2001 although it was launched in July 2000. This programme was intended to relieve tensions[6] arising over co-habitation between white farmers and occupiers, especially over agricultural production on the farms and to end the technical paralysis that the state unceremoniously found itself in. Ideally, the first phase of land allocations (during 2001-2002) could have been to legally resettle the land occupiers in line with the A1 and A2 schemes designated by government. Instead, the first phase of allocations went according to the script of those who were now in charge, i.e. the land committees. Frustrating as the move was to the NMWVA, the struggle continued as shown in section five below and the important lesson to draw from these developments is that 'no matter how a typical participant describes his/her reasons for joining the movement, or what motives may be suggested by a social scientist on the basis of deprivation, it is clear that the original decision to join required some contact with the movement' (see Freeman 1983).

Crystallising individual actions into collective actions

A number of processes generated by the occupation and the subsequent Fast Track Land Reform had an impact on transforming individual actions into collective action on one hand and the weakening of the appeal of the

NMWVA to its membership on the other hand. Snow et al (1986) refers to such a process as 'micro-mobilization' and suggests that groups must convince their members that the group's 'world-view' is worth adopting. There is little explicit elaboration of the process by which groups 'convince' people, but Gurr (1970) had earlier elaborated on the concept of appeals. He suggested that by issuing appeals, groups are able to 'convince' people to adopt the group's 'world-view' and thus mobilise them. Gurr (1970) outlined a five-point typology of appeals used to mobilise individual discontent into rebel movements. These are identified as: (i) appeals to corporate identity; (ii) appeals to an individual's sense of relative deprivation; (iii) appeals that identify the existing regime as the source of that discontent; (iv) appeals to normative justifications for taking violent action; and (v) appeals promoting the utilitarian value of rebellion.

The NMWVA effectively challenged the existing land ownership regime, by issuing two categories of appeals. One was directed at mobilising those who had not decided to risk their lives for risky collective action and the other focused on preventing the active members from leaving the movement by focusing on real issues that provided answers to the state of deprivation faced by individual members. The appeals used by the NMWVA included real issues such as the (i) shortage of land (ii) the displacement of members by non- members who were officially allocated land on the occupied farms ahead of occupiers, (iii) preferential treatment of some white farmers at the expense of occupiers by the District Land Committee, (iv) selective allocation of land to members of the association by the official land committee and (v) access to external assistance to those who were officially allocated land vis-à-vis the landless members of the same association and (vi) the preference given to A1 occupiers vis-à-vis A2 occupying members of the NMWVA. Evidence of the NMWVA's capacity to draw concessions from the District Land Committee and benefits for its members from donors, government and public enterprises buttressed the organisation's appeal to its members.

Appeals on the shortage of land

That the individuals' sense of relative deprivation in relation to land ownership was the main cause for the land occupations has never been in doubt. The same can be said about the identity of the source of that deprivation which in the Zimbabwean case was the Large Scale Commercial Farmers and the property rights laws that ring-fenced them. Table 4.4 below shows the level of deprivation. About 83 per cent of the NMWVA members, compared to

86 per cent in the AIAS baseline study, had no access to land at all. Of the four respondents who already had pieces of land, three had less than one hectare each. This compared unfavourably with the average of 2,000 hectares that each of the white Large Scale Commercial Farm owners had at their disposal. The translation of this level of individual deprivation into a sense of collective relative deprivation through micro-mobilisation by the association enabled the individual members of the NMWVA to articulate their frustrations in rebellious action and to identify with the world view expressed by the NMWVA.

Table 4.4: Land ownership prior to occupation

Ownership Status	Number	Percentage
Had a Piece of Land	4	17
Did not have Land	20	83
Total	24	100

Source: Masuko 2008

According to the sociological analysis of M. Sayles,

> Unlike individual deprivations, collective deprivation exists only among individuals who have a collective consciousness. Collective deprivation therefore describes group deprivation sensed through comparisons made between those in the group and other groups in the society in this case between the occupiers who had nothing and the large- scale commercial farmers who on average owned 2000 hectares each. There is, however, no sense of deprivation concerning one's position within the in-group itself (Sayles 1984).

There was no sense of deprivation within the group because most NMWVA members were landless.

Selective allocation of land as a source of deprivation

Official land allocations commenced in earnest in 2001 and 8 per cent of the interviewed members of the NMWVA were allocated pieces of land at this time. The bulk of the allocations, however, took place between 2002 and 2004 (Table 4.5) at a time when the NMWVA representatives had been pushed out of the District Land Committee and no longer had a direct influence on the deliberations of the land committee.

Table 4.5: Selective allocation of land by the District Land Committee

Year of allocation	Number	Percentage
2001	2	8
2002	9	38
2003	7	29
2004	6	25
Total	24	100

The leadership of the NMWVA had swiftly and deliberately been displaced from the farm they had occupied and were replaced officially by the national leadership of the ZNLWVA in a cunning move by the bureaucrats and senior politicians in Mashonaland Central aimed at paralyzing the NMWVA. Desperate times called for desperate measures. A number of embarrassing demonstrations were staged by the association at the Governor and District Land Committee offices to put pressure on the land committees (both at district and provincial levels). At this point, these committees were identified as the new source of deprivation, in their method of prioritising the occupiers when allocating land. It was during one of these demonstrations in 2002, that the Member of Parliament for Mazowe West was forced to emerge from a District Land Committee meeting; he promptly offered three 'holding' farms for allocation to the members of the association who qualified for A1 (common property) scheme plots. While there were no concessions made to A2s, the offer was enough to temporarily pacify the membership.

However, the land committee's decision to allocate land to A1 members of the association marked the beginning of the selective allocation strategy and emergence of divisions within the ranks of the association. In a way, bureaucrats and politicians were using their hegemony and newly-found ownership of the land reform to get back at the association. While the selective allocations were positive news to the membership, they had the effect of weakening the association because more than 60 per cent of the association membership was made up of the occupiers whose only wish was to own no more than an A1 plot. With the allocation of land, came the threats of land withdrawal from those who continued to engage in collective activities such as demonstrations, evictions of white farmers and Saturday meetings. A few identified NMWVA members also received offer letters for A2 farms during the course of 2001 and these members were exclusively those who originated from Mashonaland Central Province.

The NMWVA had to redesign its strategies and tactics in the face of these new challenges. The first tactic was to appeal to the A1 members, who had been provisionally allocated land by the officials, but had not received letters conferring ownership of the land to them individually. Threats of eviction by the land committee were a real possibility to the A1 members (see Abuja Agreement 2001 to evict invaders) and the association used this card to its advantage. The fact that the holding farms[7] had not been surveyed and pegged into individual plots gave the leadership of the association ammunition to appeal to the prospective A1 members to safeguard themselves against evictions by maintaining their membership with the association. Information about evictions that had taken place on some farms and how they were returned to former owners on technical grounds were used as examples. To leverage against the possible evictions, the A1 members of the association decided to stick with the association and participate in its struggles.

Displacement, preferential treatment of white farmers and lack of access to resources

Prospective A2 farm beneficiaries who had not been allocated land had a strong reason to continue with the land struggle through the NMWVA. The practice where the Provincial Land Committee issued offer letters to non-members, who in turn went on to displace NMWVA member occupiers, at the recommendation of the District Land Committee bound them together. Several factors did not make the task of the association any easier. These included: preferential treatment by the land committee of some white farmers at the expense of occupiers; the lack of access to cheap inputs from the Grain Marketing Board and the Ministry of Energy; and the lack of access to loans from AGRIBANK by the members of the association who had no offer letters. The association, between 2002 and 2004, had to respond in ways that would increase its appeals to the members who by now had a choice: either to follow the District Land Committee's promises or to continue struggling from within the ranks of the association.

New strategies that were more action-oriented, accompanied by varied tactics, were employed by the NMWVA to pre-empt those of the land committees. It went about destroying properties belonging to those farmers who would have been officially allocated land on occupied farms ahead of the occupiers[8] and confiscated their offer letters. Mackay Farm near Nyabira is a case in point. White farmers who enjoyed the protection of the committee and senior politicians had their farms raided and looted and they were forced to pack and leave the occupied farms.[9] Both the prospective A1 and A2 members executed these new tactics with a lot of enthusiasm and zeal.

Besides sending a clear message to the authorities that the occupiers were there to stay, the new tactics had concrete material benefits for the participants in the form of access to diesel, chickens and other small items that they secured during the forced evictions of the white farmers. The association refused to protect the farmers where the land committee would have shown bias, such as in the case of Saint Hurst Farm. To further legitimatise itself in the eyes of its membership, the association managed to negotiate with AGRIBANK and GMB for special facilities for loans and inputs, respectively, through group lending schemes, which is a form of collective responsibility and was accepted as collateral. The association acted as guarantor against default by the members in their loan repayments. Members started receiving inputs and loans without producing offer letters for the 2001-2002 farming season and this relationship continued until 2004, when all occupiers had officially been allocated their pieces of land.

The above actions were meant to provoke reaction from the Provincial and District Land Committees, as much as they were meant to consolidate group cohesion. The reaction from these committees was swift. Joint teams comprising the provincial and district committees, ZANU-PF structures (province and district), war veterans, security organs and uniformed forces visited the occupied areas to identify plots that were vacant and to recommend their allocation to the NMWVA occupiers. The association coordinated the visits and organised its members to be present on each and every farm that was being assessed. The handover of offer letters to the A1 farmers was followed by the movement of these farmers to four farms, namely Adhora Farm, Kruger Farm, etc., that had been officially demarcated. An olive branch was extended to the association's leadership, who had been displaced by the land committee. An alternative farm of their choice was identified and allocated to them. As events unfolded, one of the two leaders ended up being resettled on the same farm and only one was moved to the other farm of his choice.

The attention to the land allocation grievances by the land committee drained life out of the association. Upon receiving offer letters for their plots, A1 members of the association were advised by the District Land Committee and local politicians not to engage in demonstrations or any other activities of protest. Over and above the land, A1 farmers could now use the offer letters to access inputs (fertilizers, seed, chemicals, etc) from the GMB depots closest to them. They also qualified for loans from AGRIBANK and/or their individual banks. The same applied to the A2 beneficiaries. The resultant effect was non-attendance of Saturday meetings and contributions to the coffers of the association dwindled drastically (see Table 4.6). Over 71 per cent of the members last attended the associations' meetings in 2003. By this year, 75 per cent of the members of the NMWVA had legally acquired land.

Table 4.6: Year when last meeting of the association was attended

Year	Numbers	Percentage
2001		
2002	1	4
2003	17	71
2004	5	21
Other	1	4
Total	24	100

Source: Masuko 2008

The association's Mazowe wing collapsed, leaving only the Nyabira wing active. This was simply because land was not delivered at the same pace in Mazowe ((Mashonaland Central) and in Nyabira ((Mashonaland West). Nyabira lagged behind to the extent that the only active members left in the association were those from Mashonaland West Province. As a result, the NMWVA merged with another occupiers' organisation from Mashonaland West to form a new organisation called Gwebi-Hunyani War Veterans Association. The chair of the NMWVA continued to lead the new organisation.

The point, advanced by Masiiwa and Chipungu (2004), Raftopoulos and Phimister (2004) and Hammar (2003), that the land movement was focused on a single issue was indeed relevant at this juncture. What the argument missed is that the social capital realised through the formation of groups and networks developed during the land invasions and land occupation process laid the foundation for future development not of the same organisation, but of new organisations with new objectives and abilities. Section six below focuses on the development of this social capital and what it meant for the continuation of the transformation of the reform agenda beyond the single- issue discourse.

The development of new social networks in farming areas

It can be argued that a given group's level of mobilisation is not simply a function of the cat-net structures[10] of society, but a function of that group's ability to exploit the cat-net structures present in society and to forge new ones (Moore and Jaggers 1990). With this realisation in mind and operating in an area where structures had not existed, the NMWVA had started from scratch. The areas that were occupied by the NMWVA were basically Large Scale Commercial Farming areas, where trespassing attracted prosecution. Farm workers were the 'property' of the farmer just like the land and other immovables and movables. These were 'liberated zones' for white farmers who had unbridled authority over everything within them. Active politics was non-existent, except at the behest of the white farmers. In other words, not even ZANU-PF or the ZNLWVA had structures in these areas.

But this is in no way to suggest that both the farmers and the workers lived in isolation. The farmers had their clubs where they would meet and share farming ideas, talk politics and socialise. They had a very strong mobile radio communication system, so effective that they were just one call away from each other. Over and above these, the white farmers had the Commercial Farmers' Union that represented them at national level. More importantly, they had established very strong networks with senior politicians, who had, over the years, acquired land and shared the same interests of protecting the Large Scale Commercial Farms as discussed earlier in this chapter. On the other hand, farm workers had their own networks within compounds and between compounds. These were basically kinship relations that kept their world rotating. Although independence had sought to democratise the work place through workers' committees at the farm level, the attitude of farmers frustrated any development towards this direction (Loewenson 1992; Amanor-Wilks 1995; Kanyenze 2001; Chambati and Moyo 2004). The NMWVA, like other groups elsewhere around the country, had to contend with these networks: networks that could be turned against them and their objective of land redistribution.

It, therefore, became prudent that the survival of the NMWVA depended on its ability to at least tap into the network of the farm workers (who had their grievances against the white farmers) and to create new networks with other groups that shared the same objectives. Its first strategy was to establish ZANU-PF cell and ward structures within the farm compounds. This was a new and exciting development to the farm workers, who felt useful and recognised at the same time. All positions at the cell level were filled by farm workers, who were voted in by their own and the same was encouraged at ward level. Fear of reprisal from the 'Boss' was a major block for many farm workers, who thought the arrival of war veterans was temporary and likely to upset the applecart. Temporary structures that were put up at the occupied farms and the permanent presence of the youth at the farms played a very important role in assuring the farm workers that the white farmers' reign at the farms was under check.

The gospel of appeal was focused on issues that most concerned farm workers. The first was on what would be their fate in the event that the farmer left, which by then looked inevitable. Second, farm workers were worried about their compound houses, employment and whether they would be resettled close to their work places in the event that some wanted their own pieces of land. During the mobilisation meetings which were also attended by the farmer at their respective farms, war veterans would assure the farm workers that their terminal packages would be settled before the farmer removed his/her moveable property and they would continue to reside in their houses and new farmers would provide them with jobs. While war veterans assured farm workers of pieces of land, the decision on the location lay with the land committees. Sandringham Farm is a living example of the success of this approach to winning the farm workers' support.

The setting up of district-level ZANU-PF and NMWVA structures had a direct appeal to politicians and war veterans in the Districts of Mazowe and Zvimba. They viewed these as providing an opportunity to penetrate what had been the white farmers' fortress. In the process, the new district was integrated into the District Coordination Committees (DCC) of the respective districts. Senior provincial party officials were invited to address the workers and occupiers, legitimatising the structure and presence of the occupiers in the areas in the process.

To survive future onslaught, the NMWVA strengthened its relationships and networks with the other groups occupying farms adjacent to their area. All but one (which was led by a spirit medium) of these groups were led by war veterans. However, competition for membership always existed among the occupying groups, depending on the benefits that each group extracted from the politicians and land committees for its membership, such as recognition, land allocation and support, whether in cash or kind. The NMWVA lost some of its members to one neighbouring group that had, through its chairperson (who was wife to a national hero), used influence to good effect to get A2 official allocations for their members ahead of the other groups. However, the working relations were not affected by these defections as the objective of the groups was the same and they collaborated during demonstrations at the land committee offices and even in the evictions of white farmers.

More critical for the future sustainable reform programme are the social networks that had already been developed between the occupiers themselves and between the occupiers and the farm workers. All occupiers who qualified for A2 were resettled on the farms they occupied, depending, of course, on the holding capacity of the farm. What this means is that all 'base commanders' were resettled on the farms they commanded. Besides the ZANU-PF and war veteran structures, new production-focused organisations also emerged in the post-FTLRP period. These included farmers' clubs, input group support schemes, input collection groups, irrigation syndicates, etc. Murisa (Chapter 7) supports and elaborates on this point. The organisation's ability to mutate was clearly demonstrated when, in 2004, the NMWVA was reconstituted as the Gwebi-Hunyani Farmers' Association, involving members from Mashonaland West Province. The consistent interaction of people who share the same traits is assured within these new categories (groups/organisations) and networks.

War veterans continue to channel information through the new organisations that have and continue to emerge, through their own structures and through those of ZANU-PF. Indeed, a new community of people with important shared traits who interact frequently now characterises the social, economic and political environment in the resettled areas. Such qualities are a prerequisite to the formation of social movements and show the existence of a latent capacity to take collective action. More importantly, war veterans and former members of the NMWVA have shown their ability to exploit the cat-net structures that existed before and during occupation and to forge new ones with different objectives to their predecessors. This ability is indeed the social capital that would take land reform to its higher and transformative stage of agrarian reform and, with it, to long-term political, social and economic development of the area and the nation at large.

Conclusion

This chapter explored the sustainability of Zimbabwe's land reform from a social movement perspective, using the Nyabira-Mazowe War Veterans Association (NMWVA) as a case study. Two issues were examined. First is the allegation that the occupation movement was a top-down movement conceived of and directed by a beleaguered state to spruce up the image of ZANU-PF. Second is the idea that the occupation movement was a one-issue movement, not interested in long-term social, political and economic development.

After a critical examination of the NMWVA's evolution, activities and networks, it is clear that land occupations and reform in Zimbabwe will remain a contested terrain, much more so if productivity levels in resettled farms remain subdued. However, the discussion throughout this chapter should put to rest the notion of a top down movement which in the main was a creation of the state. It has been demonstrated that not one, but many micro-organisations executed the invasions and occupations, operated independently of the state and which, in many cases, were in conflict with the state. By way of networking, they were woven into a national movement that saw through a national programme that has been characterised as the 'first radical shift in agrarian property rights in the post Cold War world' (Moyo and Yeros 2005).

While the top- down approach precludes active and voluntary participation by the generality of the aggrieved, evidence throughout the chapter shows that, whenever there is relative individual deprivation and there are groups of people with shared traits who interact consistently, a movement will thrive. If the movement does not emerge, it is because the political establishment provides adequate channels to pursue solutions. This truism that the NMWVA and other occupation movements have amply demonstrated has escaped many Zimbabwe crisis theorists. It has been shown that the original decision by an individual to join a movement does not come from the state or political party that seeks to enhance its waning popularity, but that this action requires some contact with the movement.

This chapter has shown that a number of actors with different interests were either passive or active participants during the invasions and occupations and subsequent resettlement programme. These included the white farmers themselves, the occupation movements, the state and ZANU-PF as institutions, politicians and state managers as interested parties, the opposition party MDC and NGOs. While the state and ZANU-PF as institutions might have had common objectives with the occupation movements, class interest and contradictions within these institutions led some bureaucrats and politicians, white farmers and those who sympathised with them, to go against the tide. But because the balance of class forces within ZANU-PF and society in general was tipped in favour of radical nationalist solutions, the objective of reforming the skewed land ownership, itself part of the broader land question, was realised, such that by 2003, households numbering 134,452 had accessed land (Presidential Land Review Committee 2003).

Ownership on its own does not guarantee the sustainability of the land reform. However, emerging new social structures conducive for movement activity are defined in the new groups and networks that are developing in all resettled areas. These new structures are capable of demanding solutions to the other outstanding components of the land question, including insecure land tenure and unsustainable and sub-optimal land use. The fact that the land occupation movements' composition was diverse in terms of class composition, gender and age makes the new communities much stronger than the occupation movements before them. They are, therefore, organised to deal with the last two issues of the land question. The NMWVA movement received strong community support, redefining their constituencies to include the broader working class (including the farm workers), the youth, the intellectual, the petty bourgeois, etc, into a broader movement for fundamental social and economic change. The emphasis on the intentions and ideologies of revolutionaries in explaining the cause and energy of the revolution tend to be overstated; if ways to maintain constant sound interaction exist, there is always potential for revolution.

Notes

1. Urban workers, youth and women saw an opportunity to extricate themselves from poverty, social prejudices, etc., by owning their own piece of land. In other words, the participation of these urban groups is an indication that land reform should go beyond decongestion of communal areas and satisfying only rural peasant demands.
2. Class diversity shows the national character of the grievance and the nationalist character of the movement.
3. War veterans were a very small component of the total membership. The majority of the members of the NMWVA was made up of non- war veterans of all ages. However, it should be made clear that the youth (anyone below the age of 21 years) were not required to pay joining and subscription fees to be members because they were the foot soldiers of the organization.
4. Strategies included how to occupy farms with minimum violence or injury to both the farmers and the occupiers; how to deal with those that were issued with offer letters ahead of occupiers; how to extract benefits from responsible offices (e.g., the District Administrator, District and Provincial Lands Committees) and from politicians and law enforcement agents (e.g., the police); and how to keep the membership interested and involved. Tactics changed depending on the situation at hand, but the

basket of tactics included rallies; erections of temporary shelters, tilling of land and planting of crops on the occupied farms; evictions; micro-mobilization; and demonstrations, particularly against politicians and office holders, etc.

5. Relations between ZANU-PF and the war veterans during the struggle for independence were that of mother and child and such has been the case in the post- independence era. Now, war veterans have their own agendas, different from those of ZANU-PF, although overlapping in the main.
6. Production activities were disrupted during invasions and occupations. Occupiers would plant where the white farmer would have prepared without agreement, forcing complete or partial work stoppages.
7. Holding farms were farms where occupiers were accommodated temporarily before they were officially allocated lands by the land committee. From the point of view of the war veterans, holding farms were strategically placed as launch pads for invading other farms.
8. A number of people were officially given offer letters ahead of the occupiers. To put pressure on the official structures, these were confiscated by war veterans who instructed the offer letter holders to go and collect those for the occupiers before they could be allowed to settle. In the event that they put up structures, these were razed by the group to emphasize the point.
9. Evictions were spontaneous, fast and did not give the farmer time to organize; farmers were only allowed to take with them the household goods. For assistance, many farmers would either go the ZNLWVA or the presidium, where they would also receive some political orientation and lessons on co-existence. Farmers were only allowed back if they were accompanied by the leaders of the association or the police.
10. Cat-net structures refer to categories (groups) and the networks they establish over time.

References

AIAS, 2005, Baseline Study, Harare: AIAS.

AIAS, 2007, Baseline Study, Harare: AIAS.

Amanor-Wilks, D., 1995, *In Search of Hope for Zimbabwe's Farm Workers*, Harare: Panos Institute.

Bonano, A., 1988, 'Theories of the State: The Case of Land Reform in Italy, 1944-61', The Sociological Quarterly, Vol. 29, No. 1, pp. 131-147.

Chambati, W. and Moyo, S., 2004, Impacts of Land Reform on Farm Workers and Farm Labour Processes, Harare: AIAS.

Chambati, W. and Moyo, S., 2007, *Key Debates on Zimbabwe's Contemporary Land Reform*, Harare: AIAS.

Freeman, J., 1983, *Social Movements of the Sixties and Seventies,* New York: Longman.

Garreton, A.M., 1997, 'Social Movements and Democratization', in S. Lindberg and A. Sverrisson, eds, *Social Movements and Development: The Challenge of Globalization and Democratization,* New York: McMillan Press Ltd.

GoZ, 2001, Rural Land Occupiers (Protection from Eviction) Act, 2001.

GoZ, 2001, Abuja Agreement of 2001, A Commonwealth Initiative to solve the land impasse.

GoZ, 2003, Mazowe District Land Committee Minutes 2003, Mazowe Rural District Council District Administrator's Office.

GoZ, 2003, Presidential Land Review Committee, Harare.

Gurr, T.R., 1970, *Why Men Rebel*, Princeton: Princeton University Press.

Hammar, A., B. Raftopoulos and S. Jensen, eds, 2003, *Zimbabwe's Unfinished Business: Rethinking Land, State and Nation in the Context of Crisis*, Harare: Weaver Press.

Kanyenze, G., 2001, 'Zimbabwe's labour relations policies and implication for workers', in F. Banda, C. Gavi, eds, Zimbabwe's Farm Workers: Policy Dimension, Harare: Panos Southern Africa.

Loewenson, R., 1992, *Modern Plantation Agriculture: Corporate Wealth and Labour Squalor*, London: Zed Books.

McCarthy, J. D, 1987, 'Pro-Life and Pro-Choice Mobilisation: Infrastructure Deficits and New Technologies', in *Social Movements in an Organisational Society*, M.N. Zald and J.D. McCarthy, eds, New Brunswick, NJ: Transaction Books.

Masuko, L., 2004, *The Role of the State in Land Reform: The Case of Land Reform in Zimbabwe, 1998-2002,* Dakar: CODESRIA.

Masuko, L., 2008, 'War Veterans and the Re-emergence of Housing Cooperatives', in S. Moyo, K. Helliker, T. Murisa, eds., *Contested Terrain: Land Reform and Civil Society in Contemporary Zimbabwe*, Pietermaritzburg: S&S Publishing.

Masuko, L., forthcoming, 'China in Africa: Case of Zimbabwe's Extractive Industry', OSISA, RSA.

Masiiwa, M. and Chipungu, L. 2004, 'Land reform Programme in Zimbabwe: Disparity between Policy Design and Implementation', in M. Masiiwa, ed, *Post Independence Land Reform in Zimbabwe: Controversies and Impact on the Economy*, Harare: FES and IDS-University of Zimbabwe.

Masuko, L., 2008, NMWVA Fieldwork Interviews, IDS-UZ.

Moore, H.W. and K. Jaggers, 1990, 'Deprivation, Mobilisation and the State: A Synthetic Model of Rebellion', Journal of Developing Societies, Vol. 6, No. 1, pp. 17-36.

Moyo, S., 2001, 'The land occupation movement and democratisation in Zimbabwe: contradictions of neoliberalism', Millennium Journal of International Studies, Vol. 30, No. 2.

Moyo, S. and P. Yeros, 2005, 'Land occupations and land reform in Zimbabwe: towards the National Democratic Revolution', in S. Moyo, P. Yeros, eds., *Reclaiming the land: the resurgence of rural movements in Africa, Asia and Latin America*, London: Zed Books, pp. 165–208.

Moyo, S. and Yeros, P. 2005a, 'The Resurgence of Rural Movements under Neoliberalism', in S. Moyo, P. Yeros, eds., *Reclaiming the land: the resurgence of rural movements in Africa, Asia and Latin America*, London: Zed Books, pp. 165–208.

Moyo, S., Murisa T. and Helliker, K. eds, 2008, *Contested Terrain: Land Reform and Civil Society in Contemporary Zimbabwe*, Pietermaritzburg: S&S Publishing.

NMWVA, 2000, Membership Register.

NMWVA, 2000-2003, Executive Committee Meeting Minutes.

NMWVA, 2000-2004, Minutes of the NMWVA general meetings.

Poulantzas, N, 1975, *Classes in Contemporary Capitalism*, London: New Left Books.

Raftopoulos, B., 2003, 'The State in Crisis: Authoritarian Nationalism, Selective Citizenship and Distortions of Democracy in Zimbabwe', in Hammar, A. Raftopoulos, B. Jensen, S. eds, *Zimbabwe's Unfinished Business: Rethinking Land, State and Nation in the Context of Crisis,* Harare: Weaver Press.

Raftopoulos, B. and Phimister, I., 2004, 'Zimbabwe now: the political economy of crisis and coercion', *Historical Materialism*, Vol. 12, pp. 355–82.

Tilly, C., 1978, *From Mobilisation to Revolution*, Reading, MA: Addison-Wesley.

Sachikonye, L., 2003, 'From Growth with Equity to Fast Track Reform: Zimbabwe's Land Question', *Review of African Political Economy*, Vol. 30, No. 96, pp. 227–240.

Sadomba, W., 2008, 'Movements within a Movement: Complexities within the Land Occupations', in S. Moyo, Helliker, K. Murisa, T., eds, *Contested Terrain: Land Reform and Civil Society in Contemporary Zimbabwe*, Pieter maritzburg: S&S Publishing.

Sayles, M., 1984, 'Relative Deprivation and Collective Protest: An Impoverished Theory?', *Sociological Inquiry*, Vol. 54, No. 4, pp. 449-465.

Skocpol, T., 1979, *States and Social Revolutions*, Cambridge: Cambridge University Press.

Snow, D.A., R.B. Rochford, Jr., S.K. Worden and R.D. Benford, 1986, 'Frame Alignment Processes, Micromobilization and Movement Participation', American Sociological Review, Vol. 51: pp. 464-481.

Truman, D., 1951, *The Governmental Process: Political Interests and Public Opinion*, New York: Knopf.

5

Changing Agrarian Labour Relations after Land Reform in Zimbabwe

Walter Chambati

Introduction

The agrarian labour relations generated after the 'Fast Track' Land Reform Programme (FTLRP) tend to be neglected in most literature after 2000. This neglect largely resulted from the dismissal of the redistributive nature of the FTLRP and changing patterns of agricultural production by some studies (see Marongwe 2009; Masiiwa and Chipungu 2004; Hellum and Derman 2004; Sachikonye 2004; Davies 2004; Richardson 2005). The distributional outcomes of the FTLRP, which in turn shape a restructured agrarian labour regime, have, however, been acknowledged in a few empirical studies (Moyo et al 2009; Scoones et al 2010).

Most studies analysing agrarian labour relations after 2000 have adopted modernisation perspectives, in which formal wage labour in the capitalist LSCFs is treated as superior to self-employed forms of labour in the 'backward' peasant sector (Freund 1984), assuming that returns to wage labour are greater than those of self-employed peasants. The self-employment of peasants is neglected because it also does not fit the formal employment criteria used by neo-classical economists (see Leavy and White 2001). The concern has been on the 'displacement' of former farm workers from their LSCF jobs[1] and residency in the farm compounds (see Sachikonye 2003; Hartnack 2005; NRC 2003; Magaramombe 2003; Rutherford 2004).

The former farm workers who lost their jobs, but are still resident in the farm compounds, tend to be addressed as 'displaced in situ' (Hartnack 2005; Magaramombe 2010), meaning they are out of 'work' regardless of their new livelihoods in farm and non-farm work. In this respect, the redistribution of LSCFs to peasants is thus equated with the 'end of modernity' (Worby 2003), with unemployment as the sole consequence. On the other hand, the analysis of the physical displacement from the farm compounds does not adequately examine the insecure residential tenure that farm workers faced before 2000. The notion of 'physical displacement' does not acknowledge the extent to which the FTLRP has re-established self-employed peasant jobs that were displaced by colonial land dispossession or investigate whether this change represents losses or gains in the overall scheme of social life.

Moreover, the agency of former farm workers is not examined as they are largely considered as passive victims of violence and/or human rights abuses (narrowly limited to political and civil rights) at the hands of war veterans and peasants during the FTLRP (Hellum and Derman 2004; ZHRF and JAG 2007; JAG/RAU 2008; Ridderboos 2009). The struggles they waged to improve their material conditions, including their alliances with war veterans and peasants in the land occupations (Sadomba 2008), is not assessed. Their mobilisation against land reform by land owners and their trade union, the General Plantation and Agriculture Workers Union of Zimbabwe (GAPWUZ), emphasising job protection rather than a new livelihood after redistribution, is also not considered.

The dynamic process that entailed the FTLRP in terms of the differentiated land acquisitions and allocations over the last decade, alongside changing agrarian (labour) policies, which in turn influenced the agrarian labour relations outcomes, is also not adequately acknowledged in the literature. Indeed, much of the commentaries based on the earlier phases of the land reform did not envisage changes in the outcomes with the continued implementation of the FTLRP and shifting agricultural production conditions (see, for example, Alexander 2003; Sachikonye 2003, 2004; Magaramombe 2003).

The emphasis of most contributions after 2000 on what happened to former farm workers sidelines analysis on what kind of agrarian employment structure is emerging and the social relations of production that this entails.

Agrarian labour relations need to be understood in their historical context and, in former settler colonies such as Zimbabwe, these were based on specific land-labour utilisation relations created by land dispossession. The ownership of land, though not the only decisive factor, is central in the emergence of agrarian labour relations, influencing who sells or hires labour power (see Patnaik 1997; Mafeje 2003; Bernstein 2010; Moyo 2011a). By eschewing the historical evolution of agrarian labour relations, the debate after the FTLRP misses the critical linkage between land and labour relations. Thus the comprehension of the new agrarian labour relations requires analysis that relates labour utilisation with the new land access patterns, changing land use patterns, land tenure reforms, agricultural resource flows and their effects on different classes and segments of society.

The reversal of the monopoly in land ownership via extensive land redistribution affects agrarian labour relations by generating a new agrarian employment structure in which the political power is diffused amongst a broad base of many smaller capitalist farms and peasants. Furthermore, the state tenures allocated to land beneficiaries through public leases and permits undermine the residential labour tenancy by reducing their authority to compel wage labour in return for residency and, in turn, their control over labour (Moyo 2011a). Land redistribution also provides peasants wider access to means of production for autonomous social reproduction through self-employment.

This chapter examines the transformation of agrarian labour relations after the radical restructuring of agrarian property relations since 2000. It begins with an examination of how colonial land dispossession and policy shifts after 1980 shaped agrarian labour relations in Zimbabwe, which provides the basis for evaluating the changes generated by the FTLRP. It then explores whether a new agrarian labour regime has emerged to replace those that were structured around land alienation, social relations of labour residential tenancy[2] and private property. The different types of farm and non-farm sources of employment that have emerged in relation to the new land use patterns and the social and economic conditions under which this labour is employed are assessed on the basis of detailed empirical studies conducted by the AIAS since 2000 and other secondary sources. The assessments of the struggles being waged by labourers to improve their material conditions precede our conclusions.

Agrarian structure, production and labour relations

The (agrarian) labour relations in Zimbabwe developed from a historical process structured around land alienation and private property during colonial conquest (Arrighi 1970; Clarke 1977). A semi-proletarianised African labour force economically and extra-economically subordinated to labour markets in the European farms, mines and urban industries emerged (Palmer 1977; Mafeje 2003). This labour force existed alongside a dispossessed under-employed and unemployed reserve army of labour in Communal Areas governed by customary tenure and reserved for Africans engaging in petty commodity production, mostly via self-employed family labourers (Bush and Cliffe 1984).

In the LSCFs, a master-servant relationship between white farmers and black workers was generated on the basis of a labour residential tenancy and part- time labour supplies from the Communal Areas (Clarke 1977; Rubert 1997). These exploitative labour relations were institutionalised through various repressive racial legislations such as the *Masters and Servants Act of 1899* that treated black workers as the property of white land owners (Amanor-Wilks 1995). The conditions of agrarian labour prior to 1980 were pathetic, characterised by paltry wages, intimidation of workers, racial abuse, arbitrary dismissals and physical violence (Rubert 1997; Rutherford 2001).

The colonial skewed land redistribution patterns were inherited at independence (see Moyo and Chambati, Chapter 1; Moyo, Chapter 2). However, since the 1950s, farm workers were incorporated into the national labour relations framework that covered the rest of the working class (Kanyenze 2001). This incorporation occurred concurrently with the abolition of the *Masters and Servants Act of 1899,* which governed labour relations in the LSCFs (ibid). The state regulated the labour markets, including the requirement for state consent in worker dismissals and the introduction of minimum wages that increased the real wages of farm workers by over 50 per cent in theearly 1980s (Amanor-Wilks 1995). Previously banned trade unionism amongst farm workers and its associated structures such as farm- level workers' committees also emerged, but their impact on improving the material conditions of workers was limited (Loewenson 1992; Rutherford 2001). In particular, trespass laws restricted trade union access to LSCFs, such that after two decades of independence, only a third of the permanent farm workers were members of the largest farm labour union, GAPWUZ (Kibble and Vanlerberghe 2000).

By 1980, the LSCFs had diversified from the colonial labour- intensive maize/ tobacco production systems to include other outputs, utilising high levels of modern technologies (such as hybrid seeds, pesticides and fertilisers) and capital-intensive production systems mainly geared towards exports (Loewenson 1992; Muir 1994). The new outputs included sugar, cotton, wheat, soybeans, coffee, tea, beef and dairy. The diversification of LSCF commodities intensified after ESAP, as they moved away from traditional exports such as cotton to new high- value exports such as horticulture and wildlife ranching, but tobacco was retained as a key export (Amanor-Wilks 1995; Moyo 2000).

The transformation of LSCF agricultural production was also reflected in the growth of agrarian wage employment from 218, 817 workers in 1983 to a peak of 334, 521 workers in 1996 (CSO 1984; 1997) and by 2000 there were 313,879 workers (CSO 2001). Simultaneously, the structure of the labour force was shifting towards more casualisation of labour, which was less protected by the labour regulations and poorly remunerated, as the share of permanent workers declined from 76 per cent in 1983 to 50 per cent in 1996 (CSO 1998). Women constituted

65 per cent of the total casual LSCF labour force and less than 10 per cent of the permanent employment (Chambati and Moyo 2004). The LSCF labour force was initially dominated by foreign workers brought in from Malawi, Mozambique and Zambia in the 1950s, but over the years their share decreased from over 70 per cent to around 30 per cent by 2000 (Magaramombe 2001), as more locals enlisted in the wage labour market to offset the effects of land alienation.

Most of this LSCF wage farm labour (65 per cent) was concentrated in the Mashonaland Provinces (Central, East and West) located in the high potential agro-ecological regions, whilst Matabeleland North and South and Midlands in the drier regions had the least share and the predominant plantation estates in Manicaland and Masvingo employed 16 per cent and 10 per cent of the workers (CSO 2001).

The ESAP, which encompassed new flexible labour arrangements that largely favoured employers, eroded the earlier gains that accumulated to workers via minimum wages and job protection (UNDP 1999), such that, by 1999, farm wages were only 24 per cent of the rural PDL (Kanyenze 2001). Even the high-income export oriented land use patterns did not result in improved social reproduction of agricultural workers, as the profits benefitted mostly the LSCF land owners (UNDP 1999; Davies 2000). Farm workers supplemented their wages with other income generating activities (e.g., petty trading and gold panning) and petty commodity production on small gardens provided by 20 per cent of LSCFs and in Communal Areas (Vhurumuku et al 1998).

Although social policy shifts after 1980 improved the lot of farm workers, they did not significantly alter agrarian labour relations. The LSCF agricultural workers, who constituted the largest share of formal employment (26 per cent) (CSO 2001), were the least paid amongst the working class (MPSLSW 2001; Tandon 2001). The colonial farm compound or labour reserve on the LSCFs persisted as an institution tying residency to employment. Many studies done before the FTLRP have exposed the appalling living conditions (housing, health, education, malnutrition, poverty) of farm workers (see Loewenson 1992; Amanor-Wilks 1995; Tandon 2001; Rutherford 2001). Many white farmers continued to institute some of the elements of the master-servant relationship to repress labour. This approach has been termed 'domestic government' (Rutherford 2001).[3] The social and economic conditions of farm workers were out of the public view, as trespass laws restricted access to them, whilst spatial dispersion of the LSCFs made it difficult for trade union organisation (Chambati and Magaramombe 2008). The spatial dispersion of the LSCFs also limited the reach of state labour officials (Loewenson 1992). Moreover, government provided minimal social services to farm workers as they largely considered them the responsibility of farmers (ibid).

In the Communal Areas, agriculture production on small land sizes was the basis of social reproduction, utilising labour- intensive technologies of mainly cheap food crops (dominated by maize) for subsistence and surplus for sale in domestic markets (Muir 1994; Moyo 1995; Mehretu 1994). Agrarian labour in the Communal Areas was mostly provided by self-employed family workers, though hiring wage labour was limited to a few rich households (Adams 1991). These self-employed workers, who were mostly women (Potts 2000; Muchena 1994), accounted for 84 per cent (or 2,018,808 workers) of the total agricultural labour force in Zimbabwe, with the remainder on LSCF employment (CSO 2000). Despite the growth in agricultural production and productivity (maize and cotton) in the Communal Areas after 1980, the declining farm sizes and land quality caused by demographic growth meant that many could not meet their social reproduction needs (Moyo 1995). Agricultural productivity declined from the 1990s as the state reduced input subsidies, forcing many households into cheap wage labour (Oni 1997). By 1999, migrants constituted around 30 per cent of the LSCF workers (Sachikonye 2003). Deepening poverty,[4] growing landlessness and the massive retrenchment of urban workers after ESAP fuelled demands for land redistribution by 1997 (Moyo 2000; Moyo a nd Yeros 2005).

The Communal Areas were, in fact, socially differentiated. There were households that commanded more land, superior land quality in prime agro-ecological regions and/or better access to economic and social resources than others. These households hired wage labour and performed better in agricultural production (see Adams 1991; Cousins 1992; Moyo 1995). However, the social reproduction of most self-employed and LSCF agrarian wage labour was fragile and these households could not meet their basic food and social requirements.

New agrarian labour relations after land reform

Restructuring of agrarian labour relations

The transformation of agrarian labour relations during the FTLRP was not a one-off event, but a dynamic process that was shaped, since 2000, by differentiated land acquisition and redistribution and by changing agrarian policies. Moyo periodised the FTLRP into four distinct phases (see Table 2.1, Periodisation of the FTLRP 2000-2012, Chapter 2). These four phases of land reform also entailed various changes in agrarian (labour) policies (Table 5.1). The land reform policy did not initially adequately address what would happen to the former farm workers who were employed and resident on the LSCF private properties under insecure tenure. Numerous policy measures were, however, instituted to address their needs over the course of the last decade. These included the retrenchment and re-employment policy; land access and residential tenancy reforms; amendments of the citizenship laws; repatriation; social services; and wage rates determination policy (Table 5.1). Alongside these, macro-economic conditions and various agricultural and economic policies that shaped the conditions for agricultural production and utilisation of labour also underwent dynamic changes (Table 5.1; see also Moyo and Nyoni, Chapter 6).

Table 5.1: Changing agrarian (labour) policy regime, 2000-2011

Land Reform Phase (Period)	Key Agrarian Labour Policy Events	Key Agrarian Labour Processes
March 2000 to June 2001 Revolutionary situation	-Citizenship laws amended	-Migrant citizenship in doubt -Land occupiers/LSCFs/farm workers disputes -Mass worker evictions -Formal job losses -Expanded self-employment
July 2001 to December 2003 Rationalisation of land reform	-Land access and temporary residence -Retrenchment policy -Migrant repatriation	-Land allocations to workers -Severance packages paid -Farm wage disputes/LSCFs monopoly in collective bargaining -Wage jobs in A2, more self-employment on A1 -Non-farm livelihoods -Low land utilisation on new farms
January 2004 to June 2008 Bureaucratisation of land reform	-Migrant citizenship rights restored - Land tenure reforms - Inclusion of new farmers in collective bargaining - Subsidised farm mechanisation - Input and output market price controls - State input subsidies - Contract farming	-More land allocations to workers -Depressed farm wages -Shortages of key inputs -Food production dominant -Limited wage employment capacity -Expansion of non-farm livelihoods -Labour shortages
July 2008 to December 2011 Residual land redistribution	- Multiple foreign currencies introduced - Liberalisation of agricultural markets -Reduction in state input subsidies -Expansion of contract farming	-Dollarized wage payments -Expanded land utilisation -Increased availability of inputs -Expanded wage employment capacity -Diversification of commodities -Concentration of labour in contract commodities

Source: Moyo, Chapter 2 for land reform phases; key labour policy events compiled by author from Chambati and Moyo 2004; Moyo et al 2009; Moyo and Nyoni, Chapter 6.

The transformation of agrarian labour relations after 2000 entailed the expansion of the peasantry and creation of many new smaller capitalist farms that in turn generated various forms of self-employment on farm and non-farm activities and full and part-time farm wage jobs as discussed below. The restructuring also involved the loss of an estimated 200,000 formal farm worker jobs in the acquired LSCFs. These job losses were phased throughout the last decade, but most occurred in the earlier phases of land reform characterised by mass expropriations of LSCFs. An estimated 100,000 farm jobs (of which 50 per cent were part-time) were, however, retained in the LSCFs and plantation estates that were not acquired (also discussed below). Amongst those who lost their jobs, some have been completely displaced from the former LSCFs and others are still resident in the farm compounds. Between 30,000 and 45,000 workers are estimated to have been displaced from the former LSCFs to Communal Areas, towns and informal settlements (Chambati and Moyo 2004; Chambati and Magaramombe 2008). Most of these physical displacements occurred during the early phase of land reform that was characterised by confrontation between land occupiers and farm workers. Some former farm workers have also been reported to have migrated to neighbouring countries to work in their LSCFs, especially in South Africa (Rutherford and Addison 2007) and Mozambique (Hammar et al 2010). However, the extent of former farm worker migration to neighbouring countries is not known in the absence of systematic data collection on this question.

Nationally, it is estimated that over two- thirds of the former farm workers remained on the former LSCF land (Chambati and Moyo 2004; Moyo et al 2009; Magaramombe 2010). Nearly 69 per cent of the former farm workers indicated that the majority of their colleagues were still resident in the compounds (AIAS Farm Worker Survey 2005/06). There were variations in the displacement: For example, in Chiredzi, Chipinge, and Zvimba Districts over 60 per cent of the former farm workers reported that most of their colleagues had stayed put, compared to 29 per cent and 9 per cent in Kwekwe and Mangwe respectively (AIAS Farm Worker Survey 2005/06). Overall, there has been a net gain in livelihoods, as 45,000 farm workers and 4,000 physically displaced farmers have been replaced by 170,000 farm households (see Moyo 2011a) plus new types of employment.

The competing demands for land access during the FTLRP by different classes, including landless peasants, the urban working class, farm workers, semi-proletariats and an emerging middle class bourgeoisie interested in commercial farming, meant that many former farm workers' preferences for land resettlement[5] could not be met. Various discourses have been mobilised on whether farm workers were supposed to benefit from the FTLRP land allocations. Some argued for the preferential treatment of farm workers since the redistributed lands were tied to their livelihoods as agrarian labourers (Magaramombe 2003). Others wrongly perceived farm workers as foreigners who could not claim land in Zimbabwe (see Moyo et al 2000). The potential of farm workers as independent producers has also been questioned, as some advocated for their continued role as suppliers of wage labour to the new farms.[6] The perceived political allegiance of farm workers to the Movement for Democratic Change (MDC) and white farmers (who are viewed as anti-land reform) was also considered as a justification for their exclusion in land allocation (see Chambati 2011a).

The possibilities for farm workers to benefit from land allocation was also limited by their exclusion from the list of targeted land beneficiaries in the FTLRP policy document, which emphasised landless peasants as targeted land beneficiaries, followed by other groups in need of land and offered a 20 per cent quota of all land allocations to liberation war veterans (GoZ 2001).

Former farm workers were thus largely marginalised in allocation of land in the initial phases of the FTLRP, but their position received more attention during the GoZ land audits (Utete 2003) and the Parliamentary Portfolio Committee on Lands and Agriculture enquiries, as well as in advocacy by NGOs such as the Farm Community Trust of Zimbabwe (FCTZ) and in various academic studies (see Magaramombe 2003; Chambati and Moyo 2003; Sachikonye 2003). Since mid 2002 and in 2003, District Land Committees were enjoined by GoZ officials to increase their allocations of land to former farm workers, including the setting aside of some farms for their resettlement.

In this situation, former farm workers were not completely excluded from land access as some studies suggest (see Alexander 2003; ZHRF and JAG 2007); a few of them were accommodated. Former farm workers constituted only 8.1 per cent of the land beneficiaries (Moyo et al 2009: 22).[7] They explored multiple avenues to gain access to land, including through registration with traditional leaders in Communal Areas, participation in land occupations and official applications with District Land Committees (without revealing their status as former farm workers) (Chambati and Moyo 2004). This form of application was done to evade victimisation and being left out of the resettlement exercise as perceived "anti-land reform" reactionaries and MDC supporters (ibid), as some of their constituencies were mobilised to oppose land reform and land occupations as discussed below. Only 38 of the 70 former farm workers who got land disclosed their status when they registered for land (AIAS Farm Worker Survey 2005/06).

There was variation in the distribution of land to former farm workers during the FTLRP. Chipinge had the highest proportion of former farm workers among land beneficiaries (17.7%), followed by Mangwe (12.4%), whilst they were only 6 per cent and 7 per cent in Kwekwe and Zvimba respectively (Moyo et al 2009).[8] In Chiredzi, they constituted less than one per cent of the land beneficiaries (ibid). Some former farm workers also informally accessed small land pieces (between 0.04 and 2.0 hectares) as detailed below, while others (26.8%) got Communal Area plots (AIAS Farm Workers Survey 2005/06).

The former farm workers remaining in the newly redistributed lands who did not benefit from land allocations continued to sell their labour to farm and non-farm activities, as part of the new agrarian labour regime, alongside some landless people from the Communal Areas. This new agrarian labour regime is centred on the diverse sources of employment in the reformed agrarian structure, comprising the peasantry, small-to-medium capitalists, large capitalists and plantation estates and conservancies.

There exists competition for labour resources amongst a set of activities that include self-employed farming, other petty commodity production (e.g. natural resources exploitation) and hiring out labour to farm jobs. The competition for labour resources is also amongst the different classes in the new agrarian structure. This competition has resulted in shortages of farm labour across all the different farm classes. These shortages were faced by 38.4 per cent of the new land beneficiaries (Figure 5.1), as well as the remaining LSCFs and plantation estates (USAID 2010). The farm labour shortages arise because of the expansion of self-employment amongst the potential agrarian labour force from the Communal Areas, who gained access to land, poor wages and new non-farm job opportunities. Farm labour mobilisation strategies have been transformed as they now often involve more sources beyond the Communal Areas as discussed below.

Extrapolating the district surveys' farm labour utilisation rates and other secondary information nationally, we estimate that, by 2010, permanent jobs had grown to over 478,013 in comparison to 167,459 jobs in 2000 (Table 5.2).

Table 5.2: Changing Agrarian Employment Structure, 2000-2010

Farm sector	No. of employers/ farms	Average farm size	Average no. of employees per farm			Total number of employees					
			Permanent	Casual	Family	Permanent	% of total permanent	Casual	% of total casual	Family	% of total Family
2000											
Peasantry											
Communal Areas	1,050,000.0	16.0	0.1	0.6	.9	72,768.0	24.7	609,811.0	80.6	2,018,806.3	89.9
Old resettlement	75,000.0	49.0	0.1	na	2.5	9,588.0	3.3	na		87,500.0	8.3
Subtotal	**1,125,000.0**					**82,356.0**	**27.9**	**609,811.0**	**80.6**	**2,206,306.3**	**98.2**
SSCF	8,500.0	165.0	2.6	na	2.0	22,440.0	7.6	na		17,000.0	0.8
LSCFs	4,956.0	1,754.0	33.8	29.7	0.9	167,459.0	56.8	146,960.0	19.4	5,898.0	0.3
Estates	296.0	8,672.0									
Subtotal	**5,252.0**					**189,899.0**	**64.4**	**146,960.0**	**19.4**	**22,898.0**	**1.0**
Total	**1,138,752.0**	**165.0**			**2.0**	**294,695.0**	**100.0**	**756,771.0**	**100.0**	**2,246,204.3**	**100.0**
2010											
Peasantry											
Communal Areas	1,100,000.0	15.0	0.1	0.6	1.9	77,000.0	16.1	638,000.0	32.1	2,112,000.0	72.2
Old resettlement	75,000.0	49.0	0.1	na	2.5	9,750.0	2.0	na		187,500.0	6.4
A1	145,800.0	40.0	1.0	5.9	3.6	145,800.0	30.5	860,220.0	43.2	524,880.0	17.9
Subtotal	**1,320,800.0**					**232,550.0**	**48.6**	**1,498,220.0**	**75.3**	**2,824,380.0**	**96.5**
Medium capitalist											0.0
SSCF	8,500.0	165.0	2.6	na	2.0	22,440.0	4.7	na		17,000.0	0.6
Small-medium A2	22,700.0	71.0	6.6	17.2	3.6	149,820.0	31.3	390,440.0	19.6	81,720.0	2.8
Subtotal	**31,200.0**					**172,260.0**	36.0	**390,440.0**	19.6	**98,720.0**	3.4
Large capitalist											
Large A2	217.0	508.9	15.0	25.2	3.6	3,255.0	0.7	5,468.4	0.3	781.2	0.0
LSCF	577.0	574.0	33.8	29.7	2.0	19,496.6	4.1	17,108.05	0.85	1,192	0.05
Corporate estates	247.0	6,051.0	138.5	268.5	n.a	34,209.5	7.2	66,319.5	3.3	n.a	
Total	**1,353,618.0**	**6,051.0**				**478,013.2**	**100.0**	**1,989,195.6**	**100.0**	**2,925,408.0**	**100.0**
New jobs created						**183,318.2**		**1,232,424.6**		**679,203.8**	

Sources: World Bank 2006; Moyo et al 2009; Moyo 2011a; CSO 2001.

The newly redistributed areas had generated an estimated 679,203 new self-employed farm jobs, representing a ninety- fold growth from the situation in LSCFs in 2000 (Table 5.2). Despite the shortages of labour currently being experienced, the new agrarian labour structure represents an overall growth in agricultural employment (Table 5.2). The structural changes in agrarian labour relations in the four agrarian classes are detailed below.

Agrarian labour relations in the peasantry

The expansion of the peasantry after land redistribution has entailed the broadening of self-employed family labour in the agrarian labour structure. Increased land access after the FTLRP to the formerly- land short and landless peasants has allowed them to engage in petty commodity production through self-employment of family. On the basis of field and secondary data, an estimated 524,880 new self-employed farm jobs have been added via the A1 resettlement scheme (Table 5.1). New self-employed jobs have also been added to the Communal Areas on lands vacated by FTLRP beneficiaries. Out of 1,158 land beneficiary households from the Communal Areas, 864 of them had given up their landholdings to their extended family relatives (AIAS Household Baseline Survey 2005/06). Cumulatively, the peasantry now accounts for over 2,824,380 self-employed farm jobs, equating to 96.5 per cent of the self-employed farm jobs nationally (Table 5.2).

The use of family labour is most important for the peasantry in the Communal Areas, where up to 90 per cent of the labour for the major field operations[9] is provided by the family (Langyuito 2005). Even in the A1 sector, most of the labour force is composed of self-employed family labour, but provides all the farm labour in only in 26.1 per cent of the households (Moyo et al 2009). Overall, self-employed labour deployed to farming averaged 3.6 persons per household in newly redistributed areas (Table 5.2).

Self-employment also entails the reciprocal exchange of labour amongst farm households. This exchange entails the grouping of family labour from several households to carry out time- sensitive tasks such as weeding on a single plot over an agreed amount of time, which is then reciprocated to all the households participating in the group. Reciprocal labour exchanges were utilised by 20 per cent and 23 per cent of the A1 and A2 households respectively (Moyo et al 2009). This practice is also common in the Communal Areas (see Adams 1991; Muchena 1994).

Based on patriarchal gender relations, women and children continue to provide the bulk of self-employed family labour in the Communal Areas (Paradza 2010;

Mutopo 2011; see also Muchena 1994). However, in the newly redistributed areas, men accounted for 55.0 per cent of the family labour, reflecting the slightly male dominated population in these areas (Moyo et al 2009).

Some of this self-employed family labour is mobile and straddles the newly redistributed areas and the Communal Areas. Over 20 per cent of the beneficiaries that originated from the Communal Areas still practice petty commodity production in these areas (Moyo et al 2009). Family labour is exchanged between the newly redistributed and Communal Areas by these land beneficiaries.

The self-employed workers in the peasantry also hire farm wage labour. The A1 sector employed most of the farm wage jobs amongst the peasantry, amounting to an estimated 145,800 and 860,220 full and part- time wage workers (Table 5.2). In the small A1 farms, 21.3 per cent and 57.1 per cent hired permanent and casual workers respectively (Moyo et al 2009). Overall, 71.6 per cent of the A1 households hired farm wage labour (ibid). Agrarian wage labour markets also continue to exist in Communal Areas. Recent research in Svosve Communal Areas (Goromonzi District) showed that up to a third of the households surveyed by Chimhowu and Woodhouse (2008) were participating in casual agricultural labour markets (see also Paradza 2010).

It is in this context of labour hiring that some peasants also experienced shortages of farm wage labour. In the A1 farms, 38.6 per cent of the land beneficiaries reported facing them (Moyo et al 2009). These shortages were highest in Kwekwe District (36.5%), in the context of competition with gold panning activities, followed by the districts located in the higher potential agro-ecological regions (Zvimba and Goromonzi – around 30%) that hired more wage labour (ibid).

In the Communal Areas, farm labour shortages were experienced as a result of the movement of 10 per cent of the households into the newly redistributed areas and via the economic crisis context that resulted in the outmigration of mostly young people to urban and other areas and even across borders in search of wage employment. Some scholars (Kinsey 2010) have reported declines in average household sizes by as much as 42 per cent between 2001 and 2008 in selected Communal Areas in Mashonaland Central and East and in Manicaland. In a similar vein, field surveys in Mangwe Communal Areas revealed that 20 per cent of the 150 households interviewed had a member who had left for South Africa between 2001 and 2003 (Maphosa 2009). Paradza (2010) also found that non-farm job opportunities created in the former LSCFs such as gold panning were resulting in labour shortages as unemployed school leavers from Goromonzi Communal Areas were flocking to these areas.

Agrarian labour relations in the small-to-medium and large capitalist farms

The small-to-medium and large capitalist farms, dominated by the new A2 land beneficiaries, are defined by the reliance on hired wage labour for their agricultural production activities. Specifically, this sector has diversified the sources of farm wage employment by increasing the potential number of employers from around 4,500 to over 31,000 (Table 5.2). Cumulatively, they employ a total of 211,253 and 424,656 full and part- time farm jobs respectively (Table 5.1). The family labour in the middle and large capitalist farms accounted for only 3.5 per cent of the total self-employment nationally. In contrast, permanent farm wage labour constituted 44.2 per cent of the total, yet they constitute less than 4 per cent of the farm units.

Figure 5.1: Levels of agrarian wage labour hiring

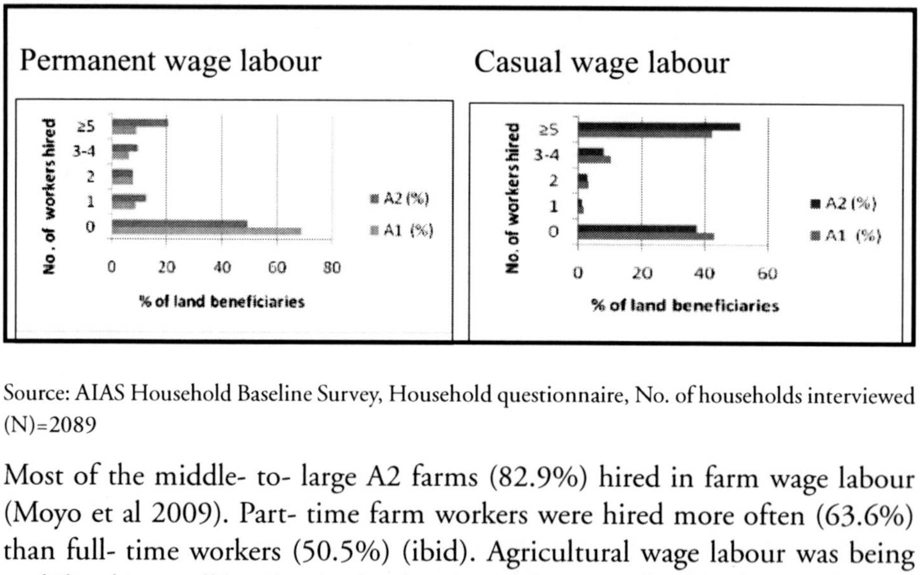

Source: AIAS Household Baseline Survey, Household questionnaire, No. of households interviewed (N)=2089

Most of the middle- to- large A2 farms (82.9%) hired in farm wage labour (Moyo et al 2009). Part- time farm workers were hired more often (63.6%) than full- time workers (50.5%) (ibid). Agricultural wage labour was being mobilised in small batches in the downsized farms, as few hired more than 10 workers (Moyo et al 2009), compared to an average of 50 workers engaged in the LSCFs prior to 2000 (CSO 2001). The capitalist A2 farms tended to hire larger amounts of labour in comparison to the small A1 farms (Figure 5.1). Farm wage labour use tended to be higher in the high agro-ecological potential districts of Chipinge, Goromonzi and Zvimba, where over 75 per cent used it, compared to around 60 per cent in the drier Mangwe and Kwekwe Districts (Moyo et al 2009).

The reliance of the new capitalist farms exposes them to shortages of farm labour, which are characteristic of the new agrarian structure. These were faced by 37.4 per cent of the capitalist farms in the newly redistributed areas (Figure 5.2). The capitalist farms located in the higher agro-ecological potential districts that require more labour (such as Goromonzi) had the highest proportion of households (47.6%) that reported facing labour shortages, followed by Kwekwe district (42.5%), which is endowed with high- wage alluvial gold that has been attracting labour. In Chiredzi, Mangwe and Zvimba Districts, farm labour shortages were reported by 31 per cent, 16.7 per cent and 27.8 per cent of capitalist farms surveyed, respectively.

Figure 5.2: Land beneficiaries experiencing farm labour shortages

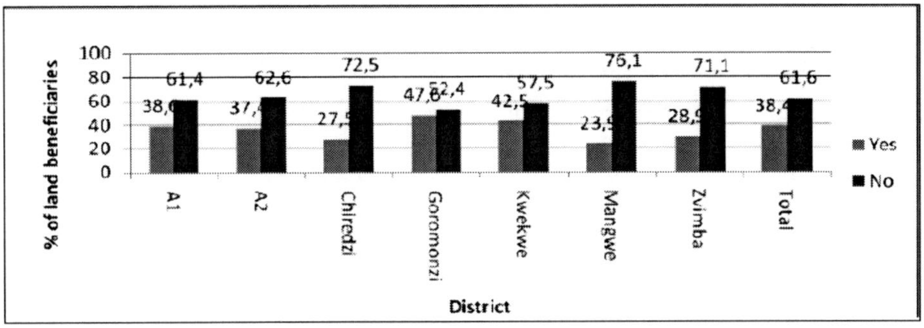

Source: AIAS Household Baseline Survey (2005/6) N=2089

Thus, the new agrarian employers increasingly have to search for farm labour, unlike in the past when landless people from the Communal Areas would actively seek employment in the LSCFs. Field surveys in Mazowe and Zvimba districts, for instance, showed that some new capitalist farms were trucking unemployed youths from Bindura and Banket towns to work as day casual labourers (Chambati 2009). The recruitment of unemployed urban youths from the nearby high-density suburbs of Epworth and Mabvuku was also observed in Goromonzi District. Former farm workers were employed by 36 per cent of the land beneficiaries (Moyo et al 2009). The employment of former farm workers was higher in Chipinge, Goromonzi and Zvimba districts, which also had a higher concentration of former farm workers before 2000 (ibid). The majority of the former farm workers (76%) who have been re-absorbed into farm employment have been engaged in A2 capitalist farms (AIAS Farm Worker Survey 2005/06), whilst 79 per cent of the A1 farms that hired wage labour engaged new farm workers from communal and urban areas (ibid). Overall, 42 per cent of the land beneficiaries recruited wage labour from the Communal Areas (AIAS Household Baseline Survey 2005/06).

Landless relatives from the extended family were also being incorporated into the wage agrarian labour force. This development introduced new forms of social patronage in which work relations are defined by kinship ties between the employer and the employee (Chambati and Moyo 2004). These social patronage labour relations entail qualitatively different social relations than the remnants of the master-servant relationship in LSCFs, as discussed later. The social patronage labour relations were existent in 11.5 per cent and 16 per cent of the A1 and A2 farms respectively (AIAS Household Baseline Survey 2005/06). Children are also being exploited as underpaid farm labourers to reduce the costs of labour and alleviate shortages.[10]

The gender composition of the labour forces recruited by the new land beneficiaries imitates that of the LSCF, as men still dominate the permanent forms of employment, comprising 70 per cent of the full- time work force, but women's share has increased significantly to 30 per cent (Moyo et al 2009) from the 10 per cent they occupied in the LSCF sector before 2000 (CSO 2001).

Agrarian labour relations in the plantation estates

As has been noted above (Moyo 2011b; Moyo, Chapter 2), foreign- owned large agro-industrial plantation estates involved in sugar, tea, coffee and timber in the Eastern Highlands and Lowveld were not acquired under the FTLRP. On their expansive land sizes, they still employ large batches of farm labour, estimated at 34,209 and 66,319 full and part- time workers, respectively (Table 5.1). The hired permanent labour in the agro-industrial plantation estates constitutes 7.2 per cent of the total full- time agricultural wage employment nationally. The plantation estates are exclusively reliant on wage labour, some of which still operates under the labour residential tenancy. They command the largest concentration of wage labour per farm unit, averaging in excess of 200 workers per estate (Table 5.1).

A key development in the agrarian labour relations in the plantation estates during the FTLRP has been the shortages of farm labour. Since the plantation estates relied to a large extent on part-time labour from the Communal Areas, particularly during harvesting periods (Mtisi 2003), they also face labour shortages as some of the potential labour force were allocated land during the FTLRP. For instance, Tanganda Tea Company (the largest tea producer in the country) attributed reduced output in 2006 to harvesting labour shortages (USAID 2010). Similar labour shortages were also encountered by sugar plantation estates in the Lowveld (Parliament of Zimbabwe 2006; Scoones et al 2010). The poor wages eroded by inflation after 2000 also forced some farm labourers in the plantation estates to seek farm work in nearby South African and Mozambican LSCFs (Hammar et al 2010; Rutherford and Addison 2007).

Non-farm sources of rural employment

Land redistribution and tenure reforms after 2000 have also opened up self-employment opportunities in natural resources exploitation to land beneficiaries, to those in nearby Communal Areas and to former farm workers. The LSCF freehold property rights excluded other population segments from accessing natural resources through their security protection systems in the form of farm guards and other security people (see Moyo 1995). Land beneficiaries with state tenures are less equipped to exclude others from natural resources on their lands. In Kwekwe District, for example, over 20 per cent of the land beneficiaries confirmed the presence of informal gold miners on their lands (Moyo et al 2009). Land beneficiaries blame former farm workers for the "destruction" of natural resources (Chambati and Moyo 2004). Many former farm workers (81 per cent) confirmed accessing natural resources in new farms (AIAS Farm Worker Survey 2005/06).

About 21 per cent of the former farm workers were involved in the commoditisation of natural resources on a full time basis (AIAS Farm Worker Survey 2005/06). Alluvial mining (especially gold and diamonds) has created an alternative source of income for farm workers and others in various provinces. In Goromonzi and Kwekwe districts, 30.7 per cent and 21 per cent were self-employed in gold panning, while 27 per cent in Chipinge panned for diamonds (ibid). In some instances, these activities have led to the abandonment of FTLRP plots by, for example, 300 former farm workers in Zvimba North (Magaramombe 2003). Firewood sales also represent one of the major sources of non-farm employment as firewood is in high demand, being the major source of energy for 81.3 per cent of the land beneficiaries, as well as for urban dwellers in the context of nation-wide electricity shortages (Moyo et al 2009). Other natural resource extraction livelihoods include fish and thatching grass sales (Chambati 2009), as well as wildlife hunting for meat consumption in the former LSCFs in districts such as Mwenezi (Scoones et al 2010).

New rural jobs also arose out of the demand for housing and other productive infrastructures by land beneficiaries. A wide array of construction jobs and material supplies (bricks, thatching grass and poles) were created as over 73 per cent of the land beneficiaries had constructed homesteads by 2007, whilst other sources of rural construction jobs included storage facilities (7.3 per cent) and worker housing (8.9%) (Moyo et al 2009). Over 5 per cent of the small A1 land beneficiaries were involved in supplying bricklaying services, whilst 2.1 per cent provided carpentry services (Moyo et al 2009). Farm equipment repairs and services also provided non-farm employment. In Goromonzi District, one A2 farmer who previously worked for an engineering firm prior to 2000 was operating a workshop at his homestead for water pumps and tractor repairs for other land beneficiaries. Such new economic opportunities were also observed in the other districts surveyed (Moyo et al 2009).

Petty trading activities such as the vending of clothes and value- added farm products have also grown as result of the unrestricted movement of people and goods with over 5 per cent of the land beneficiaries involved (ibid). Within the newly redistributed areas surveyed, we found some land beneficiaries involved in the pressing of oil from oilseeds (soybeans and sunflower seeds) and the grinding of groundnuts into peanut butter using hand- operated machines in preparation for sales in the local markets. Others have also taken advantage of the absence of facilities such as retail shops[11] through the operation of small informal shops at their homesteads: they buy daily consumer goods and services such as foodstuffs (salt, cooking oil, sugar, flour, etc.) from nearby urban centres for resale to others. About 5 per cent of the land beneficiaries were found to be involved in these types of activities (AIAS Household Baseline Survey 2005/06).

Changing agricultural production relations and labour utilisation

The utilisation of the different forms of labour is differentiated according to the land sizes redistributed, input usage, asset ownership, cropping and livestock production patterns, value of crops produced and farm machinery and equipment resources distribution (Moyo et al 2009; Chambati 2011a; Scoones et al 2010). Agrarian labour utilisation is one of the critical factors in explaining broader social differentiation in newly redistributed areas (Moyo et al 2009). Amongst the peasantry in the A1 sector, 68.7 per cent were low-level labour users who relied mostly on family labour with occasional hiring of part- time workers, while 23 per cent were high- level labour users, hiring over seven full- time workers (Table 5.3). More middle- to- large capitalist A2 farms (38%) were high- level labour users compared to A1 peasantry (23%) (Table 5.3).

Table 5.3: Emergent Structure of Rural Labour in New Resettlement Areas

Level of labour use	A1						A2						% Total
	No. of HH[4]	of HH	Average labour use				No. of HH	% of HH	Average labour use				
			FT[5]	PT[6]	Family	Hired out			FT	PT	Family	Hired out	
Low[1]	1134	68.7	0.0	4.22	3.65	0.14	217	49.5	0.0	12.3	3.38	0.02	64.7
Medium[2]	140	8.5	1.0	5.95	3.56	0.17	55	12.6	1.0	12.1	3.45	0.05	9.3
High[3]	377	22.8	7.2	12.75	3.77	0.11	166	37.9	8.2	13.1	3.76	0.06	25.9
Total	1651	100.0	1.74	6.31	3.67	0.14	438	100	3.2	12.6	3.53	0.04	100

Source: AIAS Household Baseline Survey (2005/06). 1. Household utilises family labour in combination with part- time labour; 2. Household hires in one full- time worker; 3. Household hires in at least two full time workers, plus part- time; 4. HH – household; 5. FT – full- time; 6. PT – part- time.

Small farms had higher labour/land ratios (or labour employed per unit of farm area) than the middle- to -large capitalist farmers. As expected, there was a direct relationship between the number of workers hired and the areas cropped (Moyo et al 2009; Chambati 2009). The average labour/land ratios decreased as the farm size increased (Table 5.4). This relationship arises from the existence of labour substituting farm machinery in the larger farms. For instance, among the land beneficiaries owning tractors, 48.3 per cent had farm sizes over 50 hectares, in comparison to 17.0 per cent of the non-tractor owners (AIAS Household Baseline Survey 2005/06). We also saw that tractor owners had a lower number of permanent workers per cropped area of 0.4 in comparison to 0.6 amongst non-tractor owners (AIAS Household Baseline Survey 2005/06). As a result, small farms generate more employment in comparison to the larger farms, which is critical given Zimbabwe's current unemployment problem.

Table 5.4: Labour Intensities by Farm Sizes

Labour type	**Farm sizes**				
	1-19	**20-49**	**50-99**	**100+**	**Total**
					No. of workers per unit area
Permanent workers	0.33	0.88	0.33	0.16	0.19
Casual workers	1.20	0.37	0.10	0.02	0.71
Family labour	0.60	0.15	0.06	0.21	0.35
Family + permanent	0.94	0.23	0.10	0.04	0.54
N	962	620	191	208	1981

Source: AIAS Household Baseline Survey (2005/06)

Prior to the FTLRP, the LSCFs were the lowest employers of labour per unit of cropped area, which averaged 0.7 workers per cropped hectare, compared to an average of 3.5 and 5 in the old resettlement and SSCF sectors respectively during the 1990s (GoZ 2001). The labour utilisation per unit of cropped area (0.62 workers per cropped area), excluding casual workers, in the newly redistributed areas is slightly less than the combined rates that prevailed in the former LSCF sector (Moyo et al 2009), implying greater capacity for labour absorption in a situation when land utilisation reaches optimal levels.

The socio-economic context during the last decade was also largely influential in determining the trajectory of agricultural production and labour relations. The wider economic crisis, especially between 2004 and 2008, induced production constraints, but land beneficiaries managed to establish farming activities to varying degrees (see Moyo and Nyoni, Chapter 6).

The FTLRP transformed the export- oriented LSCF land use patterns towards food crops such as maize and small grains. However, from around 2008, the re-integration of the country into international capital markets and economic stabilisation through dollarisation has drawn a few larger land beneficiaries who have been able to accumulate capital into production of high value export and cash crops such as tobacco, oilseeds and horticulture (Moyo et al 2009; Scoones et al 2010; Moyo 2011c). Investments into productivity enhancement inputs, irrigation and mechanical equipment such as tractors are relatively higher amongst this group (ibid).

Most of the beneficiaries who had diversified their agricultural production to include other cash/export crops such as tobacco, oilseeds and horticulture tend to have higher proportions of representation in the medium and high level labour users than those focused on food production. For instance, amongst tobacco, oilseed and horticultural producers, over 40 per cent were medium and high- level farm labour employers in comparison to 33.9 per cent of maize- focused producers (AIAS Household Baseline Survey 2005/06). The farm wage labour employed was significantly higher in land beneficiaries involved in tobacco and oilseed production who hired an average of 3.85 and 3.60 permanent workers respectively in comparison to 1.74 workers by food crop producers (AIAS Household Baseline Survey 2005/06).

These farm labour use patterns were also reflected in input usage. The average amount of fertiliser applied to maize production, for example, was highest amongst high level farm labour employers (1,257 kilogrammes per hectare), compared to low level farm labour employers (100.2 kilogrammes per hectare) (AIAS Household Baseline Survey 2005/06). Overall, the expensive inorganic fertiliser was more commonly used by the medium and high- level farm labour employers than the low-level farm labour employers (ibid).

Access to scarce and/or expensive inputs through contract farming has also been introducing new shifts to agrarian labour relations. Since 2009, there has been an increased promotion of contract farming by merchant capital and agro-industrial processing companies that derive their inputs from agriculture in export crops (cotton and tobacco), oilseeds (soybean) and food crops (sugar beans) (Moyo and Nyoni, Chapter 6; Moyo 2011a). Contract farming, which provides inputs and technical advice on credit in return for crop sales, lessens farmers' difficulties in mobilising financial resources in a tight external financing environment.

The tobacco sector specifically registered the largest growth in contract farming (Moyo and Nyoni, Chapter 6; Moyo 2011b). This growth resulted in the concentration of agrarian labour in tobacco production. It is estimated to have employed 112,000 people or 20 per cent of the current formal employment nationally in the 2009/10 season, whereas in 1998 it employed 172,000 people (Price Waterhouse Coopers [PWC] 2010) or 12.7 per cent of the formal employment (CSO 2000). Nevertheless contract farming has entailed the super-exploitation of self-employed agrarian labour through low output prices and overpriced inputs provided by contractors, as well as via poor wages paid to labour hired by the small capitalist farms (Moyo 2011b; Moyo and Nyoni Chapter 4). Moreover, in contract farming, the bulk of the production risks (harvest failure, labour shortages and product quality) were also borne by the peasants and small capitalists (see Oya 2011).

The areas cropped and commodity choices are also influenced by the farm machinery and equipment endowments possessed by households, which in turn affect the demand for farm labour. The farm machinery and equipment endowments such as tractors that allow households to crop more land areas were not accessible to most households. For instance, only 36 per cent of large A2 land beneficiaries had access to a tractor, whilst in the small A1 farms, 52 per cent and 6.2 per cent had access to animal- drawn ploughs and tractors respectively (Moyo et al 2009). Moreover cattle, which provide low cost draught power, were owned by 42.8 per cent of the new land beneficiaries (Moyo et al 2009). The farm mechanisation programme launched by the government in 2003 to address this problem and labour shortages (GoZ 2007) was limited in its reach, as it mostly benefited a few large farmers (World Bank 2006; Chambati 2009).

There was a direct relationship between capital intensity and scale of labour establishment in the newly redistributed areas (Table 5.4). The land beneficiaries with high capital intensity also tended to be high-level farm labour employers (Table 5.5). The high-level labour users constituted the majority of the households with high capital intensity. These high capital intensity land beneficiaries also tended to have high cropped areas, as 43 per cent of them cropped over 10 hectares in comparison to 11 per cent of the low and medium capital intensity land beneficiaries (AIAS Household Baseline Survey 2005/06). Access to irrigation, which was limited to 16 per cent of the land beneficiaries, was available to more high-level farm employers (19.5 per cent) in comparison to 16.2 per cent and 14.2 per cent of the low-level and medium- level farm labour employers (ibid).

Table 5.5: Level of labour use vs. capital intensity*

Level of Labour Use	*Capital Intensity (No. and % of households)*							
	Low		*Medium*		*High*		*Total*	
	No.	*%*	*No.*	*%*	*No.*	*%*	*No.*	*%*
Low**	822	68.4	432	63.8	97	46.2	1351	64.7
Medium***	93	7.7	80	11.8	22	10.5	195	9.3
High****	287	23.9	165	24.4	91	43.3	543	26.0
Total	**1202**	**100**	**677**	**100**	**210**	**100**	**2089**	**100**

Source: AIAS Household Baseline Survey (2005/06).

* Farm machinery and equipment endowments are used as a proxy of capital intensity in newly redistributed areas; ** Low capital intensity do not own any power driven implements and mostly rely on animal drawn implements and hiring in services;

*** Medium capital intensity-own at least one of the power driven implements, not including a tractor;

****. High capital intensity -own at least three items from a set of power driven equipment that includes tractor.

The agricultural production processes and labour relations are also largely shaped by access to finance by land beneficiaries, which in turn influences the commodity choices, farm machinery and equipment employed and areas cropped. Thus medium to large capitalist farms with access to more finance through personal savings and salaried jobs in the metropolis also tended to be medium and high level employers. Most resources invested in agricultural production were from land beneficiaries' own finances as credit facilities were limited for most of the last decade (Moyo et al 2009).[12] The low level farm labour employers had the least percentage of land beneficiaries (24%) who had access to income from professional employment, compared to medium level (27.3%) and high level farm employers (37.4%) (AIAS Household Baseline Survey 2005/06).

Even the quality of formal jobs of the land beneficiaries were also differentiated, as the high level farm labour employers (32.1%) had the highest proportion of those in high-paying managerial jobs in the civil and private sectors, followed by the medium-level farm labour employers (26%) (ibid). Of the low- level farm labour employers, 23.6 per cent were employed in managerial jobs (ibid). Furthermore, high-level farm employers had the highest percentage of land beneficiaries (12.7%) who had access to savings and pensions from recent professional employment, compared to 9.5 per cent and 8.6 per cent amongst low and medium-level farm employers respectively (ibid).

The agricultural production processes and labour relations in the retained export oriented plantation estates and the peasantry in the Communal Areas also underwent transformation after 2000. Some agro-industrial estates such as Tanganda Tea Company, invested in mechanical harvesters to supplement hand plucking by human labour (Tanganda Tea Company 2006) and expanded outgrower schemes (*The Herald* 2006), which entailed the transfer of production risks and/or labour shortages to small producers.[13] Labour exploitation of small tea and sugar producers is also rife through under-grading of output and low prices (Mtisi 2003; Parliament of Zimbabwe 2006). These problems arise from the monopoly control of value chain of plantation crops by the agro-industrial estates (Moyo 2011b).

Alongside labour shortages, after 2000, low international prices for plantation crops such as coffee (USAID 2010; Moyo 2011b) also resulted in the shift to other lucrative crops. An example is that of Tanganda Tea Company, which reduced its coffee hectarage from 5,000 hectares to 3,000 hectares and replaced it with macadamia nuts, which is less labour intensive, yet commanding better international prices (*The Daily Mirror* 2006). New land beneficiaries in some of the sugar plantation estates acquired during the FTLRP have also been reported to be shifting land use to the less labour- intensive food crops (Scoones et al 2010; Moyo 2011b).

In the Communal Areas, the shortage of critical farming inputs (fertiliser, seeds, agro-chemicals) and the recurrent droughts resulted in the decline in land areas under cultivation (see Moyo and Nyoni, Chapter 6; Paradza 2010). Research conducted in Svosve Communal Areas showed that the land areas under cultivation declined from an average of 1.7 hectares in 1998 to 1.08 hectares in 2005 (Chimhowu and Woodhouse 2008), resulting in under-employment of family labour.

Economic and social conditions of agrarian labour

Land redistribution has altered the ways in which agrarian labourers realise their social reproduction. The diversified farming structure provides the agrarian labour force with broader choices for wage work than before, when the farm wage labour markets were monopolised by a few LSCFs. Agrarian political power has been diffused among many smaller employers. Quite crucially, the new dispensation *delinks* the employment rights of former farm workers from their land rights.

Nonetheless, the economic crisis during the FTLRP negatively affected farming and its associated returns (Moyo and Nyoni, Chapter 6) and the social reproduction of agrarian labourers. The wages of those in both rural and urban employment were eroded by inflation as the average real earnings index (1990 as the base year) declined from 90.7 in 2000 to 10.0 by 2004 (Kanyenze 2007). New A2 land beneficiaries have been pressing for low farm wages through their respective farming unions in the collective bargaining process. Agrarian wages have been repressed at below 10 per cent of the rural Poverty Datum Line (PDL) (Sachikonye 2003; Magaramombe 2010; Chambati 2011a) during the hyperinflationary period (between 2005 and 2008), such that part of the wages were paid in kind (maize grain, salt, cooking oil and soap). Even after the relative stabilisation of the economy, prevailing farm wages at \$30 to \$50 still remain way below the monthly rural PDL, estimated at over \$300 (ZIMSTAT 2011).[14] Beyond wage payments, other statutory benefits such as leave days and protective clothing are being received by a few farm workers (Parliament of Zimbabwe 2003; Utete 2003). There were, however, cases of non-super-exploitative wage payments, especially among export crop producers such as tobacco, which require specialised skills (Moyo et al 2009; Moyo 2011). The overall result has been under-consumption of basic food and social requirements by landless labourers.

Even with the opening up of diverse sources of farm and non-farm employment opportunities during the FTLRP, the social reproduction of many agrarian labourers was precarious. About 53 per cent of the farm worker households were managing to consume three meals per day (breakfast, lunch and supper) around 2006, whilst 4.3 per cent and 42.6 per cent had one and two meals per day respectively (AIAS Farm Worker Survey 2005/06). This rate of consumption was in contrast to that of land beneficiary households in which over 75 per cent were able to consume three meals per day during the same period (AIAS Household Baseline Survey 2005/06). Moreover, proteins were glaringly missing in farm worker diets dominated by the staple *sadza*[15] and vegetables (AIAS Farm Worker Survey 2005/06). Similar to the situation before 2000, malnutrition and food insecurity remain major challenges for farm workers, as noted by 31 per cent of them in the survey. Many farm worker households were not able to send their children to school, as only 22.2 per cent of the school-age children (6 to 18 years) were in school (AIAS Former Farm Worker Survey 2005/06). Most of the youths were employed in the new farms, with 30.9 per cent and 15.8 per cent of them as permanent and casual labour respectively. Land beneficiary households presented a different picture as 77 per cent of the school-age children were in school (AIAS Household Baseline Survey 2005/06).

There were, however, some differences in the social reproduction of agrarian labour between kinship and non-kinship work relations. Although the wages paid in kinship work relations were not significantly different from those paid to non-relatives, in the former case more workers received additional benefits than in the latter (Moyo et al 2009). For instance, 84 per cent of related employees received food rations in comparison with 54 per cent of non-related employees (AIAS Farm Worker Survey 2005/06). This was also reflected in the number of meals consumed in related employee households; 67.3 per cent of these households consumed three meals per day in comparison with 46.4 per cent of non-related employee households. Annual leave was also enjoyed by more kin employees (58.9%) compared to non-related employees (48.9%).

The social reproduction of farm workers is also being affected by the perpetuation of labour tenancy, as the larger- scale land beneficiaries still seek to link residential tenure to employment. On the contrary, government policy allows landless labourers to reside on redistributed lands irrespective of their employment status (Chambati and Moyo 2004). The implementation of this policy was varied across the districts and, in some areas, conflicts have emerged as land beneficiaries seek to evict those workers refusing to work for them.[16] About a third of the former farm workers indicated that insecure access to residency and related land conflicts had been problems for them during the FTLRP (AIAS Farm Worker Survey 2005/06). Some A2 land beneficiaries complain that they incur electricity and water charges on behalf of former farm workers who they do not employ (Utete 2003). While the residential rights of some former farm workers are as insecure as they were before 2000, the state's user land rights offer land beneficiaries significantly less authority to compel labour to work for low wages, often leading to labour shortages. Moreover, most of the new smaller capitalist farms that were allocated land with the farm compounds cannot absorb all the labour from there. This lack of absorption capacity makes it inevitable for some former farm workers to sell their labour elsewhere, de-linked from residency.

Struggles for labour access have, however, ensued among larger land beneficiaries as they seek to compel landless labourers resident in redistributed lands to work for them through the institution of labour tenancy. Some farm workers view the FLTRP as an opportunity to dismantle labour tenancy, even in the face of eviction threats, which 15.8 per cent face (Moyo et al 2009). Other former farm workers also argue that, even if they are not working on the new farms, their residency is justified since they are yet to receive retrenchment packages; only 56 per cent had received part or the full amount promised (AIAS Farm Worker Survey 2005/06).[17]

The labour tenancy is being prolonged by the recruitment of landless relatives and other new workers. Eighty-four per cent (or 292) of the new farm workers surveyed were resident in the homes provided by their employers (AIAS Farm Worker Survey 2005/06). Of these, 277 new farm workers (including relative-employees) indicated that they stayed there because of their employment. This labour tenancy arrangement is qualitatively different from the overcrowded compound which was located far away from the white farmer's luxurious mansion,[18] as we observed that new farm workers, in most cases, were housed at the land beneficiaries' homestead and shared several amenities with them, including water, sanitation and energy. Labour tenancy also continues in the remaining plantations and LSCFs, where a range of self-employed family farmers and the remaining landless continue to provide precarious farm labour, with some continuing to reside on such farms under insecure tenure (Moyo 2011b).

Although the data presented above suggests the entrenchment of functional dualism, whereby farm workers continue to subsidise farming for the emergent mid-sized and large capitalist farms through under-consumption of necessary social requirements that cannot be met by the wage incomes (Moyo and Yeros 2007), after 2000 it persists in a different form. The working class or semi-proletariat have increased their bargaining power or ability to withdraw labour from agriculture and sell it elsewhere.

Furthermore, land redistribution, by extending landholdings to the landless and/or land- short peasantry and working classes released them from the economic compulsion to engage in the wage labour markets in the LSCFs, mines and urban industries. However, in the context of the economic crisis during the FTLRP, some small producers who could not mobilise adequate resources to fully utilise their lands also engaged in farm wage labour (see

Table 5.2). This suggests that the semi-proletarianisation process that existed before 2000 with limited access to land (Moyo and Yeros 2005) has been altered as paid work is sought with autonomy of land access.

Land redistribution has also paved the way for non-racial labour relations systems in the new farms. The inhumane treatment that farm workers suffered at the hands of 'superior' white farmers (see Amanor-Wilks 1995; Rutherford 2001) declined significantly after 2000. Only 0.8 per cent of the farm workers reported the use of physical beatings by employers in the accomplishment of tasks (AIAS Farm Worker Survey, 2005/06), indicating a shift in some of the struggles farm workers face in improving their material conditions.

Politics of rural labour after the FTLRP

In a bid to improve their material conditions, farm labourers are involved in various organised and unorganised struggles. The self-organisation of farm labourers outside of farm-level worker committees prior to the FTLRP was restricted by the labour tenancy system. The former LSCFs maintained a tight grip on workers through their involvement in issues that transcended the employment contract to include resolving marriage and other social disputes in the farm compound (Rutherford 2001). These peculiar relationships in the LSCF were broken down by the FTLRP. Land redistribution and tenure reforms have opened up opportunities for farm workers to organise themselves independently of external agents such as political parties and NGOs, as well as their lead trade unions.

They are involved in the struggle to access land for residency, farming plots and natural resources. During the land occupations around 2000, some former farm workers viewed their future as independent producers and formed alliances with liberation war veterans and peasants in occupying LSCFs (Sadomba 2008). Indeed, 18 of the 70 former farm workers (25.7%) who got access to land participated in the land occupations and got their occupied lands regularised under the FTLRP (AIAS Farm Worker Survey 2005/06). They also asserted their autonomy in the struggle for land through independent land occupations (Musungwa 2001; Sadomba 2008).

Some former farm workers were also mobilised by white farmers and GAPWUZ to defend existing freehold land rights and their jobs and to oppose the new constitution that contained a clause on compulsory LSCF acquisition (Chambati 2009; Sadomba 2008). The alliances forged during the land occupations also tend to define the current relationships between former farm workers and land beneficiaries. Overall, about 18 per cent of the former farm workers reported being involved in violent confrontations with A1 land beneficiaries over farm compound land rights and natural resource access. These violent confrontations were reported by over 20 per cent in Chiredzi, Goromonzi and Zvimba, where GAPWUZ had a strong membership base and fewer farm workers participated in land occupations, in comparison to less than 7 per cent in the other districts. In Mangwe District, which had the highest proportion of former farm worker participation in the land occupations, only 3.7 per cent reported these violent confrontations (AIAS Farm Worker Survey 2005/06).

Land struggles by farm workers have continued after the land occupations and formal land allocations during the FTLRP. Some landless former farm workers are deploying their agency to informally occupy unutilised lands in the newly redistributed areas. In the areas surveyed, 9.7 per cent of the former farm workers reported that they were practising their own agricultural production of these self-provided lands in the former LSCFs where they resided (AIAS Farm Worker Survey, 2005/06). This is indeed a source of conflict between former farm workers and land beneficiaries. In general, 68 per cent of the farm workers reported that land conflicts in terms of access to residency, farming plots and natural resources for their basic needs were major problems in the newly redistributed areas (ibid). Over 13 per cent of the farm workers interviewed had experienced violence between themselves and the new land beneficiaries, owing to conflicts emanating from the farm workers' 'encroachment' into the boundaries of land beneficiaries by accessing farming plots and natural resources (Moyo et al 2009).

Former farm workers' struggles also relate to them being wrongly perceived as 'foreign' migrants (Moyo et al 2000). In fact 26.5 per cent of the former farm workers are descendants of migrant workers who were born in Zimbabwe, while 90.5 per cent were born in Zimbabwe to immigrant parents (AIAS Farm Worker Survey 2005/06). This birth citizenship is recognised by the *Citizenship Amendment Act of 2004* that entitles people born in Zimbabwe of Southern African Development Community (SADC) parents to citizenship, although state administrative capacities are weak in offering registration services (Chambati and Moyo 2004). Most of the former farm workers of foreign origin consider themselves de facto citizens of Zimbabwe as only 0.5 per cent stated a preference to return to their countries of origin after the FTLRP (Chambati and Moyo, forthcoming). As of late 2004, the Repatriation Unit of the Ministry of Public Service, Labour and Social Welfare and the International Organisation of Migration (IOM) had not handled any requests from former farm workers for repatriation (Chambati and Moyo 2004).

The organisation of farm workers through external agents such as trade unions and NGOs weakened after 2000. Land redistribution and tenure reforms created opportunities for trade unions to mobilise workers. The state tenures allow farm worker trade unions freer access to wage workers, access that was previously constrained by trespass laws. Their mobilisation base has also been expanded by the growth of farm wage employment after the FTLRP. However, the creation of a multiple range of smaller farms which hire labour in smaller batches has increased the costs of trade unions in mobilising agrarian wage labour. Farm worker trade unions have not taken advantage of these changes as they have been actively involved in MDC opposition politics (see Chambati and Magaramombe 2008; Chambati 2011a). In fact, most of the leadership of the trade unions, including the farm workers' union, have taken up leadership positions within the MDC structures. The union's alignment with the MDC hampers its mobilisation efforts as it is treated with suspicion by land beneficiaries.[19] The GAPWUZ organising secretary for Kadoma, Chegutu, Mhondoro and Selous was reported in the press complaining that their efforts to represent agrarian labourers were being hampered by accusations of being 'MDC activists' (*The Standard* 2011).

As a result, in 2006, few farm workers (4.4 per cent) were aware of the existence of a labour union in their area and only 3 per cent were paid up members of a labour union (AIAS Farm Worker Survey 2005/06). Trade unionism amongst farm workers was more common in Goromonzi (9.6%) and Chipinge (4%), compared to less than one per cent in the other districts (ibid). Workers' committees that handle worker grievances exist on only a few farms, as only 12 per cent of the workers had such structures at their places of employment (AIAS Farm Worker Survey 2005/06). The small- sized labour forces existing in the new farms seem to have also reduced the necessity of such trade union structures.

Although GAPWUZ began up-scaling worker mobilisation from around 2007 (Chambati and Magaramombe 2008), some former farm workers have been expressing reservations about its absence for the most part of the last decade. Indeed, former farm workers at one A2 farm in Zvimba District chased away GAPWUZ officers canvassing for membership as the union had 'abandoned them in their time of need' (Chambati 2011b).[20]

The absence of trade unions weakens the voice of farm workers seeking the enforcement of their labour rights. Indeed, wage bargaining of farm workers is weak at the National Employment Council for the Agricultural Industry (NECAIZ) level, as land beneficiaries have refused to endorse some collective bargaining agreements (e.g., July 2003) arguing lack of affordability, with a weak response from GAPWUZ (Chambati and Moyo 2004). At the local level, farm workers have, however, deployed their bargaining power to sell their labour in non-farm jobs. The realisation of farm worker labour and social rights is also negatively affected by the limited presence of state labour officials in the newly redistributed areas (Chambati and Moyo 2004).

Furthermore, a few NGOs still remain working in new farms (e.g., FCTZ, Kunzwana Women's Association and Farm Orphan Support Trust [FOST]). This number has been reduced from a peak of about twelve in the late 1990s (Chambati and Magaramombe 2008) due to funding constraints (Moyo et al 2009) and state perceptions of them being MDC supporters who utilise welfare projects for political campaigning (Helliker 2008).

Farm workers are increasingly organised independently at the local level in different ways to provide labour services to a range of farmers and others, with the power to withdraw labour. Some formed groups of skilled former farm workers or specialised labour to provide consultancy services to new land beneficiaries as a resistance to the low wages, as they tend to demand higher payment rates than permanent and casual workers. The specialist services being provided by these groups include overall farm planning activities, machinery operations, tobacco operations and livestock disease management and were utilised by 11.2 per cent of the land beneficiaries (AIAS Household Baseline Survey 2005/06). Unskilled former farm workers' groups also perform general tasks such as weeding, harvesting and stumping and were used by 24.4 per cent of the land beneficiaries (AIAS Household Baseline Survey 2005/06). Beyond farm labour consultancy groups, former farm workers in districts such as Zvimba have also formed groups that harvest thatching grass and firewood for sale.

They are resisting poor conditions of wage employment in the new capitalist farms by diverting their labour resources to non-farm employment, which offers higher wage rates as noted earlier. Rural labour struggles also entail 'the often invisible day-to-day acts of resistance' (Scott 1985, cited in Jha 1996: 132), such as frequent absenteeism from work by farm labourers. Some A2 farmers in Zvimba District complained that former farm workers had the habit of missing work on the pretext of being sick, yet in reality they would be using that time to do non-farm employment activities such as fishing in the nearby Darwendale Dam (Chambati 2011b).[21] Former farm workers are thus engaged in wider autonomous struggles to improve their material conditions and social reproduction.

Conclusion

The FTLRP introduced a new broad agrarian labour regime characterised by expanded self-employment in farm and non-farm jobs amongst small land owners, but farm wage labour is hired in a differentiated pattern in the middle to- large capitalist farms. This system replaced the predominantly wage labour system monopolised by a few LSCFs. The land allocations received under the FTLRP provide autonomy in social reproduction for the peasantry and have relieved their pressure to engage in cheap farm labour markets.

The state's user land rights have undermined the control agrarian employers have over workers and the degree to which they can compel them to work in return for residential rights as was the case in LSCF private properties. The exploitation of farm workers via labour residential tenancy, however, persists in the plantation estates and LSCFs that were not acquired. The political power conferred by land ownership is now diffused amongst many smaller farmers who compete for labour, while de-racialisation of the agrarian labour relations has diluted remnants of the master-servant relationship of the past. This transformation has opened up the opportunities for farm labourers to engage in a variety of struggles to improve their material conditions.

Although functional dualism persists in the new agrarian structure, the semi-proletarianisation process resulting from landlessness and land shortages before the FTLRP has been altered. The freeing of labour from residential tenancy tied to provision of farm labour means that farm workers can sell their labour in competing non-farm jobs. Farm workers resist poor working conditions in the farms through self-organisation in autonomous groups to provide high-wage demand-driven consultancy services to land beneficiaries.

Nevertheless, many agrarian labourers remain unable to meet the costs of social reproduction. For the social relations of agrarian labour to be fully equitable, there is a need to further redistribute land to former farm workers and clarify their tenure security, as well as for the state and NGOs to step up the provision of social services, the enforcement of labour rights and the implementation of agrarian labour quality and skills development programmes.

Notes

1. The analysis of job losses tends to be treated in an undifferentiated manner, i.e., whether they were full or part time.
2. The form of labour tenancy existing in Southern Africa does not entail rental payments, but the residential rights of "labour tenants" in the farm compounds were tied to the labour supply (see Neocosmos 1993).
3. The white farmers resolved disputes internally and transcended the employment contract by resolving marriage and other social disputes in the farm compound (Rutherford 2001). Many used violence in labour management (Kanyenze 2001).
4. By 1995, 62 per cent of the national population were poor, the majority of whom resided in the communal areas, where 80 per cent of the people were poor in comparison to 46% in urban areas (GoZ 1998).
5. Most former farm workers (73%) preferred land resettlement under the FTLRP to re-employment, repatriation and relocation (GoZ/IOM 2004).
6. Peter Gambara, then Director in the Vice President's Office, as quoted in the *Daily Mirror*, 4 May 2006 (Magaramombe 2010).
7. In Masvingo Province, they were 7 per cent of the land beneficiaries (Scoones et al 2010). The farm worker land allocations are also presented as a proportion of former farm workers as group. The GoZ/IOM (2004) estimates that 15 per cent of the former farm workers got land.
8. Similar findings were also reported in Masvingo Province, which had about 14.8 per cent of the land beneficiaries as former farm workers (Scoones et al 2010).
9. Land preparation, planting, weeding, fertilisation, harvesting and threshing.
10. A tragedy in 2004 in Bindura exposed the growth of child labour when a lorry carrying farm workers from work overturned, killing 22 people. The survivors included children aged between 13 and 18 years old (Independent Catholic News 2004).
11. Less than a third of the original LSCFs that were surveyed had a farm store (Moyo et al 2009).
12. By 2006, less than 10 per cent of the new land beneficiaries had accessed external finance through mostly the resource- constrained state input programmes (Moyo et al 2009; see also Moyo and Nyoni, Chapter 6).
13. The expansion of outgrower schemes was also influenced by government calls to un acquired large plantation estates to promote new land beneficiaries allocated land in some of the acquired plantation estates (Moyo 2011b).

14. The monthly rural PDL in Zimbabwe is overestimated in comparison to the World Bank international poverty line of $1.25 per person per day which translates to $225 per month for a family of six.
15. Thick porridge made from maize meal.
16. The High Court of Zimbabwe has heard several eviction cases of former farm workers (Chambati and Magaramombe 2008).
17. Permanent workers losing their jobs were entitled to retrenchment packages deducted from the LSCF compensation for land improvements (*Statutory Instrument No. 6 of 2002).* However, disagreements on the compensation process and delays in the legal aspects of land acquisition meant that few LSCFs had been valued (Moyo *et al* 2009).
18. Between 1996 and 2000, 30 per cent of the funds invested in housing in the LSCFs were for farmers when they constituted only 0.2% per cent of the population and 70 per cent was devoted to the 2,000,000 people that included farm workers and their families (CSO 2001).
19. Various senior MDC officials (including its president, Morgan Tsvangirai) have threatened to reverse the land redistribution programme on assumption of power (Moyo and Yeros 2009).
20. Field interview with farm worker, Zvimba District, 23 November 2011.
21. Field interview with A2 farmer in Zvimba District, 23 November 2011.

References

Adams, J., 1991, 'The Rural Labour Market in Zimbabwe', *Development and Change*, Vol. 22.

AIAS, 2005/06, 'Inter-District Household Baseline and Farm Worker Survey', Harare: AIAS.

Alexander, J., 2003, 'Squatters', Veterans and the State in Zimbabwe', in A. Hammar, B. Raftopolous, and S. Jensen, eds, *Zimbabwe's Unfinished Business: Rethinking Land, State and Nation in the Context of Crisis*, Harare: Weaver Press.

Amanor-Wilks, D., 1995, *In Search of Hope for Zimbabwe's Farm Workers*, Harare: Dateline Southern Africa and Panos Institute.

Arrighi, G., 1970, 'Labour Supplies in a Historical Perspective: A Study of Proletarianisation of African Peasantry in Rhodesia', *Journal of Development Studies*, Vol. 6, No. 3, pp. 197–234.

Bernstein, H., 2010, *Class Dynamics of Agrarian Change*, Canada: Fernwood, Pluto and University of Michigan Press.

Bush, R. and L. Cliffe, 1984, 'Agrarian Policy in Migrant Labour Societies', *Review of African Political Economy*, Vol. 29, pp. 77-94.

Chambati, W., 2009, 'Land Reform and Changing Agrarian Labour Processes in Zimbabwe', MM Thesis, South Africa: University of Witwatersrand.

Chambati, W., 2011a, 'Restructuring of agrarian labour relations after Fast Track Land Reform in Zimbabwe', Draft manuscript submitted to Journal of Peasant Studies.

Chambati, W., 2011b, Field interviews, Zvimba District, 23 November 2011.

Chambati, W. and Magaramombe, G., 2008, 'An Abandoned Question: Farm Workers', in S. Moyo, K. Helliker and T. Murisa, eds., *Contested Terrain: Civil Society and Land Reform in Contemporary Zimbabwe,* Johannesburg: SS Publishing.

Chambati, W. and Moyo, S., 2003, 'The Political Economy of Agricultural Labour and Land Reform', Harare: African Institute for Agrarian Studies.

Chambati, W. and Moyo, S., 2004, 'The Impact of the Fast Track Land Reform on Farm Workers and Farm Labour Processes', Harare: AIAS Monograph series.

Chambati, W. and Moyo, S., forthcoming, The Political Economy of Farm Workers in New Resettlement Areas, Harare: AIAS Monograph Series.

Chimhowu, A. and Woodhouse, P., 2008, 'Communal Tenure and Rural Poverty: Land Transactions in Svosve Communal Area, Zimbabwe', *Development and Change*, Vol. 39, No.2, pp. 285-308.

Clarke, D., 1977, *Agricultural and Plantation Workers in Rhodesia*, Gwelo: MamboPress.

Cousins, B., Weiner D. and Amin, N., 1992, 'Social differentiation in the communal lands of Zimbabwe', *Review of African Political Economy,* Vol. 1, No. 53, pp. 5 -24.

CSO (Central Statistical Office), 1984, *Indicator Monitoring Labour Force Survey*, Harare: Government Printers.

CSO, 1998, *Statistical Yearbook 1997,* Harare: Government Printers.

CSO, 2000, *1999 Indicator Monitoring Labour Survey*, Harare: Government Printers.

CSO, 2001, *Agricultural Production on Large Scale Commercial Farms,* Harare: Government Printers.

CSO, 2004, *Labour Statistics October 2004*, Harare: Government Printers.

Davies, R., 2000, 'The Impact of Globalization on Local Communities: A case study of the cut flower industry in Zimbabwe', Harare: ILO Discussion Paper No. 3.

GoZ, 1998, *Poverty Assessment Survey Main Report,* Harare: Government Printers.

GoZ, 2001, 'People First, Fostering Social Justice and Economic Growth: Zimbabwe's Land Reform Programme', Harare: Department of Information and Publicity, Office of the President and Cabinet.

GoZ, 2007, 'Agricultural Engineering, Mechanisation and Irrigation Strategy Framework: 2008 – 2058, Draft Strategy', Harare: Ministry of Agricultural Engineering, Mechanisation and Irrigation.

GoZ and IOM, 2004, 'Zimbabwe Third Farm Workers Survey Report: Labour Dimensions of the Land Reform Programme', Harare: Government Printers.

Hammar, A., McGregor J., and Landau, L., 2010, 'Introduction. Displacing Zimbabwe: Crisis and Construction in Southern Africa', *Journal of Southern African Studies*, Vol. 36, No. 2, pp. 263-283.

Hartnack, A., 2005, 'My Life Got Lost: Farm Workers and Displacement in Zimbabwe', *Journal of Contemporary African Studies*, Vol. 23, No. 2, pp. 173–93.

Helliker, K., 2008, 'Dancing on the same spot: NGOs', in S. Moyo, K. Helliker and T. Murisa, eds, *Contested Terrrain: Land Reform and Civil Society in Contemporary Zimbabwe*, Pietermaritzburg: SS Publishing.

Hellum, A. and Derman, B., 2004, 'Land reform and human rights in contemporary Zimbabwe: balancing individual and social justice through an integrated human rights framework', *World Development,* Vol. 32, No. 1, pp. 1785–805.

Independent Catholic News, 2004, 'Zimbabwe: Child labour a growing problem'. (www.independentcatholicnews.org), 6 February 2004.

JAG/RAU, 2008, 'Reckless Tragedy: Irreversible? A Survey of Human Rights Violations and Losses Suffered by Commercial Farmers and Farm Workers from January 2000 to 2008', Harare: Justice for Agriculture and Research Advocacy Unit Report.

Jha, P., 1996, *Agricultural Labour in India,* Delhi: Vikas Publishing House.

Kanyenze, G., 2001, 'Zimbabwe's Labour Relations Policies and their Implications for Farm Workers', in D. Amanor-Wilks, ed., *Zimbabwe's Farm Workers: Policy Dimensions,* Lusaka: Panos Southern Africa, pp. 86-114.

Kanyenze, G., 2007, 'The Labour Market, Sustainable Growth and Transformation in Zimbabwe', Paper prepared for the Zimbabwe Institute.

Kibble, S. and P. Vanlerberghe, 2000, *Land, Power and Poverty: Farm Workers and the Crisis in Zimbabwe,* London: CIIR.

Kinsey, B., 2010, 'Who Went Where...and Why: Patterns and Consequences of Displacement in Rural Zimbabwe after February 2000', *Journal of Southern African Studies,* Vol. 36, No. 2, pp. 339-360.

Langyuito, A., 2005, 'Maize Production Systems in Zimbabwe: Setting Indicators for Impact Assessment and Targeting', Harare: CIMMYT.

Leavy, J. and H. White, 2001, 'Rural Labour Markets and Poverty in Sub-Saharan Africa', Paper presented at 2001 Development Studies Association, Development Economics Study Group (DESG) Conference, Nottingham University, 27-29 March 2000.

Loewenson, R., 1992, *Modern Plantation Agriculture: Corporate Wealth and Labour Squalor,* London and New Jersey: Zed Books.

Mafeje, A., 2003, *The Agrarian Question, Access to Land and Peasant Responses in Sub-Saharan Africa,* Geneva: United Nations Institute of Social Research and Development.

Magaramombe, G., 2001, 'Rural Poverty: Commercial Farm Workers and Land Reform in Zimbabwe', Paper presented at the Southern African Research on Poverty Network Conference on Land Reform and Poverty Alleviation in Southern Africa, Pretoria.

Magaramombe, G., 2003, 'An Overview of Vulnerability within the Newly Resettled Former Commercial Farming Areas', Harare: Draft Report prepared for the United Nations Humanitarian Coordinator.

Magaramombe, G., 2010, 'Displaced in Place: Agrarian Displacements, Replacements and Resettlement among Farm Workers in Mazowe District', *Journal of Southern African Studies,* Vol. 36, No. 2, pp. 361-375.

Maphosa, F., 2009, 'Rural Livelihoods in Zimbabwe: Impact of Remittances from South Africa', Dakar: CODESRIA Monograph Series.

Masiiwa, M. and L. Chipungu, 2004, 'Land Reform Programme in Zimbabwe: Disparity between Policy Design and Implementation', in M. Masiiwa, ed., *Post Independence Land Reform in Zimbabwe: Controversies and Impact on the Economy*, Harare: Frederich Ebert Stiftung and University of Zimbabwe.

Mehretu, Assefa, 1994, 'Social Poverty Profile in Communal Areas', in M. Rukuni and C. Eicher, eds., *Zimbabwe's Agricultural Revolution*, Gweru: Mambo Press. pp. 56-69.

Ministry of Public Service Labour and Social Welfare (MPSLSW), 1998, 'Farm Workers Survey Report', Harare: MPSL&SW Unpublished report.

Ministry of Public Service Labour and Social Welfare (MPSLSW), 2001, 'Land Reform: a Labour Perspective, Farm Worker Survey Report', in collaboration with the International Organisation of Migration, Harare.

Moyo, S., 1995, *The Land Question in Zimbabwe*, Harare: SAPES Books.

Moyo, S., 2000, *Land Reform under Structural Adjustment in Zimbabwe: Land Use Change in the Mashonaland Provinces,* Uppsala: Nordic Africa Institute.

Moyo, S., 2007, 'The Future of Land and Agrarian Reform in Zimbabwe', Harare: AIAS Monograph Series.

Moyo, S., 2011a, 'Three decades of agrarian reform in Zimbabwe", *Journal of Peasant Studies,* Vol. 38, No. 3, pp. 493-531.

Moyo, S., 2011b, 'Land concentration and accumulation after redistributive reform in post-settler Zimbabwe', *Review of African Political Economy*, Vol. 38, No. 128, pp. 257-276.

Moyo, S., 2011c, 'Changing agrarian relations after redistributive land reform in Zimbabwe', *Journal of Peasant Studies*, Vol. 38, No. 5, pp. 939-966.

Moyo, S. and Yeros, P., 2005, 'The Resurgence of Rural Movements under Neoliberalism', in S. Moyo and P. Yeros, eds., *Reclaiming the Land: The Resurgence of Rural Movements in Africa, Asia and Latin America,* London and Cape Town: Zed Books, David Phillip. pp. 165–208.

Moyo, S. and Yeros, P., 2007, 'The Radicalised State: Zimbabwe's Interrupted Revolution', *Review of African Political Economy*, Vol. 34. No. 111, pp. 103-121.

Moyo, S., Rutherford B., and Amanor-Wilks, D., 2000, 'Land Reform and Changing Social Relations for Farm Workers in Zimbabwe', *Review of African Political Economy*, Vol. 84, No.18, pp. 181-202.

Moyo, S., Chambati, W., Murisa, T., Siziba, D., Dangwa, C., Mujeyi K., and Nyoni, N. 2009, 'Fast Track Land Reform Baseline Survey in Zimbabwe: Trends and Tendencies 2005/06', Harare: AIAS Monograph.

Mtisi, J., 2003, 'Green Harvest: the Outgrower Tea Leaf Collection System in the Honde Valley, Zimbabwe', in M. Roth and F. Gonese, eds, *Delivering Land and Securing Rural Livelihoods: Post Independence Land Reform and Resettlement in Zimbabwe,* Harare and Wisconsin: Centre for Applied Social Sciences and Land Tenure Center.

Muchena, O., 1994, 'The Changing Perceptions of Women in Agriculture', in M. Rukuni and C. Eicher, eds, *Zimbabwe's Agricultural Revolution,* Gweru: Mambo Press. pp. 348-360.

Muir, K., 1994, 'Agriculture in Zimbabwe', in M. Rukuni, and C. Eicher, eds, *Zimbabwe's Agricultural Revolution,* Gweru: Mambo Press. pp. 40-55.

Mutopo, P., 2011, 'Women's Struggles to access and control land after fast track land reform in Mwenezi District', *Journal of Peasant Studies,* Vol. 38, No. 5, pp. 1021-1046.

Neocosmos, M., 1993, 'The Agrarian Question in Southern Africa and "Accumulation from Below": Economics and Politics in the Struggle for Democracy', Uppsala: Nordic Africa Institute Research Report No. 93.

NRC, 2003, 'Displaced and Forgotten', Harare: Global Internal Displacement Project.

Oni, S.A., 1997, T*he Impact of ESAP on Communal Areas in Zimbabwe,* Harare: Friedrich Ebert and Stiftung.

Oya, C., 2011, 'Contract Farming in Sub-Saharan Africa: A Survey of Approaches, Debates and Issues', *Journal of Agrarian Change,* Vol. 12, No. 1, pp. 1-33.

Palmer, R., 1977, *Land and racial domination in Rhodesia,* London: Heinemann Educational Books.

Paradza, G., 2010, 'Single women, land and livelihood vulnerability in a communal area in Zimbabwe', PhD Thesis, Wagenigen University.

Parliament of Zimbabwe, 2003, 'Second Report of the Portfolio Committee on Public Service, Labour and Social Welfare on the Plight of Farm Workers and Newly Resettled Farmers', Harare: Government Printers.

Parliament of Zimbawe, 2006, 'Fourth Report of the Portfolio Committee on Lands, Land Reform, Agriculture, Resettlement and Water Development on the Viability of the Sugar Industry', Harare: Government Printers.

Patnaik, U., 1997, *The Long Transition,* Manohar Publishers.

Potts, D., 2000, 'Worker-Peasants and Farmer-Housewives in Africa: The Debate about 'Committed' Land beneficiaries' Access to Land and Agricultural Production', *Journal of Southern African Studies,* Vol. 26, No. 4, pp. 807-832.

Price Waterhouse Coopers (PWC), 2010, 'Zimbabwe Agricultural Sector Assessment Study', Harare: Price Waterhouse Coopers Unpublished Report.

Ridderboos, K., 2009, 'Stateless Former Farm Workers in Zimbabwe', *Forced Migration Review,* Vol. 32, pp. 73.

Rubert, S., 1997, 'Tobacco Farmers and Wage Labourers in Colonial Zimbabwe, 1904–1945', in A. Jeeves and J. Crush, eds., *White Farms, Black Labor: The State of Agrarian Change in Southern Africa, 1910–1950,* Portsmouth, Oxford and Pietermaritzburg: Heinemann, James Currey, University of Natal Press.

Rutherford, B., 2001, *Working on the Margins: Black Workers, White Land beneficiaries in Post Colonial Zimbabwe,* Harare and London: Weaver Press, Zed Books.

Rutherford, B. and L. Addison, 2007, 'Zimbabwean Farm Workers in Northern South Africa', *Review of African Political Economy,* Vol. 34, No.114, pp. 619–35.

Sachikonye, L., 2003, 'The Situation of Commercial Farm Workers after Land Reform in Zimbabwe', Harare: Unpublished Report prepared for FCTZ.

Sachikonye, L., 2004. 'Land reform and farm workers', in D. Harrold-Barry, ed., *Zimbabwe: the Past is the Future,* Harare: Weaver Press.

Sadomba, W., 2008, 'War veterans in Zimbabwe's Land Occupations: Complexities of a Liberation Movement in an African Post-Colonial Society', PhD. Thesis, Wagenigen University.

Scoones, I., Marongwe, N., Mavedzenge, B., Mahenehene, J., Murimbarimba, F. and Sukume, C., 2010, Zimbabwe's Land Reform: Myths and Realities, New York and Harare: James Currey and Weaver Press.

Tandon, Y., 2001, 'Trade Unions and Labour in the Agricultural Sector in Zimbabwe', in B. Raftopolous and L. Sachikonye, eds, *Striking Back: The Labour Movement and the Post Colonial State in Zimbabwe 1980 – 2000*, Harare: Weaver Press. pp, 221–49.

Tanganda Tea Company, 2006, *Financial Report for Year ended 31 October 2006*, Harare.

The Daily Mirror, 2006, 'Coffee output down', 22 June 2006.

The Herald, 2006, 'Tanganda posts $375,8 bn profit,' 22 June 2006.

The Standard, 2011, 'ZANU PF chefs exploiting farm workers', 2 October 2011.

UNDP, 1999, *Zimbabwe Human Development Report 1999*, Harare: UNDP.

UNDP, 2008, 'Comprehensive Economic Recovery in Zimbabwe: a Discussion Document', Harare: UNDP.

USAID, 2010, 'Zimbabwe Agricultural Sector Market', Harare: Unpublished report prepared by Weidemann Associates, Inc.

Utete Report, 2003, *Report of the Presidential Land Review Committee* under the chairmanship of Dr M.B. Charles Utete, *Volumes 1 and 2: Main report to His Excellency the President of the Republic of Zimbabwe,* Harare: Presidential Land Review Committee (PLRC).

Vhurumuku, E., McGuire M., and Hill, V., 1998, 'Survey of Farm Worker Characteristics and Living Conditions in Zimbabwe: Verification Report', Harare: Agricultural Labour Bureau, Farm Community Trust of Zimbabwe and USAID FEWSNET.

Worby, E., 2003, 'The End of Modernity in Zimbabwe? Passages from Development to Sovereignty', in A. Hammar, B. Raftopolous and S. Jensen, eds., *Zimbabwe's Unfinished Business: Rethinking Land, State and Nation in the Context of Crisis,* Harare: Weaver Press.

World Bank, 2006, *Agricultural Growth and Land Reform in Zimbabwe: Assessment and Recovery Options*, Report No. 31699-ZW, Harare: World Bank.

Zimbabwe Human Rights Forum (ZHRF) and JAG, 2007, 'Adding Insult to Injury: A Preliminary Report on Human Rights Violations on Commercial Farms, 2000-2005', Harare: ZHRF and JAG.

ZIMSTAT, 2011, 'Poverty Datum Lines – June 2011', Harare: ZIMSTAT.

6

Changing Agrarian Relations after Redistributive Land Reform in Zimbabwe

Sam Moyo and Ndabezinhle Nyoni

Introduction

Zimbabwe's Fast Track Land Reform Programme (FTLRP) initiated from 2000 extensively redistributed land, mainly to peasants and working peoples (see Moyo 2011c) and, in doing so, unravelled the labour reserve economy created over a century of settler-colonial agrarian capitalism. This change has created a broader range of prospects for progressive agrarian transformation, despite the persistence of inequalities and exploitative social relations. The dominant discourses have reflected such agrarian changes, however, by narrowly focusing their attention on the immediate consequences of the FTLRP, particularly on the decline in agricultural output and formal employment which are conceived of in a linear fashion (e.g., UNDP 2010). Moreover, this narrow view is dramatised by such erroneous claims as, for example, Zimbabwe's large-scale white farms having been the breadbasket of Southern Africa, when, in fact, they constituted an irregular food exporter and importer, as South Africa met the regional food deficits (Moyo 2010). Such narrative attributes the decline merely to the replacement of skilled large-scale white farmers with alleged 'subsistence' producers (e.g., Tupy 2007) and to the loss of private property rights (e.g., Richardson 2005).

Reluctant to recognise the new farmers, this perspective vilifies their behaviour, defining them in blanket terms as 'part-time', 'weekend', or 'cell phone' farmers, who are 'unproductive' and hold land for speculation. This definition is set in contrast to their erstwhile commercially-oriented counterparts. Anecdotal evidence is used to make vacuous claims that new farmers have destroyed the environment by wantonly cutting trees, degrading soils and silting water sources, in debates reminiscent of conservative colonial agrarian and environmental discourses (see Gore et al 1992). Ignoring Zimbabwe's agrarian history, they claim that almost all exports and 'marketed' foods were produced on a large-scale by white farmers until 1999. Yet, most small producers who supplied over 80 per cent of the nationally consumed wage-foods, such as maize and various pulses and basic meat products (see Moyo 2000), do not fit the label of 'subsistence' farmers. They also contributed significantly to exports such as cotton (80%), tobacco (10%) and paprika (over 30%) from 1980, despite the continued bias of agricultural support towards white farmers and their historical exclusion from irrigation development (see Moyo 2000). When the economy was liberalised from 1990, per capita and land productivity levels of food grains among small producers began to decline (Anderson 2007), although cotton outputs remained steady, while large-scale farmers cashed in on the export incentives and financial market de-regulation (Moyo 2000).

This pessimism, regarding the ascribed productivity deficiencies of black farmers, occludes the re-orientation of Zimbabwe's agrarian system towards improvements in the broader livelihoods of differentiated classes of farmers and its focus on social priorities. Limited research has been done on the re-composition of agricultural outputs and production relations in relation to the reconfiguration of Zimbabwe's agrarian classes, changing market relations and shifting state interventions in order to decipher the emerging accumulation trajectories. Consequently, the changing uses of land and labour by the heterogeneous farming classes with differentiated access to agrarian markets, within different agro-ecologies and contexts are glossed over, missing their diverse production outcomes. The literature fails to note that the agrarian labour regime has changed due to the creation of a diverse range of smaller capitalist farms and peasants (see also Chambati, Chapter 5), who compete with remaining large-scale farms for labour and access to agrarian markets and credit. Moreover, changing agrarian relations in a labour reserve economy which is being restructured were inevitably influenced by what was happening with urban production and labour markets.

The FTLRP and agrarian reform imposed varied changes on the workings of capital, in the domestic and external spheres, given the differentiated exposure of Zimbabwe's 15 main agricultural commodities to international and national commodity and financial markets. Few scholars have examined how Zimbabwe's agrarian markets have been reconfigured as capital adapts its strategies to the new agrarian structure, changing state policy and increased speculation and volatility on global commodity markets, especially from 2005 (Moyo 2011b). Those who explore the restructuring of 'commodity value chains' (e.g., Scoones et al 2010), do not adequately capture the national and international reconfiguration of agricultural inputs, outputs and financial markets, nor the shifting logic and orientation of state interventions in agriculture in response to capital.

In particular, the nuances of state agrarian intervention are clouded by perspectives which assume that 'chaos' and 'politicisation' pervaded the agrarian reform, missing the substantive class and regional dynamics in which state action is embedded. Media-based reports highlighted corruption and patronage in favour of ZANU-PF- aligned elites based on populist perspectives which ground all state action and social agency in the ruling political party (ZANU-PF). The state is narrowly conceptualised as being intrinsically neopatrimonial, allegedly driven by unproductive 'rent seeking' and consumptive distributional behaviour. For instance, a mass of small farmers are seen to engage in production without state support, conceived in terms of direct inputs donation (see Scoones et al 2010), despite the evidence which suggests otherwise. Moreover, it is argued that public extension services collapsed and failed to promote progressive agronomic practises, despite the increased presence of the state locally (see Murisa, Chapter 7). These perspectives under study the emerging state-capital relations in agrarian markets and the new forms of surplus value creation and extraction, as well as their implications for the wider politics of agrarian reform.

The Government of Zimbabwe (GoZ) has insisted that the root cause of the decline in output and productivity has been the external sanctions imposed on Zimbabwe, alongside the effects of three droughts between 2001 and 2011, as well as 'sabotage' by capital. Allegedly, the white farmers dismantled key farm infrastructures and commercial banks were reluctant to fund new farmers, ostensibly for lack of title, while input suppliers and commodity merchants, who tended to go on a 'capital strike', were apparently more interested in profiteering and externalisation of earnings (Reserve Bank of Zimbabwe [RBZ] 2007a). Moreover, the government argues that it supported farmers and intermediaries in a non-partisan manner, in the face of droughts, structural constraints on credit and fiscal capacity and political isolation and sanctions imposed by western nations (RBZ 2007a). Indeed, little empirical work has been undertaken to understand the constraints facing newly settled farmers, particularly their limited access to inputs and credit.

Various constraints existed, but policy implementation was also riddled by inconsistencies and class contradictions (Moyo and Yeros 2007, 2009). The dilemma was how to finance agrarian reform in favour of peasants, at a time when the political and class struggles evoked by radical land reform were highly polarised, while securing state autonomy in the face of political isolation, economic sanctions and other external interventions in domestic politics.

Moreover, the state's confrontation with western powers, whose strategic interests in the settler economies and security architecture of Southern Africa were being challenged, not only led to debilitating sanctions, but also unravelled historic regional economic cooperation. Together, these engendered a hostile external environment, rather than support for progressive agrarian reforms. Indeed, the manner in which the commodity, food and financial crisis of global capitalism significantly influenced agrarian change in Zimbabwe (as elsewhere) at the height of land reform is under-examined.

Consequently, the policy alternatives which are presently being proposed by key political actors, donors and think tanks focus narrowly on dispossessing some (if not most) of the land beneficiaries. They assume that the land beneficiaries are inherently incapable of producing commercially and promote privatising land tenure, ostensibly to improve access to credit. Such green revolution-type reforms are intended to integrate small farmers into dominant foreign agribusiness and to obviate state agrarian intervention (e.g. USAID 2010). Moreover, some international agencies (e.g. World Bank 2012; BBC 2011) are 'surprised' by the current scale of agricultural recovery, largely due to the empirical and analytic weaknesses of dominant discourses on agrarian change since the FTLRP.

These questions require attention in the context of a progressive vision of transforming settler-colonial agrarian relations, recognising that land redistribution on its own cannot address all pre-existing agrarian inequities. Yet the egalitarian landholding structure and the relief against the super-exploitation of labour which emerged represent social progress which cannot be understated because of the present failure to institute fully socialised agrarian relations (such as collective and/or state farms), as some imply, or because of the loss of capital accumulation at scale as others imply (see Sender and Johnson 2004). Redistributive land reform responds to the political and social imperatives of addressing the historical social injustices and debunks the presumed inevitability of an economic and agricultural 'development' system created through a functional dualism in favour of a settler dominated capitalist transition and accumulation from above.

Nonetheless, Zimbabwe's agrarian class composition and the social orientation of farming have changed substantially enough to restructure the technical organisation of agricultural production, largely around family labour processes, while enabling capital to re-orient its strategies of agro-industrial inputs supply and to adjust the market mechanisms used by agrarian merchants and finance capital to extract surpluses. Contrary to perspectives which overemphasise the (in)capacity of new and smaller farmers in recovering agricultural output, capital continues to play a critical role in shaping agricultural production relations.

Progressive agrarian reforms in Zimbabwe ought to promote increased productivity among small producers to increase food sovereignty and other supplies to home markets and to enhance industrial diversification and employment. This improvement requires transformation of the agro-industrial system to adapt to the new technical and social relations of production, through equitable forms of inputs production and distribution and democratic systems of generating knowledge and controlling intellectual property rights. Agrarian reform requires an articulated national development strategy, which emphasises accumulation from below. Critical is the protection of producers, through trade policy and subsidies, to insulate them from the highly protected, subsidised, speculative and volatile world markets (Chang 2009; Ghosh 2008) and to balance the interests of producers and consumers (see Patnaik 2008). Such a vision requires stronger producer cooperation and activism, in alliance with working class consumers, against dominant capital which prioritises externally-oriented production and markets, while depressing commodity prices (Moyo and Yeros 2005). Realising this vision ultimately depends on the correlation of social forces, the nature of formal and informal social struggles and the substantive content and politics of struggles for democratisation.

This chapter explores the macro-processes of agrarian change that emerged during the FTLRP. After outlining Zimbabwe's agrarian history and the new agrarian structure, section three examines the agrarian policies instituted since 2000. The chapter then interrogates the emerging composition and trends of agricultural outputs and productivity, within their class and regional contexts, while identifying the accumulation trajectory underway. Patterns of access to agrarian markets, including the role of the state, are then explored, highlighting their reconfiguration, the re-insertion of diverse foreign capitals and the socially differentiated access to and utilisation of inputs. Farming contracts tied to inputs supplies intended largely for export increasingly entrench differentiated investment and productivity. Finally, the chapter examines the way agrarian politics are re-oriented by farmers' reorganisation for state support and access to markets, while defending their land (see also Murisa, Chapter 7). This examination highlights the waning agrarian radicalism within a state with limited fiscal capacity and the renewed dominance of a multi-racial and foreign capital. Western donors simultaneously use limited aid and sanctions to influence the orientation of agrarian policy as Zimbabwe is saliently re-integrated into more diverse world markets.[1]

Agrarian history and land reform

Equitable agrarian reform in Zimbabwe was compromised during independence negotiations in 1979 in favour of power transfer and liberal democratic reform (Habib 2011). Settler-colonial accumulation by dispossession from 1890 created a labour reserve economy (Amin 1972), dependent on cheap domestic and foreign migrant labour (Arrighi 1973). Peasant farming, rural small-scale industry and commerce were repressed through extra-economic regulations and taxes, but this did not create full-scale proletarianisation (Bush and Cliffe 1984; Yeros 2002). Racial and class inequalities in the agrarian relations were consolidated by discriminatory subsidies to large-scale farmers (Moyana 2002) and narrow import substitution export-led strategies. The consequent rise and fall of the peasantry is well documented (Weiner 1988).

Rhodesia's agricultural transformation strategy entailed state support for large-scale farming, including individual white settlers with an average land size of 2,000 hectares and foreign and domestic estates, with average landholdings well above 5,000 hectares. From 1966, state support to large-scale irrigated estate farming, through dams, rural electrification and other infrastructures, was increased (see Rukuni et al 2006), to expand exports and reduce sugar and wheat imports (see Stoneman 1988). By the 1970s, state-owned farm estates were created, including through the Agricultural and Rural Development Authority (ARDA), which succeeded the Tribal Trust Lands Development Authority,[2] the Cold Storage Commission (CSC) and other parastatals. Large scale private estates before the FTLRP were largely owned by South African based transnational corporations, such as Triangle Sugar Corporation and Hippo Valley (Sugar) Estate (see EU 2007) and European and domestic white capital (e.g., Meikles, Tanganda Tea, Liebigs, Mazoe Estates, Ariston Holdings).

Domestic agribusiness conglomerates and estates included pioneer white family owners, some of which held mining exploitation licenses.[3] The sugar estates had promoted the creation of white large-scale outgrower farmers called Independent Commercial Growers (ICGs), largely through Mauritian and South African immigrants, with average landholdings of 217 hectares (EU 2007). By 1971, Mkwasine Estate, owned by Triangle and Hippo Valley Estates, created black sugar outgrowers with 10 hectares (ibid). The tea estates had also created about 1,000 white and black outgrowers (see USAID 2010).

Thus, independent Zimbabwe inherited a racially skewed agrarian structure and discriminatory land tenures dominated by 6,000 white farmers and a few foreign and nationally- owned agro-industrial estates, alongside 700,000 peasant families and 8,000 small-scale black commercial farmers (Table 6.1). This tri-modal agrarian structure was dominated by the large- scale capitalist farmers and secondarily by the estates at the expense of the peasantry. From 1980, Zimbabwe pursued a market-based land reform programme whose outcome by 1999 was limited, but relatively large in scale (e.g., compared to Kenya), while successfully meeting the limited production and livelihood targets it sought (Cusworth and Walker 1988; Moyo 1995; Cliffe 2000). This objective left the prevailing settler-colonial agrarian structure and labour regime generally intact.

The adoption of the Economic Structural Adjustment Programme (ESAP) in 1990 further encouraged renewed land concentration and foreign ownership. Furthermore, it exacerbated agrarian polarisation. This polarisation was also fuelled by increased export-oriented production on large-farms and the creation of extensive conservancies for eco-tourism (Moyo 2000). State agrarian subsidies and social welfare transfers to peasants were reduced, undermining the production gains that had been realised by the top 20 per cent of the peasantry and leading to a longer term decline in maize yields from 1991 (Anderson 2007). ESAP exposed farmers to volatile and monopolistic world markets and reinforced unequal production relations (Moyo 2000), while fuelling wider social dislocations as fiscal capacity dwindled (Bochwey et al 1998).

Increased rural landlessness and retrenchment of urban workers extended land hunger (Yeros 2002). Wage repression led to extensive strikes and protests by industrial and agricultural workers between 1994 and 1998 (Sachikonye 2003; Rutherford 2003). Unprecedented political conflicts emerged within and outside the ruling ZANU-PF, while external intervention in domestic politics escalated (Moyo and Yeros 2007). Elections, involving the newly- formed MDC, were bitterly contested from 2000, leading to increased electoral violence, authoritarian rule (Raftopoulos and Mlambo 2009) and the imposition of western sanctions. These contradictions ignited popular land occupations from 1997, affecting 20 per cent of the LSCF farms and many of these were led by liberation war veterans (Moyo 2001; see Sadomba and Masuko, Chapters 3 and 4).

By 2010, only about 300 white farmers remained in agriculture, alongside some agro-industrial estates. The government co-opted the process and gained control of it through state-led land expropriations and official allocations of land to over 150,000 families in two types of schemes under the Fast Track Land Reform Programme (GoZ 2001).[4] Consequently, a more broadly based tri-modal agrarian structure representing competing models of accumulation has emerged (see Moyo, Chapter 2 for details), based on relatively distinct landholding size, forms of land tenure, social status of landholders and the dominant forms of labour used (Table 6.1).

Table 6.1 Agrarian structure: estimated landholdings from 1980 to 2010

Farm category	Farms/households (000s)						Area held (000 ha)						Average farm size (ha)		
	1980		2000		2010		1980*		2000*		2010*		1980	2000	2010
	No	%	No	%	No	%	Ha	%	ha	%	ha	%			
Peasantry	700	98	1,125	99	1,321	98	16,400	49	20,067	61	25,826	79	23	18	20
Mid-sized farms	8.5	1	8.5	1	30.9	2	1,400	4	1,400	4	4,400	13	165	165	142
Large farms	5.4	1	4.956	0.4	1.371	0.1	13,000	39	8,691.6	27	1,156.9	4	2,407	1,754	844
Agro-Estates	0.296	0.1	0.296	0.02	0.247	0.02	2,567	8	2,567	8	1,494.6	5	8,672	8,672	6,051
Total	**714**	**100**	**1,139**	**100**	**1,353**	**100**	**33,367**	**100**	**32,726**	**100**	**32,878**	**100**	**46.7**	**28.7**	**24.3**

Source: Adapted from Moyo 2011a (Table 1). NB: The average farm size of the peasantry includes common grazing lands.

The enlarged peasantry is now dominant in terms of number of landholdings, covering over 75 per cent of all farming land. Agrarian relations among the socially differentiated peasantry continue to be defined mainly by self-employment of family labour towards producing foods for auto-consumption and selling some surpluses, as well as various non-farm work and short-term wage labour. Some peasants hire limited labour, while others provide labour services. Peasant families hold customary rights to cropping and homestead plots and common grazing areas (in Communal Areas and A1 areas), although the latter hold perpetual state permits for such land rights.

The number of middle-sized capitalist farmers who are more dependent on hired labour than on family labour has been tripled and these hold a wide range in size of landholdings (see Moyo, Chapter 2). The FTLRP substantially downsized the number, farm sizes and area of the large-scale individual or corporate owned capitalist farms. Together, these capitalist farms hold 40 per cent of the redistributed land, mostly on the leasehold tenure provided by the state. They mainly comprise former and new 'middle class' people with relatively higher levels of education, better access to jobs and more connections to the state and markets.

The main agro-industrial estates were retained, but on a smaller area as both the state-owned and privately held estates and conservancies were reduced in area (Table 6.1). The white independent sugar and tea producers were almost eliminated, while the number of black outgrowers was substantially increased. The level of foreign ownership of land was substantially reduced, but was retained among the sugar, tea and timber holdings, while the shareholdings of the conservancies were partly transferred to some black elites. The overall range of actors involved in outgrower farming, conservancies and forestry has been diversified in terms of race, nationality and class. Thus, some concentration of land, water, wildlife and woodlands resources was retained to preserve large scale, specialised and integrated enterprises, preserving some elements of the colonial land grab.

Thus, landlessness was not fully reversed, especially among farm labourers and in some overcrowded Communal Areas, creating a platform for new processes of labour exploitation and wider class and social struggles such that the exclusion of some potential land beneficiaries evokes persistent 'illegal' land occupations. While the policy of limiting access to redistributed land by former farm workers was partially motivated by the desire to ensure the availability of cheap labour supplies, the number of full-time hired labourers has declined, but short-term hired labour expanded. Such labour is provided on diverse types of farms and in diverse rural non-farm activities.

Thus, both re-peasantisation and semi-proletarianisation are simultaneous outcomes of the agrarian reforms since 2000 (Moyo and Yeros 2005). One cannot foreclose the trajectory (see O'Laughlin 2002), for unequal land and labour relations are being consolidated through new but limited mechanisms of land concentration (e.g., informal land rental markets), as well as capitalist farmers' advocacy for the commodification of land and efforts to evict smaller landholders which are actively resisted. Social differentiation is on-going among all farming classes based on differential access to means of production (e.g., through sub-contracts), non-farm incomes, credit and state support (discussed below), while unequal gender relations and ethno-regional identity continue to influence agrarian struggles. However, expanded agrarian petty-commodity production imposes new structural conditions for capital accumulation with competing demands for state interventions within the tri-modal agrarian structure, while the space for accumulation from below remains contingent on the nature of agrarian struggles and mediation by the state.

Agrarian reform policies since the FTLRP

The FTLRP marked a major policy departure from neoliberal prescriptions on land and agrarian reform by eliminating private land ownership and land markets, while the land redistribution itself undermined the supply of cheap labour from large landless reserves. Redistribution also called into question the prevailing wisdom that agricultural growth and 'viability' required large-scale farms, which in Zimbabwe were pegged at a minimum of 500 hectares (Moyo 2002; Cousins and Scoones 2010). However, since redistribution provided black commercial farmers with relatively large individual plots ranging in size on average from 50 to 300 hectares, the idea of large-scale farms and the reliance on cheap labour were partially retained, presumably expecting abundant labour supplies from landless workers. Moreover, agrarian labour policy continued to be largely based on collective bargaining and flexible hiring rules, which has allowed for persistently low wages (Chambati 2011), although limited output growth constrains our measurement of the precise shares of wages and profits. Nonetheless, the FTLRP process rowed against the current of escalating land alienation intended to create larger-scale capitalist estates in Africa, drawing political opposition and western sanctions.

The new tri-modal agrarian structure necessitated agrarian policy reforms, reversing the liberalisation adopted between 1990 and 2001, given its failure to stabilise supplies and prices, particularly for poor producers and consumers. Heterodox economic and agrarian policies were re-introduced within a dirigiste framework between 2001 and 2007, while negotiating 'normalisation' with capital by allowing it to operate in controlled markets and subsidising some agro-industrial estates and agri-businesses (Moyo 2011b). The wider goal was to promote auto-centred development, alongside addressing fuel, food and inputs shortages and price hikes. The agrarian reform strategy, particularly its specific policy instruments, did not begin as one holistic and coherent plan, but rather evolved in response to changing social and production conditions and struggles on the ground, especially as output fell and inputs shortages grew, in the face of increasing sanctions. By 2003, the state had adopted various economic plans and in 2006 issued the National Economic Development Priority Programme (NEDPP), in which it partially relaxed some market restrictions and escalated the state subsidies.

Significant coordination and implementation inconsistencies emerged from the start and a strategy of 'Command Agriculture' was introduced to direct agricultural production towards set output targets, using subsidised inputs and credit (GoZ 2003). Agricultural commodity market controls and trade protection were introduced in 2001, while inputs and food prices were regulated. The parastatal Grain Marketing Board (GMB) monopolised grain buying (GoZ 2001). Genetically Modified Organism (GMO) seeds were actively prohibited and open-pollinated seed was encouraged. Cheap foreign currency for targeted imports was provided to agro-industry. Subsidies also targeted distressed industries, including agro-industries, to improve the supply of inputs to farmers and state farms, as well as to agricultural merchants to enable crop purchases. The variety of state support schemes intended to support the new and existing farmers, state farms, agro-industries and merchants were loosely coordinated in an evolving agrarian reform programme (Table 6.2).

Inflation escalated beyond 400 per cent by 2005 and then hyper-inflation (at over 50 per cent a month) emerged from late 2006. This led to aggressive price controls which fuelled underground markets, at a time when sanctions were escalated. The price controls between 2005 and 2007 now included arresting managers of non-compliant firms. However, capital withdrew goods from formal markets, while private supplies of agricultural inputs and credit continued dwindling and informal markets proliferated.

Table 6.2: Coverage of RBZ agricultural financing schemes

Scheme/Years	Objectives of scheme	Support provided	Targeted beneficiaries	Comments on beneficiaries
Free Government Inputs 2002 -2011	Support peasant production	Seed and fertiliser packs	Communal/A1 farmers	
Productive Sector Financing 2004	Provide agricultural credit when private finance declined	Subsidised loans at 25% interest vs. 300% private banks	A2 Farmers	
ASPEF 2005	Enhance food export production	Cheap credit	A2 farms, agro-industry, merchants, state farms	80% of funds targeted A2 farmers
Operation Maguta 2005	Boost food security through command agriculture	Inputs and ploughing support (maize/wheat)	A2 farms and A1 and CA in 2005/06	
Champion Farmer 2008-2009	Boost food security through capable farmers	Inputs subsidy	A2 farms	
Farm Mechanisation 2003 – 2008	Address labour shortages and expand cropped area	Machinery and equipment for free and on cheap credit	A2, A1 and state farms	Small proportion of large farmers benefitted
Seed supply recovery 2002–2008	Increase area and number of seed producers	Cheap credit; Subsidised forex; Output contracts	Seed producing firms to contract A2 farmers	Relied mostly on new larger-scale farms
Irrigation rehabilitation and development 2004 – 2011	To resuscitate and expand irrigation	Cheap credit for equipment; Subsidised water and electricity	A2 and state farms	Mostly benefited A2 farms
ARDA recovery 2003 -2006	Increase ARDA cropped areas	Cheap credit; Seasonal land leases	ARDA farms Agribusiness	Agribusiness did not invest cash

Source: Adapted from AIAS research, World Bank (2006); Scoones et al (2010)

From 2005, contract farming was being formally encouraged by the Ministry of Agriculture and the RBZ, which respectively facilitated the tobacco and soybeans contracting and the emergence of new black agricultural merchants by allowing them to retain more of the foreign earnings from exports. When the Look East Policy was escalated from 2005 to diversify sources of foreign loans and markets, at a time when western agrarian merchants had retreated and as liquidity was declining, some concessional loans were secured for imported inputs and machinery. Mechanisation subsidies to counter labour shortages and to expand the areas cropped were escalated from 2006 through such loans and subsidies in local currency.

As food shortages increased and foreign currency earnings declined, import capacity fell and inflation and interest rates escalated, expansionary fiscal interventions escalated through excessive printing of money and the opaque use of parallel currency markets by the state and businesses and hyperinflation skyrocketed. Farming increasingly depended on GoZ finance and credit, although its capacity to subsidise inputs and outputs marketing was limited (ibid).

The GoZ increasingly compelled and persuaded capital through various policy measures to increase their production and to support the new farmers. It appealed to new farmers' patriotism to prioritise food production for self-sufficiency (RBZ 2007a), including by introducing a clause in the A2 land lease requiring them to put 20 per cent of their land to food grains and/or beef, depending on their agro ecological location. The state also turned to large agricultural estates, justifying their retention on the grounds of their alleged superior scale economies, productivity and technological advantage and the need to preserve the 'bulky investments' (e.g. irrigation and agro-processing infrastructures) already 'sunk' into the estates (see Sukume and Moyo 2003). Sub dividing the estates and their 'integrated infrastructures' was considered dysfunctional and a source of disputes among new farmers (see Utete 2003). Land tenure insecurity on the estates apparently discouraged investment (EU 2009). Thus, the GoZ sought to encourage production on the large agricultural estates, towards expanding food and agro-fuel production, partly by "allowing" them to retain their land and requiring them to incorporate more black 'outgrowers' into their enterprises. Some remaining estates responded positively to this, inter-alia to avoid being perceived as undermining land reform and supporting 'regime change'.

The state had attempted to recover and expand agricultural production on public estates held by ARDA and the CSC estates from 2002, through their own efforts of increasing the area under various seeds, wheat and maize using cheap credit and foreign currency supplied by the Reserve Bank of Zimbabwe (RBZ).

Then, in 'alliance' with the Nuanetsi Ranch owned by the Development Trust of Zimbabwe (DTZ), Masvingo Province authorities and the state began in 2003 to clear some DTZ lands for maize production through a state- contracted Chinese firm. These initiatives floundered due to inadequate financing. In 2005, the RBZ was contracting locally-based food-processing and inputs supply agribusinesses (e.g., National Foods, INNSCOR, Chemco, Seed Co etc.) to produce seed, wheat and oilseed on ARDA estates. This approach also floundered due to disagreements over product pricing and profit sharing, since the agribusinesses invested little of their own cash, but made profits from 'free' state land and financial subsidies (NERC 2006). In addition, these GoZ leases and sub-contractual production relations did not succeed because most estates required substantial repairs and construction of new dams and irrigation infrastructure, for which the businesses could not secure local financing.

But the agro-industrial estates had always been considered by the state as critical to export growth, employment promotion and agro-industrial development (GoZ 1998). In practice, the decision to preserve them was influenced by declining food and agricultural export production and rising imports, particularly after the 2002/3 drought and as world prices for food and fuel spiked. The large state-owned estates were now expected to fill the production gaps, as national import cover declined. By 2006, a renewed Import Substitution Industrialisation strategy, aimed at reducing fuel imports through local agro-fuels and cutting food imports and dependence on food aid, was emerging. This strategy was also in defence against economic sanctions from and political isolation by the West.

The agrarian reform strategy of resurrecting production on state-owned enterprises was by 2007 shifting towards attracting foreign investment from the East and South (Moyo 2011b). But the reconstitution of large- scale estate farming through the FTLRP and agrarian reform policies evoked competing accumulation strategies among various elements in the new landholding regime, while struggles over the natural resources controlled by estates escalated. A scramble over access to water for irrigation, which the remaining large estates currently dominate, raised wider regional and institutional struggles, involving the private transnational and public estates against the new sugar outgrowers (see EU 2007) and new farmers up stream. This situation called for new forms of state regulation of access to water and the expansion of dams, with the latter recently being led by foreign investors, including through partnership with the public estates. Furthermore, while the central state saw large estates as critical to meeting the import substitution and expanded forex earnings required for national development, contradictions emerged as regional politicians, bureaucrats and farmers' organisations wanted such resources redistributed under their leadership.

Thus these simultaneously defensive and proactive agrarian policies did not extensively oust capital nor fully socialise production relations. Full change was blocked partly because of the embedded positions of some ruling party leaders and other politically influential actors in capital and because of the rootedness of some commodity production relations. The plan was opposed ideologically and in practise by big business (see Moyo 2011b). Due to conflicting class and political interests, factions of the ruling party elite and other factions of domestic capital and international capital (represented by black nationals) clashed over the allocation of thinly spread and limited public resources and subsidies. These measures affected the interests of opposing classes and politicians, with some seeking to evade or benefit from them.

Planning deficiencies were evident as the heterodox plan was not adequately coordinated, especially in relation to food production and exports. It soon faced implementation problems, including the general evasion of some controls (e.g., of maize marketing) and the countervailing tendencies of 'underground' or 'informalised' trade and petty businesses. Clearly, less of the state's inputs support went to the peasantry as new capitalist farmers were more influential in the distribution process, but a wide spectrum of farmers in various provinces did gain access. Corruption emerged within and outside ZANU-PF as various classes and actors in general competed for access to the subsidies and rents. Patronage often included or excluded both political opponents and supporters. These events widened the fractures within ZANU-PF and fuelled the violent contestation of the mid 2008 elections (Moyo and Yeros 2009).

Facing dramatic economic collapse and a political stalemate over the 2008 elections, political players engaged in negotiations over 'power-sharing' and various policy reforms ensued through SADC mediation. Agreement was reached in September 2008 (GoZ GPA 2008) and an inclusive government was formed in February 2009. By mid 2008, in the midst of negotiations, the economy was being liberalised. Controls on agricultural markets, the capital account and trade and off-budget subsidies were abandoned. Most critically, the economy was 'dollarized' by December 2008 (RBZ 2009). A new recovery plan called Short Term Emergency Recovery Programme (STERP) was issued in May 2009, following two new cash budgets in February and March 2009. But the state's fiscal capacity remained low as it operated a foreign currency cash budget with limited revenue collection and concessional loans, although some new revenues came from diamonds, now mined in joint ventures with the government.

Thus, state- subsidised inputs directed mainly at peasants were retained in late 2009 and this now involved some donors, supplemented by ZANU-PF's 'Well-Wishers Fund'. Subsidised credit through Agribank was resuscitated in 2011. Public and private financing of agriculture remained inadequate, as only 50 per cent of the estimated $2 billion required for full scale production (Ministry of Agriculture 2011) was being met by the market and the state. Contract farm lending dominated the supply. Securing finance was increasingly being seen as a problem of 'unsecuritised' land tenure, leading some to push for privatised land tenure. Eventually, however, the political parties agreed on tradable leases. Agricultural bonds were by 2010 being raised on the market by state institutions (e.g. Agricultural Marketing Authority) and banks, but for limited amounts (ca. $20 million) (*Zimbabwe Independent* 2011). Credit provision to salaried farmers was also being considered for A2 farmers, while the mortgaging of cattle for credit was being considered by one bank (TN Bank). However, it was evident that the supply of farm credit would remain insufficient, as the financial system was considered to be 'illiquid'.

By mid-2011, trade protection for grain and oilseed milling firms increased as imports outcompeted local industry. The liberalisation of GMO seed use, ostensibly to improve yields and the competitiveness of grain and soybean producers, was being extensively pushed by capital, some politicians and scientists, while some agro-industries called for protection from imports. A commodity-exchange market was being mooted by large-scale farmers representing blacks and the Ministry of Trade and Commerce.

Foreign investors were now being more readily entertained, but within the 'Indigenisation Policy' framework, requiring domestic control of majority shares (GoZ 2011). Some privatisation of parastatals was initiated (e.g., the ZISCO steel works sold 54 per cent of its shares to the transnational ESSAR from India), but the agricultural parastatals were starved of public funds. Foreign loans were, in 2011, being revived primarily from the east and south, mainly for agricultural machinery and other imports. The West maintained sanctions by restricting International Financial Institution (IFI) loans, while slightly increasing aid, mainly for social services, HIV/AIDS and farm inputs for vulnerable groups.

These policy shifts reflected the changing agency of diverse and farming classes, the renewed influence of capital and new public contestations over agrarian policy which was reverting to more state agricultural financing and trade protection (e.g., see GoZ 2010, Budget Speech). This changing and contradiction-riddled agrarian reform policy- making process shaped various dimensions of the agrarian relations which had been fundamentally restructured by land redistribution.

Changing agrarian production relations

Overall production pattern

Extensive land redistribution is expected to alter the structure and orientation of agricultural production, while passing through a transitional decline in output, as new farmers mobilise resources to establish themselves on the land and as markets and state interventions (particularly the financing mechanisms) adapt to change. Indeed, the output of Zimbabwe's main agricultural commodities started declining in varying degrees from 2002, with some export crops and dairy falling the most, while outputs began to rise selectively from 2006 (Table 6.3). Output declines among those commodities which had been predominantly grown by small-scale farmers, who were not affected by the land transfers, were lower than the declines among commodities grown mainly on large-scale farms and plantations, as the plantations had not been totally transferred. Moreover, the outputs of some commodities, produced predominantly for export with external financing, initially declined, but they recovered after dollarization as more merchants returned. Outputs of peasant-produced food grains targeting the controlled domestic markets were affected by their limited financing throughout the decade, as they depended mainly on the state and faced numerous droughts.

The number of farmers producing diverse commodities and the overall cropped area expanded substantially, although yields generally declined. A new uneven class and regional structure of production had emerged, since the FTLRP restructured Zimbabwe's fundamental agrarian and social relations of production, while consolidating others. However, the dominant discourse has been productionist in focus and pessimistic in its projection of future output. It teleologically expects a successful agrarian reform to emulate the output composition and 'productivity' of the former large-scale white farmers, notwithstanding the heterogeneous nature and interests of the new farming population, producing within diverse agro-ecological conditions. The dominant perspective celebrates the mono-cultural and extroverted large-scale agricultural production system created by the settler-colonial and immediate post-independence state (e.g., UNDP 2008). By underplaying the social context of production, it attributes the changing output mix mainly to the transfers of land from white large farmers to black small farmers, while its understanding of agrarian change narrowly and deterministically focuses on the subjective characterisation of the relative behaviour of former and new farmers.

Table 6.3: Agricultural production trends (crops 000 tonnes): 1990s average versus 2000s

Crop	1990 savg	2002/3	2003/4	2004/5	2005/6	2006/7	2007/8	2008/9	2009/10	2010/11	2011/12
Main foods											
Maize *(% change compared to 1990s)*	1,684	930 (-44.8)	1686.1 (0.0)	915.4 (-45.7)	1,485 (-11.8)	953 (43.4)	575.0 (-65.8)	1242.6 (-26.3)	1,327.6 (-21.2)	1,451.6 (-13.9)	968.0 (-42.5)
Wheat	248	122 (-50.8)	247.0 (-0.6)	229.1 (-7.8)	242 (-2.4)	147 (-40.7)	75.0 (-69.8)	57.9 (-80.7)	41.5 (-83.3)	53.1 (-78.5)	
Small grains	167	373 (123.4)	196.1 (19.0)	65.8 (-60.1)	164 (-1.8)	120 (-28.1)	93.2 (-44.2)	270.2 (64.0)	193.9 (16.1)	156.08 (-6.5)	108.7 (-34.9)
Edible dry beans	5.3	7.1 (34.0)-	56.8 (971.7)	21.5 (305.7)	21.5 (305.7)	30 (471.7)	3.8 (-28.3)	37.3 (6.13)	17.2 (224.5)	13.1 (147.2)	10.8 (103.8)
Groundnuts (shelled)	86	86 (0)	64.2 (-24.5)	57.8 (-32.1)	83 (-3.5)	125 (45.3)	131.5 (53.5)	216.6 (154.8)	186.2 (116.5)	230.5 (168.0)	120.0 (39.5)
Oilseeds											
Soya beans	98	41 (-58.2)	85.8 (-12.4)	56.7 (-42.1)	71 (-27.5)	112 (14.3)	48.3 (-50.7)	115.8 (18.2)	70.3 (-28.3)	84.2 (-9.3)	70.5 (-28.1)
Sunflower	43	17 (-60.5)	20.2 (-50.9)	7.4 (-82.0)	17 (-60.5)	26 (-39.5)	5.5 (-87.2)	39.0 (-5.3)	14 (-67.4)	11.5 (-72.1)	6.9 (-84.0)
Key Export											
Tobacco	198	82 (-58.6)	78.3 (-60.5)	83.2 (-58.0)	55 (-72.2)	79 (-60.1)	69.8 (-64.7)	63.6 (-67.9)	103.9 (-47.5)	132.4 (-33.1)	120 (-39.3)
Cotton	201	132 (-34.3)	364.3 (81.2)	196.3 (-2.3)	153 (-23.9)	235 (16.9)	226.4 (12.6)	246.8 (22.8)	172.1 (-14.4)	249.9 (24.3)	254.9 (26.8)
Estate crops											
Sugar	438.9	502 (14.4)-	422 (-3.9)	429 (-2.3)	446 (1.6)	349 (-20.5)	259 (-41.0)	259 (-41.0)	350 (-20.3)	332 (-24.4)	372 (-15.2)
Tea	10.6	22.0 (107.5)	22.0 (107.5)	22.0 (107.5)	22.0 (107.5)	13.5 (27.4)	8.3 (-21.7)	11.5 (8.5)	14 (34.9)	20* (88.6)	
Coffee	8.4	8 (-4.8)	5.8 (-31.0)	3.5 (-58.3)	1.3 (-84.5)	0.7 (-91.7)	0.8 (-90.5)	0.5 (-94.0	0.3 (-96.4)	3* (-64.3)	
Other crops											
Citrus	90	130 (44.4)	130 (44.4)	123 (36.7	123 (36.7)	123 (36.7)	-		-	-	
Vegetables & Melons	149	180 (20.8)	181 (21.5)	161 (8.1)	161 (8.5)	162 (8.7)	-		-	-	

Source: Moyo (2011a) derived from Ministry of Agriculture, Mechanisation and Irrigation Development (2010a, 2010b, 2011, 2012), FAO (2008, 2009), World Bank (2006), GAIN Report (2010; 2012), Zimbabwe Tea Growers Association (2010) data, *RBZ (2011) projections.

This a-historical and voluntaristic understanding of social agency has privileged the loss of skilled (white) farmers in explaining production declines, while pointing to presumed deficiencies, such as commercial motivation and organisational traits among new farmers. This is reinforced by a parochial settler-colonial stereotype of what a 'viable' agricultural production model means (Moyo 2002; Cousins and Scoones 2010). More generally, this perspective rests on the dubious teleology that a Junker path of agrarian accumulation, which organises production on a large-scale, is historically the only effective trajectory of development (see Moyo and Yeros 2005); This perspective also overlooks the crisis of social reproduction that has been provoked by this settler-colonial accumulation model (Mafeje 2003).

Most analysts underplay the long-run production trends, while non-linear changes in the magnitude and rate of change (decline, stagnation and growth) of the recent outputs of various commodities are not adequately examined, such that the structural factors which underlie the recent changes are ill-understood. There is an empirically dubious tendency to use historical data on outputs based only on formal markets, despite their neglect of peasant (or) informal sector outputs (e.g., Robertson 2011). Furthermore, historical output peaks, rather than average trends, are selectively counterpoised against current 'official' marketed output data, despite the fact that food markets became even more informalised since 2004. Moreover, the systematic changes in the production of various groups of commodities and enterprise mixes, as well as the overall outputs are not sufficiently tracked in relation to changing markets and financing, let alone of their class basis of production.

This tendency arises partly due to the presumed inferiority of small-scale production and commerce and due to lack of data. In fact, few studies (including the state statistical offices with their limited capacity) capture data on the 'informal markets' and associated disaggregated production activities. Indeed, the 'informalisation' of Zimbabwean markets is poorly conceptualised by most analysts, who treat them as an aberration reflecting deviant behaviour (e.g., *Kiya kiya*), which allegedly is deliberately created by the 'destructive economic policies of ZANU-PF' for their patronage- based accumulation (Raftopolous 2010). These reductionist approaches also fail to periodise dynamic changes in production and policy context, limiting their ability to decipher the forces which have shaped the actual output trends.

Expansion of food production among the peasantry

The output of maize, Zimbabwe's main staple grain produced mostly by peasants, declined severely in an erratic long- run pattern associated with droughts (Fig. 6.1). National maize yields per hectare fell to about 50 per cent of the 1990s average (Fig. 6.2), while the average yields of A1 maize producers were half those realised by A2 producers and land beneficiaries in wetter agro-potential areas realised twice the yields of those in drier regions (AIAS 2007).

Figure 6.1: Sub-sectoral maize production trends (1980–2010) in Zimbabwe

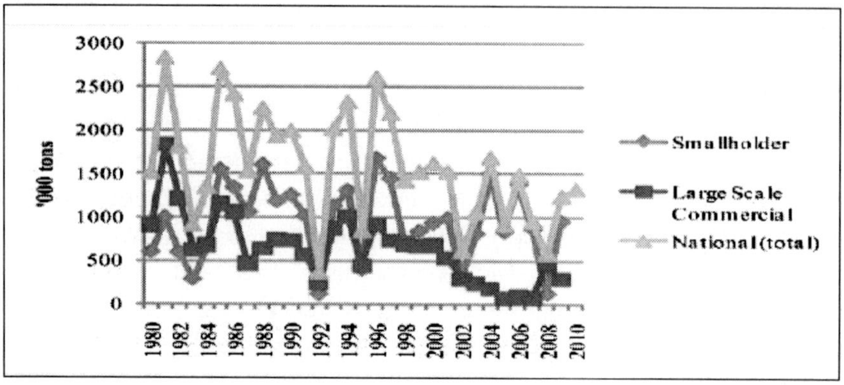

Source: MAMID (2010a, 2010b)

However, the output of small food grains such as sorghum and millet increased (Fig. 6.3), although their average yields fell 40 per cent below the 1990s average (MAMID 2010a). The output of wheat, which before 2000 was predominantly grown by large-scale farmers, declined dramatically (Fig. 6.2) and the area cropped to wheat fell from an average of 58,000 hectares in the 1990s to 18,200 hectares in 2010. Wheat output had already started rebounding in the 2005/6 season, as a consequence of concentrated provision of subsidised inputs by the government, only to deteriorate dramatically from the 2006/07 to the 2010/11 season, on account of loss of input subsidies and a sharp deterioration of electricity supply for irrigation.

Thus, the long run per capita production of cereals and maize (per capita and in absolute volumes) has been declining since the mid-1980s. Per capita cereal production on average ranged from 300kg/person during the 1980s, only to dip to 60 kg/person and 85 kg/person in 1992 and 1995 respectively. Actual national and per capita production of cereals has not been able to satisfy the needs, particularly of the recommended levels of per capita calorific requirements. Per capita maize production in Zimbabwe has never reached the 1982 level, when the country first attained a bumper crop.

Figure 6.2: Yield trends for main cereal crops

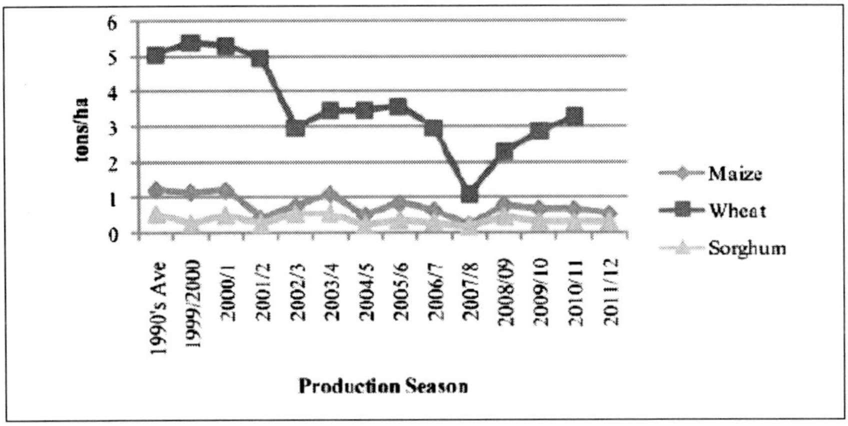

Source: MAMID (2010a, 2010b, 2011, 2012)

However, the numbers of food grain producers expanded after the Fast Track Land Redistribution Programme, leading to a major increase in the national cropped area dedicated to food grains from 1,794,527 hectares in 1999 to 2,655,687 hectares by 2011 (MAMID 2010a). Land reform beneficiaries dedicated 78 per cent of their cropped land to food grains (Moyo et al 2009). This shifted the orientation of production and use of prime lands away from exports to the staple grains prioritised by peasants.

Figure 6.3: Cereal crops output: 1990s vs. 2000s average

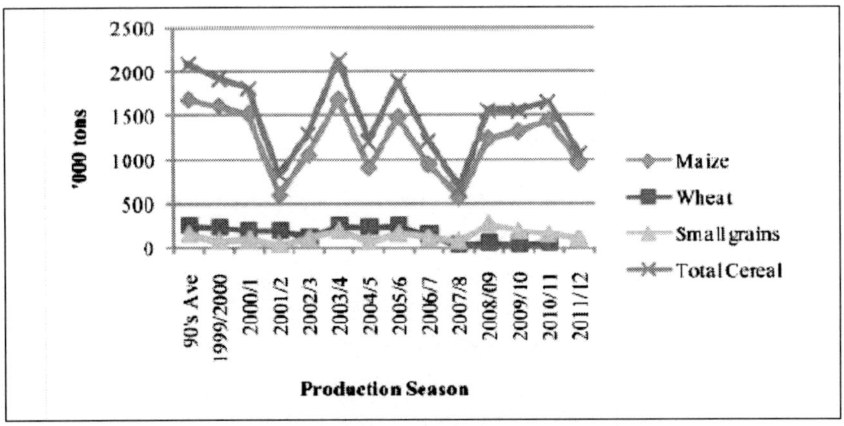

Source: FAO (2009), MAMID (2010a, 2010b, 2011, 2012)

The outputs and cropped areas of oilseeds such as soyabeans and sunflowers, which target the home market, but were mainly produced by large-scale farmers before 2000, also declined, but later they experienced a limited up-turn (Fig. 6.4). The output of groundnuts and edible beans, however, increased as a result of growth in area planted and these crops continued to be grown mainly by peasants. However, the average yield of the groundnuts has remained relatively unchanged at 0.488 tonnes, whereas that of soyabeans declined from 1,746 tonnes to 1,514 tonnes at the end of the 1990s and 2000s respectively.

Figure 6.4: Oilseed and dry beans output: 1990s vs. 2000s average

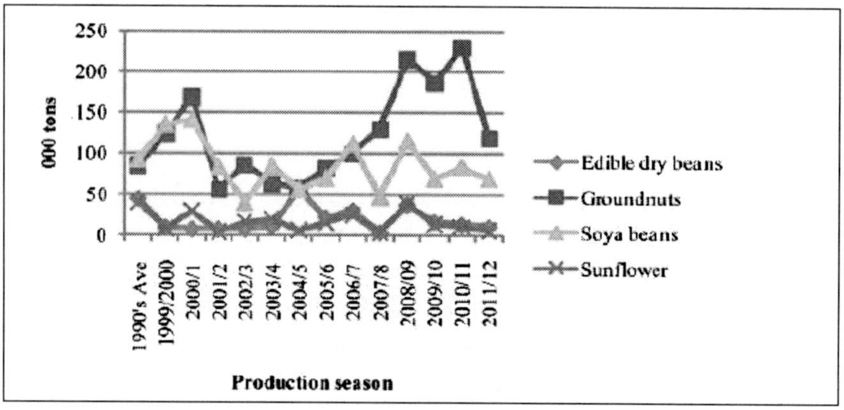

Source: FAO (2009), MAMID (2010a, 2010b, 2011, 2012)

About 21 per cent of the land beneficiaries grew groundnuts, while around 6 per cent of them grew beans and soyabeans (Moyo et al 2009). This suggests that the numbers of higher- value food producers has expanded and that peasant production is diversifying.

Resilience of export oriented production?

In terms of financial value, the largest share of agricultural output decline occurred among export commodities which had been predominantly produced on large-scale landholdings. In 2004, tobacco output had fallen by 72 per cent from the 1990s average (Fig. 6.4), but by 2011 its output was rising substantially (Fig. 6.5). Its yields and planted area declined by 55 and 20 per cent respectively during the period and fertiliser utilisation per hectare was halved by 2004 (TIMB 2010, MAMID 2011). But over 50,000 farmers, including 30,000 peasants, were now producing tobacco on smaller cropped areas, compared to about 700 large-scale and 3,500 small producers in 2002 (TIMB 2010, MAMID 2011).

Figure 6.5: Key exports' output trends: 1990s average vs. 2000s

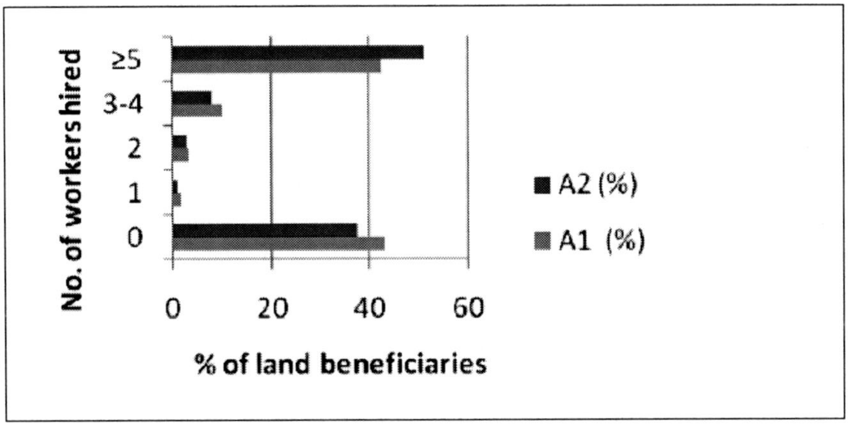

Source: TIMB Statistical data (2010), Cotton Ginners Association (2010), MAMID (2011, 2012)

The A2 farmers were realising higher yields as they used more inputs (Moyo et al 2009). Some remaining white farmers continued to produce tobacco. The output of cotton, which was largely exported and was predominantly produced by small farmers, surpassed the 1990s average by 49 per cent by 2011 (Fig. 6.5). This reflects a sizeable expansion of small-scale producers in a commodity, whose production was already entrenched and which was attractive since cotton has high tolerance of drought and since there was a continuity of contract-based inputs supplied by capital. On average, 4 per cent of the land beneficiaries grew-cotton, while 21 per cent grew it in Chiredzi District, despite its limited cotton-growing history (Moyo et al 2009). Established agro-industrial production structures and technocratic wisdom were being challenged by farmers in the new milieu (see also Scoones et al 2010).

The outputs of plantation export commodities only began to decelerate in 2004, but these were rising by 2011. Sugar output fell by 20 per cent in 2006 from the 1990s' average levels and then by 50 per cent during the hyperinflationary conditions between 2007 and 2008, only for the rate of decline to decelerate by 2011 (Fig. 6.6) (EU 2009; RBZ 2011). The structure of sugar production barely changed as the area cropped by the estates was hardly reduced, while the outgrowers' cropped area declined substantially (EU 2009; RBZ 2011).

The sugar production decline was allegedly due 'to the effects of the Land Reform process and to a lesser extent the effects associated with the present economic and inflationary conditions in Zimbabwe' (EU 2007: ii). Since production had declined mostly among outgrowers, the volatile currency markets and inflation had led to reduced supplies of inputs subcontracted to them by the estates. Sugar export prices were also deteriorating with the EU reforms leading to a 67.5 per cent reduction of raw sugar prices between 2005/6 and 2010 (ibid), while the prices of imported fertilisers and transport fuel rose during the same period. There was a slight shift in land use by the black outgrowers away from sugar towards food production for family consumption and sale and by 2006 they increasingly focused on maize, millet, beans, vegetables and cotton (Scoones et al 2010; EU 2007).

Large- scale estate production of sugarcane was mainly undertaken by Hippo Valley Estates and Triangle Limited, which were established over five decades ago.[5] Together they produced over 70 per cent of the country's sugar cane, while two groups of middle-scale farmers, white large-scale commercial and newly-resettled black farmers, produced the remaining 20 per cent of the country's sugarcane (GAIN Report 2010). There were about 47 whites with an average of 147 hectares each in 2000, whose landholdings were parcelled out to about 560 black farmers by 2007 with average hectarages of 10 to 30 hectares (Moyo 2011b). The sugar milling capacity of these two estates is 600,000 tonnes per year which was realised before 2006, while there are plans to upgrade the two mills to increase production capacity to 820,000 tonnes per year and to restore production levels from just below 300,000 tonnes per year to the past peak and expand this to the planned capacity. Triangle Sugar Limited produced under 300,000 tonnes of raw sugar between 2007 and 2010 and employed over 9,000 people. Approximately 90,000 tonnes are refined for local consumption (EU 2009), while the bulk of the output is exported as sugar and alcohol products.

Sugar production is set to increase as a result of expansion in the area planted from 2011 due to new investments by Hippo Valley estates, targeted at rehabilitating the outgrower schemes and also by Green Fuels, targeted at producing agro-fuels (Table 6.4). Large-scale agro-fuel (ethanol) production from sugarcane is concentrated in the south-eastern lowveld of the country, where the long established estates revived its production in 2007, while the recently set up sugarcane plantations on the ARDA and DTZ were expected to be in production by 2011. Together, these will account for over 80 per cent of the total agro-fuels produced in Zimbabwe in terms of land in production and the targeted area. Thus, the planned area to be allocated to sugarcane for ethanol is over 150,000 hectares, comprising over 60 per cent of all the estates' land with Zimbabwe's sugar industry considered to be the lowest cost producers in the world, due to the 'excellent growing conditions in the lowveld and highly efficient processing performance by its factories' (ibid.: ii).

Table 6.4: Estate agro-fuel production

Model Type	Project Name	Location	Production					
			Crops	**Confirmed Area (ha)**	**%**	**Targeted Area (ha)**	**%**
Private Sector Estates	Triangle Ethanol	Chiredzi	Sugarcane	40,912[6]	18.3	40,912	12.7
Parastatals	ARDA Biofuels	Chipinge	Sugarcane	40,000	17.9	50,000	15.6
Public/Private Estates	Zimbabwe Bio Energy	Nuanetsi Ranch	Sugarcane	100,000	44.8	100,000	31.2
Sugarcane Out-growers	Triangle Ethanol	Chiredzi	Sugarcane	7,200	3.2	10,000	3.1
Total				**188,112**	**84.3**	**200,112**	**62.4**

Source: Compiled by AIAS from various sources

The number and area of large-scale farm and estate producers of coffee and tea had also declined by 2006, but tea output patterns were anomalous compared to other crops. Over 86 per cent of the tea was produced by the large-scale and estate producers, with the rest produced by 100 white and black outgrowers and the state estate (ARDA Katiyo). By 2010, tea output levels were well above the 1990s averages, but this level was lower than the peak reached prior to 2007 (Fig. 6.6).

Meanwhile coffee output had declined by over 90 per cent in 2010, with its cropped area falling by over 30 per cent, mainly among the outgrowers, whose numbers almost doubled.[7] Remaining tea estates were now diversifying production towards macadamia nuts, pineapples and passion fruit, ostensibly due to labour shortages and lower prices, while outgrowers in tea and sugar growing areas were also producing some foods due to input shortages (EU 2009; USAID 2010).

Figure 6.6: Tea and coffee output trends in Zimbabwe

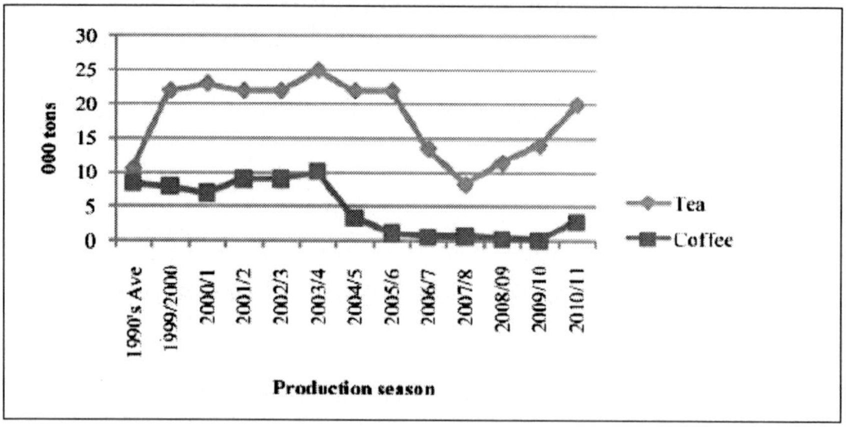

Source: IMF (2005), Zimbabwe Tea Growers Association data (October 2010), Zimbabwe Coffee Mills data

Soon after 2001, the exotic timber- producing companies (e.g. Border Timbers) stopped planting new trees, apparently because the 'illegal land occupations supported by big politicians' had brought uncertainty to their land tenure (Abu-Basutu 2010). Timber (sawn) outputs decreased from 374,779m3 in 1998/99 to about 194,181m3 in 2008/09, with over 90 per cent of this production coming from the core estates (Timber Producers Federation 2009).

The FTLRP did not directly lead to a substantial loss of formal waged agrarian labour among the estates, because most of their core production land had not been redistributed and they also retained most of their permanent workforce. When the high inflation conditions reigned, however, estate labour wages deteriorated further as plantation commodity prices fell. As a result, labour shortages increased (see Chambati and Moyo 2009) and labour emigration ensued. Agrarian reform had not fully overhauled the exploitative large-scale agrarian labour relations.

The production of formally-marketed dairy and beef was previously dominated by large-scale farms. The national cattle herd fell by 19 per cent by 2009 (Table 6.5). Most cattle are now held by many smaller-scale and middle-scale farmers practicing mixed farming and provide various foods, draught power and manure. Yet 58 per cent of the land beneficiaries had no cattle, 36 per cent had three and above and the rest had one to two beasts (AIAS 2007). However, 70 per cent of the surveyed households which had cattle were in drier agro-ecological regions, officially considered ranching areas (AIAS 2007).

Table 6.5: Cattle numbers by farming sector: 2001 – 2012

Sector	2001	2003	2005	2009	2011	2012
A2/LSCF	1,291,110	453,418	519,028	442,080	453,385	509,455
Communal	4,398,081	3,994,830	3,604,361	3,692,196	3,529,739	3,633,777
Resettlement & A1	505,360	717,969	844,800	919,616	1,020,070	899,608
Small scale	23,565	180,648	219,424	167,828	147,559	198,352
Total	**6,418,116**	**5,296,865**	**5,187,613**	**5,221,720**	**5,156,753**	**5,241,192**

Source: MAMID (2010a, 2010b, 2011, 2012), FAO (2009)

Dairy herds had already declined from their peak of 191,000 cows in 1991 to about a third by 2001, on account of the reduced profitability of dairy farming. The decline had accelerated dramatically by 2004 with breeding stock losses having been high and herd replacement slow. Milk output also fell by 66 per cent from 2001 to 2010 (MAMID 2011). The actual number of dairy cows declined continuously (Fig. 6.7) and sharply from 1990, such that the decline during the first decade of the twenty-first century was relatively slower in intensity. However, this decline is significant in effect since much of the breeding stock and the size of improved breeds also declined. Consequently, while the trends in the production of milk between 1995 and 2001 were relatively minor, between 2002 and 2010 the volume of milk output fell almost three-fold from above 200 million litres in 1995 to less than 50 million litres in 2010. Production in the dairy sub-sector is thus one of the worst performing sectors.

Figure 6.7: Dairy cows, 1990 – 2011

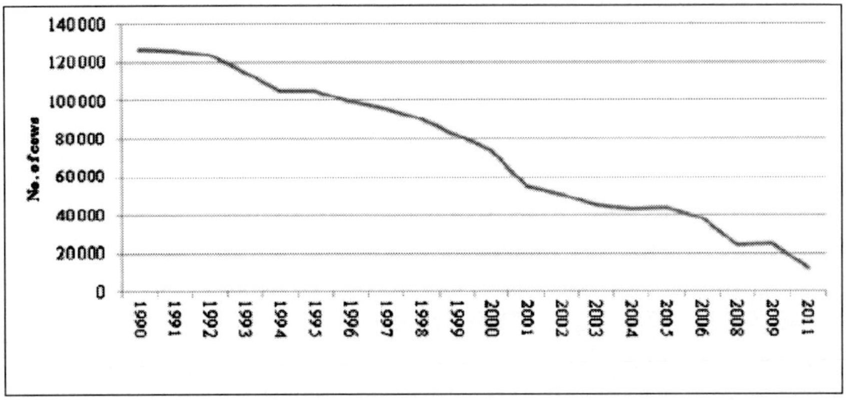

Source: Derived from MAMID (2011, 2012)

There were rapid declines in pork production by 2005, but pork output had risen substantially by 2009, while the numbers of goats were stable compared to sheep, which are not a common food (Fig. 6.8). Less than 25 per cent of the land beneficiaries had small ruminant livestock, while 2 per cent of them had piggeries by 2006 (MAMID 2011). Formally-marketed pork production was previously dominated by over 100 LSCF producers and agro-industrial plants, but by 2010 it involved over 250 smaller producers.[8] Overall, the diversification of livestock producers was accompanied by lower quality, lower breeding stocks and lower calving rates, since investments in breeding, animal health and pen-feeding had declined (FAO 2009).

Figure 6.8: Small stock production

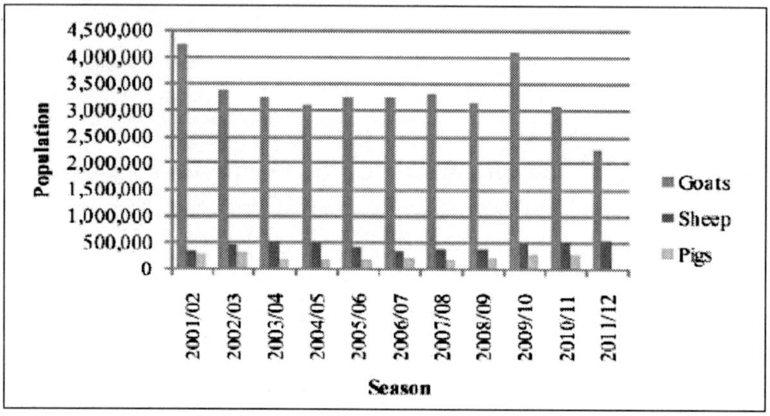

Source: MAMID (2010b, 2011, 2012)

Poultry production, both for egg and meat, expanded during the FTLRP period until 2007, when the production of chicks by key producers (Hubbard Zimbabwe, Irvines) declined, resulting in their importation from South Africa. The shortage of stock feeds due to limited availability of raw materials also affected poultry production. However, chick production (Fig. 6.9) and feed availability recovered in 2009 due to the removal of exchange rate and commodity controls in 2008 (e.g., maize, a key raw material for stock feeds).

Figure 6.9: Poultry chicks production trends

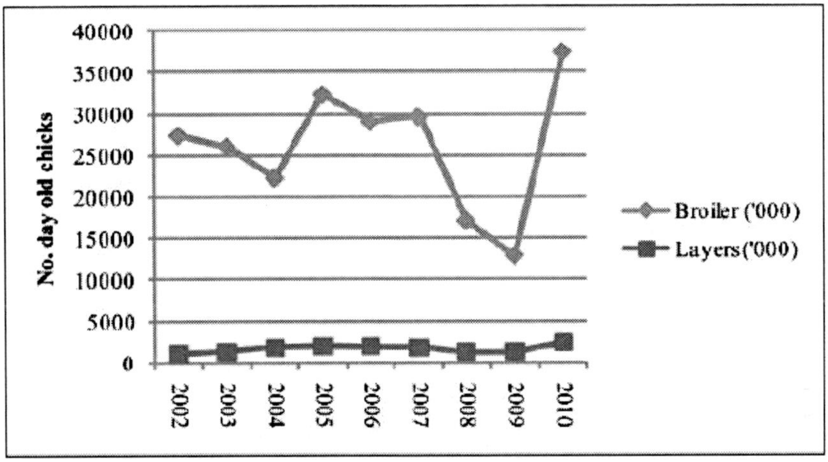

Source: Derived from the Zimbabwe Poultry Association (2010)

Diversification of the agricultural and non-farm livelihoods production base

In general, agricultural production patterns during the 2000s became more differentiated in class and regional terms, while export-oriented output rose faster than food. The production base was restructured by introducing more producers into all commodities and expanding the overall cropped area substantially, despite the decline of yields for most crops and livestock. Numerous producers earned farming incomes and provided their own food. A process of income re-distribution was underway, although this favoured an expanded range of middle-to larger-scale farmers, mainly in wetter regions. While more farmers were now producing exports, they were, however, well below 15 per cent of all the farmers.

Average land utilisation rates among land beneficiaries were at 40 per cent (AIAS 2007), comparing favourably with former large-scale farming areas (World Bank 1991). By 2006, about 54 per cent of the beneficiaries cropped less than three hectares, while only 14 per cent cropped more than 10 hectares (Table 6.6). Thus middle-scale and larger-scale landholders cropped proportionately less land than the peasants. Their pre-2000 counterparts usually cropped below 700,000 hectares, despite employing more formal labourers (World Bank 1991).

Table 6.6: Total cropped area by farm size in selected new resettlement areas

Cropped area (ha)	Farm sizes (ha)											
	1-19		20-49		50-99		100-299		300+		Total	
	No.	%	No.	%	No.	%	No.	%	No.	%	No.	%
0	177	18.8	122	19.8	30	16.3	49	27.5	11	37.9	389	19.9
0.1-1	90	9.5	44	7.2	13	7.1	27	15.2	-	-	174	8.9
1.1-3	285	30.2	145	23.6	27	14.7	37	20.8	3	10.3	497	25.5
3.1-5	222	23.5	105	17.1	24	13.0	23	12.9	3	10.3	377	19.3
5.1-10	97	10.3	92	15.0	40	21.7	12	6.7	-	-	241	12.4
10+	73	7.7	107	17.4	50	27.2	30	16.9	12	41.4	272	13.9
Total	944	100.0	615	100.0	184	100.0	178	100.0	29	100.0	1950	100.0

Source: AIAS Household Baseline Survey (2007)

While before 2000, agricultural production was predominantly export oriented, the incomes realised were concentrated among a few large farmers, alongside domestic and foreign capital. Peasants cropped much more land and used more labour towards producing various foods largely for auto-consumption. The latter trajectory has been consolidated, but land and labour productivity continue to be low. Consequently, domestic food production has, on average, been 35 per cent short of requirements, with wheat, soyabeans and dairy faring worst, while the output of pulses has expanded. Zimbabwe continues to have some relatively more secure-food enclaves, reflecting uneven production and productivity patterns. Small producers in the southern districts continue to face regular grain deficits, while those in the wetter regions increased their production of pulses and cash crops, particularly tobacco and various vegetables.

Class and regional biases in the capacities of various farmers to produce relatively larger areas of high-value crops and export commodities are being reproduced. A broader base of capitalist farmers has emerged, while the plantations are consolidating their vertical integration into world markets using more outgrowers. More peasants and middle-scale producers are now slowly expanding the supply of more diverse foods and raw materials to the home and export markets. This expansion also entails the reinsertion of large foreign capital into Zimbabwe's restructured agrarian markets.

In addition, various non-farm production and employment activities have emerged as a result of the opening up of the previously monopolised LSCF land. These non-agricultural land uses centred around farm tourism, small-scale mining, petty entrepreneurial activities, small trading businesses and industrial activities. The provision of over-night accommodation in lodges or chalets was found on 11.7 per cent of the farms surveyed, with 58 per cent of those in Chiredzi, endowed with wildlife conservancies which attracted more tourists prior to the FTLRP (AIAS 2007). In the Goromonzi and Zvimba districts, overnight accommodation facilities were found on 7.4 and 1.3 per cent of the surveyed farms respectively. Crocodile farming was another non-agricultural land use reported by key informant interviews, albeit on a low scale, as only one farm in Kwekwe was reported to be involved in crocodile farming.

Gold panning, which is associated with higher income rewards, was the most common natural resource exploitation activity for monetary gains reported by 5.8 per cent of the land beneficiaries and was more common in districts endowed with alluvial gold resources such as Kwekwe (11.3%) and Mangwe (46.9%). Districts such as Chipinge and Zvimba had fewer (0.3%) households involved in panning (ibid).

Besides natural resources exploitation activities, households were also involved in other petty entrepreneurial activities such as vending of new and second-hand clothes (5.1%), bricklaying (4.5%), tailoring (4.3%), repair works (2.6%), carpentry (2.1%), brewing of traditional beer for sale (1.7%), basketry (1.6%) and pottery (1%). Operation of small tuck-shop businesses was reported by an insignificant proportion (0.3%) (Fig. 6.10). Despite the low level of participation by the newly resettled farmers (less than 6%), all of these non-agricultural income- generating activities were more common in the A1 sector than in the more commercially oriented A2 farming sector (ibid).

These activities were spatially distributed in terms of availability of either a market or raw materials. In Chipinge, 5 per cent of land beneficiaries were involved in the vending of new and second-hand clothes. Basketry dominated in Chiredzi (7.2%) due to the availability of raw materials (ilala palm), which grow well in the hotter and dry lowveld conditions. Beer brewing is common and most pronounced in Chiredzi (9.4%), owing to the availability of raw materials (sugar, sorghum and ilala palm fruit), when compared to all the other districts, where less than 1.5 per cent of the households are involved in the activity.

Figure 6.10: Non-agricultural activities by resettled farmers

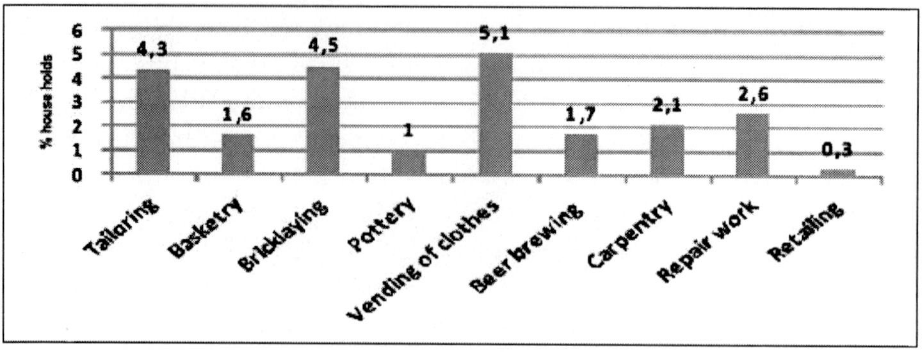

Source: AIAS Baseline Survey 2005/06, household questionnaire, N=2089

Small trading businesses were mostly found in areas nearer city centres such as Goromonzi and Kwekwe districts (which had 0.7 and 0.3 per cent of the households operating these). Industrial activities such as tractor and motor vehicle repairs were practiced by 3.5 per cent of the land beneficiaries and were more common in the A2 scheme (5.7%) than in the A1 scheme (2.9%) (ibid). Chiredzi District had the highest proportion of land beneficiaries engaged in industrial activities on the farms (13.6%), whilst participation in other districts tended to be below 2 per cent.

Differentiated access to agricultural inputs and markets

Overall productivity trends

Agricultural productivity generally declined due to reduced and uneven access to inputs and output markets. This poor access particularly affected smaller producers, who nonetheless deployed their labour to expand cropped areas. Access to inputs was also constrained by reduced public and private agricultural finance, leading to the diversification of input supply and commodity marketing arrangements by capital.

Limited and uneven access to agricultural inputs

Maize seed, particularly-locally produced seed, was in short supply during 2003 and 2009 and seed imports were required to meet the optimal national seed requirements of around 60,000 tonnes (Fig. 6.11). However, maize seed production had recovered to the full domestic requirements by the 2010/11 season (Table 6.7). Three transnational companies (Seed Co., Pannar and

Pioneer) had dominated hybrid maize seed production by 1999 through contracts with about 200 larger-scale growers whose land was redistributed. By 2010, numerous medium and large-scale farmers were being contracted to produce seed, unravelling the previous oligopoly. The seed companies prefer to deal with larger growers, ostensibly on account of their better ability to provide the required spacing and deal with the complex processing requirements of seeds.

Figure 6.11: Maize seed supply (2003-2012)

Source: Seed Co (2011) data; AMA (2012) data

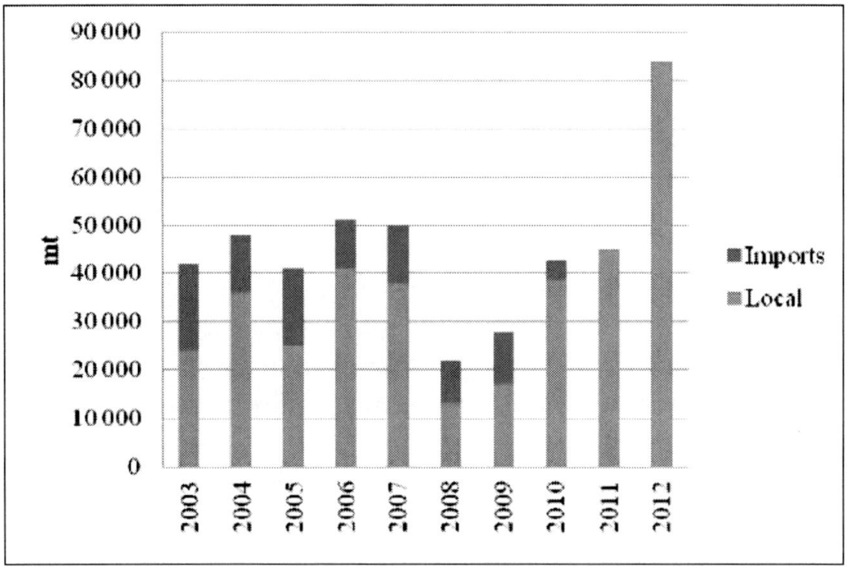

Similarly, many more middle-scale farmers were producing tobacco seedlings and meeting current demand (TIMB 2010). Shortages of potatoes and vegetable seeds persisted, leading to supplementary imports. By 2011, over 80 per cent of all the farmers were using commercial hybrid seeds now grown by more farmers contracted to capital, which retained dominance in the privatised bio-genetic industry.

Table 6.7 Maize seed delivery patterns, 2000 - 2011

Year	Maize seed delivery patterns (tonnes)		
	Hybrid seed	OPV seed	**Total**
2000/01	28187	0	**28187**
2001/02	17677	1132	**18809**
2002/03	14811	1693	**16504**
2003/04	23722	1555	**25277**
2004/05	14512	428	**14940**
2005/06	21726	315	**22041**
2006/07	16393.2	9.1	**16402.3**
2007/08	6283.4	34.4	**6317.8**
2008/09	7678.2	128.5	**7806.7**
2009/10	23460.4	641	**24101.4**
2010/11 Estimate	31887	317	**32204**

Source: Seed Company of Zimbabwe data (2011)

Fertiliser consumption in Zimbabwe has been declining since 2000 (Fig. 6.12). Its application in the 1990s averaged 30 kg/ha. This average was halved by 2004 (FAO 2009) and applied to larger cropped areas. By 2006, 50 per cent of the land beneficiaries were utilising inorganic fertilisers, mostly for maize, tobacco and cotton production, with relatively more A2 beneficiaries using fertilisers, while 20 per cent of these used pesticides (Moyo et al 2009). The use of fertiliser in the other crops (wheat, soyabeans and sunflowers) was generally limited to below 25 per cent of the producers (ibid). The wetter agro-ecological regions used more fertilisers.

Figure 6.12: Fertiliser consumption trends

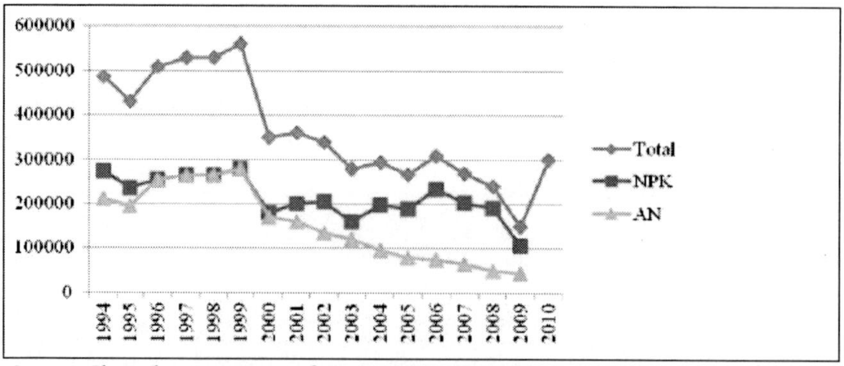

Source: Chemplex Corporation from MAMID (2010c)

Nationally, less fertiliser was applied by small farmers than by the larger farmers (Fig. 6.13). Export crops used the largest share (Tripathy et al 2007). Most A1 farmers were using animal manure rather than fertilisers (Moyo et al 2009), bearing in mind the rise of inputs prices globally (Moyo 2010). Less than 10 per cent of farmers also adapted to the rising price of inputs and access problems by adopting conservation farming to optimise absorption of water and fertiliser (FAO 2011).

Figure 6.13: Sub-sectoral utilisation of locally produced fertilizer

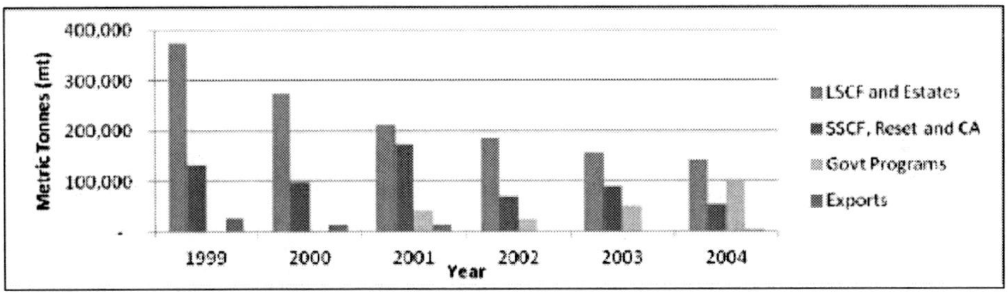

Source: Triparthy et al (2007)

The use of productivity- enhancing inputs on livestock was mainly found in the drier southern provinces. Of the land beneficiaries, 12 and 21 per cent used stock feeds and veterinary chemicals, respectively, although 34 per cent of the pork producers used them (Moyo et al 2009). Only 35 per cent of the cattle producers used public dipping and veterinary services, with slightly more A1 beneficiaries using public dips (Moyo et al 2009). This low usage limited livestock productivity. The re-insertion of large capital in the livestock sector was relatively limited because of the continued and dispersed control of breeding stock by former large-scale farmers and the pervasiveness of uneven regional investment.

Agricultural production continues to depend mainly on rain-fed farming. Only 5 per cent of the national cropped lands are irrigated and plantations control 57 per cent of this amount, while small producers control 30 per cent (World Bank 2006). About 17 per cent of the land beneficiaries had one form of irrigation facilities, while 28 per cent of the A2 farmers had irrigated crops compared to 14 per cent of A1 farmers and only 10 per cent of both groups had invested in irrigation (Moyo et al 2009). Irrigation was slightly more common in Chipinge and Chiredzi (Moyo et al 2009), where more irrigation facilities were pre-existing on plantations. Some irrigation facilities were disabled by departing landowners and land occupiers and numerous dams remain underutilised (World Bank 2006). This uneven class and regional distribution of irrigation facilities is also associated with the decrease in export production.

Agricultural productivity is also constrained by low and uneven access to farm machinery. Most peasants still depend on labour-intensive ox-drawn traction and hand weeding. Only 49 per cent of the land beneficiaries had access to animal-driven ploughs, while less than 20 per cent had access to power-driven equipment, with only 6 per cent of the A1 beneficiaries having access to tractors, compared to 36 per cent of the A2 farmers (Moyo et al 2009). Over 70 per cent of the A2 farmers who used tractors owned them (ibid). Small landholders owned less than 22 per cent of the national tractor fleet, farm equipment and machinery (MAEMI 2009). Public or private draught power hire services are limited and the government's mechanisation programme added only 3,217 tractors to the national stock (MoF 2010). During the early 2000s, limited fuel subsidies were provided, but these mainly benefited A2 farmers.

About 30 per cent of the land beneficiaries had on-farm and off-farm infrastructure, including some left on the redistributed farms (Table 6.8). Some A1 farmers shared farm houses and stores for social services, while A2 farmers gained these individually. Given the limited availability of credit, these investments are significant. Investments made were also regionally differentiated (Table 6.9). While access to subsidised inputs during the 2000s reached diverse farmers, the outcome favoured larger-scale export farmers with better access to markets (AIAS 2007). Thus, only some resource- rich farmers had access to inputs and this limited the capacity of most to hire labour (see Chambati 2011) and invest.

Table 6.8: Productive investment in newly- resettled areas

Type of investment	A1 model		A2 model		Total	
	No	%	No	%	No	%
Homestead	1089	66.0	206	47.0	1295	62.0
Irrigation equipment	168	10.2	48	11.0	216	10.3
Farm equipment & machinery	111	6.7	39	8.9	150	7.2
Storage facilities	123	7.5	30	6.8	153	7.3
Livestock	200	12.1	79	18.0	279	13.4
Tobacco barns	22	1.3	6	1.4	28	1.3
Electricity	5	0.3	2	0.5	7	0.3
Worker housing	123	7.3	62	14.2	185	8.9
Plantations & orchards	12	0.7	2	0.5	14	0.7
Environmental works	18	1.1	5	1.1	23	1.1

Source: AIAS District Household Baseline Survey (2005/06); N=2089

As of 2011, the capital intensity of farming was uneven and influenced wider agrarian relations. Those few larger-scale farmers using motorised traction included both high and low-intensity labour hirers, while the majority hired little labour and were poorly capitalised (see Chambati 2011). Class biases in the control of land, labour and access to markets have no doubt also been shaped by unequal political connections and social status, as well as the re-configuration of such markets.

Table 6.9: Investments made by farmers – Qualified gross table (excluding shelter)

Investments made by-farmers	District of study													
	Chipinge		Chiredzi		Goromonzi		Kwekwe		Mangwe		Zvimba		Total	
	No.	%	No.	%	No.	%	No.	%	No.	%	No.	%	*No.*	%*
Water & irrigation facilities	48	2.3	14	0.7	109	5.2	24	1.1	3	0.1	7	0.3	205	9.8
Farm structures	94	4.5	46	2.2	71	3.4	116	5.6	33	1.6	99	4.7	459	22.0
Farm equipment & machinery	29	1.4	7	0.3	26	1.2	45	2.1	8	0.4	22	1.1	137	6.6
Plantations & orchards	5	0.2	2	0.1	-	-	3	0.1	3	0.1	-	-	13	0.6
Environmental works	4	0.2	-	-	1	-	16	0.8	-	-	2	0.1	23	1.1

Source: AIAS Baseline Survey (2007); Notes: * N=2089

The reconfiguration of agrarian markets

The production of farm inputs such as fertilisers by domestic industry was also falling by 2003, as were imports (Fig. 6.14). Many agro-industries did not adapt to the changing demand structure. Agrarian merchants and fertiliser producers reduced local operations and increased operations in neighbouring countries. As interest rates rose, agri-business increasingly depended on subsidised foreign currency and credit in exchange for reduced prices and turned to manufacturing inputs on pre-paid contracts for large-scale exporters.

Figure 6.14: National fertilizer supply (2001 – 2012)

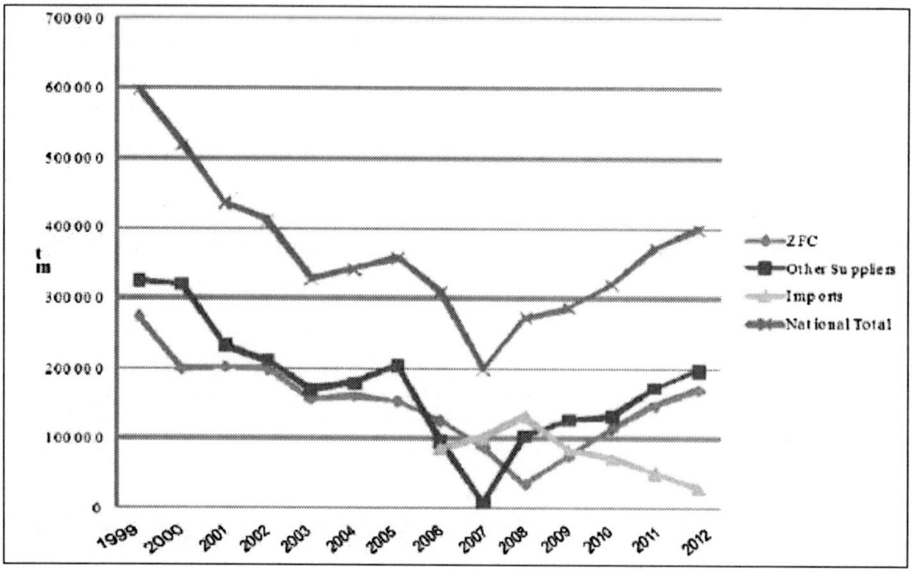

Source: Compiled by AIAS from Zimbabwe Fertilizer Company (ZFC) 2010 data.

More fertiliser and tractors were imported from China and Iran through concessional loans and from South Africa and elsewhere in cash. New indigenous importers were contracted to supply government input schemes. Similarly, domestic producers and suppliers of machinery and implements increased supplies when contracted by the government (See RBZ 2007a). This led to the recovery of some agro-industrial capacity (CZI 2010). Often, however, these schemes were alleged to have fiddled with prices, quality and distribution.

By 2006, 30 per cent of the A1 land beneficiaries relied on subsidised seed, compared to 20 per cent of A2 beneficiaries, while in high potential districts such as Chipinge, 50 per cent relied on subsidised seeds, compared to 9 per cent in remote Chiredzi (Moyo et al 2009). Between 68 and 87 per cent of the land beneficiaries, depending on the district, purchased their inputs in markets (Moyo et al 2009). A marginally larger proportion of A2 compared to A1 farmers benefited from government inputs. Over 96 per cent of the few land beneficiaries who used agro-chemicals for cotton and tobacco production bought them in markets. Similarly, less than one per cent of the land beneficiaries received government livestock inputs, aside from dipping services. International donors played a limited role in subsidising inputs until 2009. However, while such state subsidies were limited in scale, the level of support substantially surpassed that provided during the 1990s.

Formal cattle markets, whose exportation had substantially declined by 2004 due to reduced slaughters and livestock disease, became spatially and socially disaggregated. By 2010, meat trade increasingly involved direct sales between producers and new abattoirs, retailers and consumers (Scoones et al 2010). Livestock production on input sub-contracts and leased grazing created new tied markets (Zvimba Field Observation 2008[9]).

Government support for grain marketing also increased as compared to the ESAP era, but its monopoly became unsustainable. Grain procurement relied on limited subsidies and expensive credit, while the grain sold to millers was highly subsidised. Small millers mushroomed in tandem, competing and colluding with established large agribusiness to hike prices (News24 2003). Meanwhile, farmers received low prices for their grain and the payments were often delayed (FAO 2009). Many producers, consumers, millers and traders circumvented the controlled market, leading to high prices in parallel markets and unstable supplies. Few A2 producers 'obeyed' the marketing regulations, acting with patriotic solidarity, while expecting future subsidies (Mhondoro District Interview 2009[10]). This non-alignment between government subsidies and parallel market prices limited the potential welfare transfers to deficit areas, despite the substantial fiscal outlays and their supplementation by humanitarian aid.

Around 55 per cent of land beneficiaries sold their maize to the GMB, while the rest used local markets. Over 22 per cent of all the beneficiaries sold their edible beans to the GMB, with slightly more A1 beneficiaries doing so (Moyo et al 2009). Over 35 and 57 per cent of the soyabeans produced by the A1 and A2 land beneficiaries, respectively, were sold to the GMB, which used soyabean trading to generate income, while around 31 per cent of this crop was retained for own use (Moyo et al 2009). The rest was sold to agro-processing firms, contractors and local agro-dealers, including some which had secured subsidies. Distance from markets, limited state capacity and manipulative trading practises by the growing and variegated merchant classes generated various contradictions around state interventions in markets.

Most of the tobacco and cotton was being sold by land beneficiaries to contractors and private buyers in 2006, while over 65 and 55 per cent sold their sugar and tea in Chiredzi and Chipinge, respectively, to agro-processing plantations. Between 22 and 40 per cent of the A2 producers of cotton and tobacco had kept their output that year because prices were poor, while 12 and 24 per cent of the A1 and A2 farmers could not secure 'independent' markets for their sugar, allegedly because buyers rejected it on grounds of poor quality (Moyo et al 2009).

In 2006, the RBZ partially relaxed the policy of withholding large proportions of foreign currency earned by exporters (RBZ 2006). Contract farming benefited from this policy change and became central to the marketing of most commodities, excluding grains, building upon previous experiences with cotton and barley. A few black tobacco merchants piloted such contracting and later sold the businesses to transnationals. The number of tobacco contractors grew to 12 in 2010, from fewer than 3 in 2003, including four black firms, four new multi-racial contractors and four foreign contractors from China and elsewhere (TIMB 2010). Some of these were subcontractors of western transnationals such as British American Tobacco. Cotton contractors and buyers also increased and included Indians and Chinese, with the latter up scaling tobacco contracting.

Understanding the role of contract farming in class formation processes is complicated by its unclear association with land sizes and assets owned. Some contract financiers prefer peasants and middle-scale producers because they are less able to resist lower price margins, compared to larger-scale producers, who generally have higher social standing and fare better in procuring inputs using their own income, credit and subsidies. Conflicts arose between farmers and contractors over the depressed prices on offer and many 'resisted' this by side marketing their contracted commodities to other merchants who had not provided contract inputs to them. But they bought such outputs at marginally higher prices. New sugar outgrowers fought with plantation managers over transportation charges and the pricing of inputs and outputs (*The Herald* 2011b).

After market liberalisation and dollarization, the limited contract production of foods such as soyabeans faced competition from subsidised imports from the West and GMO-based grain and oilseeds imports from South Africa. This competition drove farmgate prices down, reducing local producer incomes. To some extent, this reduction propelled many small producers needing cash to reinvest in tobacco contracts. This explains the second temporal decline in the soyabean output levels in 2009.

The government revived the Agricultural Marketing Authority (AMA) in 2004 to better regulate markets and introduced new contract farming regulations in 2009 (USAID 2010). By 2011, however, the large-scale Zimbabwe Commercial Farmers Union and the Commercial Bank of Zimbabwe (CBZ), backed by the Ministry of Trade and Commerce and funded by some donors, were establishing the Commodity Exchange in Zimbabwe (COMEZ). The AMA feared this would drive speculation on food and increase foreign influence (*The Herald* 2011c). Class-based struggles over these changing agrarian markets remain potent and open with finance being critical.

Access to finance for farming

Throughout the 2000s, the volume of agricultural finance from domestic sources and external concessional loans and aid from western donors fell sharply, as compared to provisions during the 1990s (RBZ 2007b). Private agricultural credit declined from over $315 million in 1998 to about $6 million in 2008 (MAEMI 2009). European trade credit and the agricultural commodity bonds market had virtually disappeared by 2005. Government credit through Agribank had averaged around $25 million per annum between 2000 and 2007 and had peaked at $104 million in 2004, but it declined to below $3 million in 2007 (MAEMI 2009). Declining revenues limited budgetary allocations and led to larger-scale money printing. This credit constraint fuelled the diversification of the forms and sources of agriculture finance from the mid-2000s.

During this time, subsidised credit, inputs and foreign currency supplies increased (RBZ 2005). Foreign currency was being secured on parallel markets by citizens and the government, which fuelled speculative pricing, shortages of goods and hyperinflation. By 2004, the government attempted to plug the wheat deficit through sub-contracted production and other partnerships with domestic agro-industrial capital on the ARDA estates, using subsidised funds (Moyo 2011b). But this effort floundered in over pricing and profit-sharing disagreements.

Hyperinflation made agricultural credit less 'competitive' than short-term trading (Matshe 2004). The World Bank (2006) attributed decreased financing to the undermining of profitability and investment incentives by market controls. Others argued that the land user rights policy and land disputes created uncertainties for investors and limited credit supplies (Richardson 2005; Rukuni et al 2009).

During 2006, over 78 per cent of the A1 land beneficiaries used their savings, remittances from home and abroad and non-farm incomes to finance farming operations, compared to 83 per cent of the A2 farmers. Only 10 per cent of both A1 and A2 received external financing for production and 2 per cent had access to credit (Moyo et al 2009). But rising food and input prices after the 2002 drought and the global price hikes in 2005 undermined the real incomes of peasants (Wiggins 2005).

Not surprisingly, contract farming became central to the financing of smaller and middle-scale farmers (see *The Standard* 2009), who joined export production to gain access to inputs and increase their earnings. This move shifted pre-2000 agrarian relations from the dominance of private credit relationships between large-scale farmers and banks towards bonding more farmers with contracting intermediaries. Before 1986, the government had been the major lender (Moyo 1995). When foreign currency and agricultural markets were re-liberalised, agricultural sub-contractors escalated such pre-financing arrangements. Private bank credit to agriculture increased to over $300 million in 2010 (MoF 2011), but over 60 per cent of this amount went to contractors (USAID 2010).

China played a leading role in financing agriculture through loans for imported fertiliser, agro-chemicals and tractors and in contracting tobacco and cotton from 2006 (Edinger and Burke 2008). By 2009, the state had lured 'foreign investors' in partnership with domestic capital to produce and process sugarcane (for ethanol) and increase beef exports, using 20-year Build, Operate and Transfer (BOT) and land-lease arrangements on parastatal lands (Moyo 2011b). Large agribusiness was regaining dominance in agricultural input and output markets and new agrarian capital from the East and South was seeking to invest in agricultural exports and to supply inputs. Zimbabwe remained a net exporter of capital to the West during the decade (UNCTAD 2008).

In this situation, the recovery of agricultural production on the large private and public farming estates increasingly relied on foreign investments (see Moyo 2011b). The export-oriented production of the foreign-owned estates lingered on as they planned to triple sugarcane outputs on 30,000 more hectares to meet Zimbabwe's EU quota and other markets, in the context of the Economic Partnership Agreements (EPAs) under the ACP-EU Lome Convention. The European Commission aid towards a National Sugar Adaptation Strategy proposed to leave the home market with 28 per cent of such output. 'Dollarization' in 2008 created better 'incentives' for increased external financing of the foreign estates' sugar export plan, including from the EU aid and from domestic private bank credit. This increased sugar production promised to consume much of the scarce national water supplies, vis-à-vis other production needs.

Foreign financing for agriculture since 2002 has relied on new small loans and barter deals, while negotiations on the forward sales of mining concessions (ensued largely to import agricultural inputs and machinery) failed to materialise (Moyo 2010). Foreign 'investors' were by 2004 being encouraged, under the Look East Policy (focused on the Chinese), to sub-contractually buy tobacco and cotton. Substantial Chinese state trade credit to import fertiliser, agricultural chemicals, tractors, generators and pumps was secured from 2006, while foreign financing from Russia, Indonesia and Malaysia was being brokered by white domestic capital, especially to invest in the public estates.

By 2009, ARDA had signed a 20-year joint-venture agreement with private white Zimbabwean owned companies (Rating Investments Ltd and Macdom Investments Ltd) to lease over 50,000 hectares of ARDA's Middle Sabi and Chipinge estates, in a Build, Operate and Transfer Scheme (BOT). This was intended to establish 40,000 hectares of sugarcane and revive irrigation infrastructures within eight years and later to develop 10,000 more hectares (*The Herald* 2010). A two-year, rent-free grace period was provided, ostensibly to allow the sugarcane to gestate, while some sugar outgrowers were to be contracted. Construction of the $600 million sugarcane-to-ethanol distillery plant with a capacity to produce 35,000 to 40,000 litres per day (GAIN 2010) through another foreign 'investor' (Green Fuels (Pvt) Limited) was completed in early 2011 and 3,000 new hectares of sugar were being reaped for processing at ARDA's ethanol plant, with ethanol being supplied as fuel by November. No share was provided for the peasants from the adjacent Garahwa Communal Lands who originally owned the land, although by mid-2011 they were being incorporated as sugar outgrowers.

In 2008, the DTZ leased over 140,000 hectares of its land to a joint firm between DTZ and Custa (Pvt) Ltd, called the Zimbabwe Bio Energy (ZBE) project. Custa (Pvt) Ltd is owned by a white Zimbabwean large-scale capitalist (Billy Rautenbach) and foreign investors (from Russia and Spain), holding 70 per cent of the shares and investing $15 million (Moyo 2011c). About 100,000 hectares are dedicated to sugarcane production towards producing 500 million litres of ethanol per year. The rest of the land is intended to increase the cattle from 5,000 to 25,000 head, as well as to increase 100,000 crocodiles to 300,000 by 2012 and over 2,000 people were employed by DTZ. This deal led to the non-renewal of the DTZ's grazing leases with black elites and unsuccessful attempts to evict 'illegal' land occupiers, since the central government pressed the DTZ to allow 263 settlers to retain some land, dissociating itself from dispossessing this constituency.

Unlike the private sugar estates, the GoZ sought to resuscitate and expand ethanol and agro-industrial raw materials production.[11] The new foreign investments were meant to triple these industrial inputs. The inputs produced include ethanol for industrial, potable and pharmaceutical requirements; other sugar by-products, including molasses and bagasse, ingredients for yeast, carbon dioxide, livestock feedstock and fertilizer substitutes from vinesse; while generating more electricity at the mills. Sugarcane production for agro-fuel may soon dominate foreign investments in the south-eastern region's estate lands, which are expected to produce 90 per cent of Zimbabwe's agro-fuels on over 150,000 hectares by 2012.[12] This re-orientation of estate production towards substituting domestic transport fuel imports with agro-fuels runs counter to the EU's extroverted strategy, although the ecological benefits of this plan are to yet be calculated. However, it promises to reduce fuel imports and raise local agro-industrial capacity, creating scope for some national sufficiency.

The gradual recovery of production on the estates and their outgrowers indicates the premium placed on neoliberal policies by foreign 'investors', who continued to rely on borrowing locally while maintaining their control of the sugar and tea 'commodity chains'. The GoZ agrarian policy on the estates was intended to simultaneously counter agricultural production deficits and import dependence, while extending the state's capacity to direct development towards an articulated trajectory, including increased local beneficiation of agricultural raw materials and agro-industrial growth. The scale of social and 'developmental' benefits that can be expected from these foreign investments is, as yet, unclear. Thus, deeper land redistribution was being traded-off against shoring up the sanctioned objective of the state's relative autonomy from 'western' capital and the Bretton Woods institutions (various interviews).[13] This vision of state accumulation and autonomy, alongside nurturing a national 'bourgeoisie', was also initially opposed by less influential provincial officials and politicians, reflecting their desire to control sub-national accumulation processes, but these acquiesced from 2011 when the Indigenisation Policy promised to offer them shareholdings.

Moreover, political stabilisation and the liberalisation of financial markets from 2009 was leading to increased domestic agro-industrial processing capacity utilisation levels from below 20 per cent in 2008 to over 50 per cent in 2011 (MoF 2011). More foreign export crop merchants were returning. But by 2011, formal cross-border trade had increased the importation of foods, beverages and farm inputs, giving local agricultural producers and processors stiff 'competition' from cheaper (GMO and duty free) imports from South Africa, Brazil, China and the 'West'. This increase led to new demands for trade protection, which the GoZ responded to by increasing tariffs on processed and packaged foods (MoF 2011). Speedier agricultural recovery, however, remained constrained by limited access to private credit related to the low levels of liquidity and associated high interest rates (ibid).

These reconfigurations of the agrarian markets and state interventions fuelled contradictions within the state apparatus over its autonomy and the uneven benefits realised by various farming classes. The re-financialisation of Zimbabwe's economy was increasingly orienting agricultural production to exports, but diversifying its global integration. The class dynamics of the emerging agrarian relations reinforced the policy shift that enabled the dominance of large foreign and domestic capital, which, as of 2011, was only peripherally engaged directly in production.

Changing farmer organisation

These shifts in the agrarian relations altered the political landscape and re-oriented the politics of agrarian reform as new forms of farmer organisation and protest emerged at the local and national levels, through various types of farmer and commodity associations (Moyo et al 2009; Moyo 2011c; Murisa, Chapter 7). International capital, in alliance with new large-scale farmers and the established white and emerging black bourgeoisie led the reconfiguration of agrarian markets towards an increasingly neoliberal regime. New forms of social differentiation among the peasantry and emergent capitalist farmers, which shaped new forms of accumulation, gradually began to polarise agrarian reform policy and to animate the rural political constituency. Continued western sanctions and new forms of aid to vulnerable farmers also began to neutralise the political advantages derived from state subsidies. By 2010, the promise of new fruits from the indigenisation of mining and other businesses were generating new intra-class and multi-racial conflicts, slightly shifting the political heat from agrarian reform.

The former white farmers' unions had reconstituted their lobbies into two: the Commercial Farmers' Union (CFU), which retained a few members and the new Justice for Agriculture (JAG). They sought to influence national policy and international support over compensation for their lost land and the rationalisation of land use through the Land Audit proposed by the Inclusive Government and proposals to privatise land tenure. Elitist auditing perspectives also implied that poorer 'unproductive land users' could be evicted, while raising hopes among sections of the elite of gaining access to land. Some scholars and political actors sought to use the land audit to reverse the alleged ZANU-PF patronage in the land allocations (see Zamchiya 2011), while white farmers saw it as a route to revive the compensation debate.

Yet a more broadly- based class and social process defined the land allocations and the range of beneficiaries, despite the ruling party's overall dominance in the process. By 2010, many land beneficiary groups were mobilising networks to defend their land, while seeking to expand state input subsidies. Indeed, informal struggles over access (outside the official processes) persist, given the popular demand (see Moyo, Chapter 2) and competing claims over some land and natural resources. Local state authorities, including the bureaucracy and traditional leadership and ZANU-PF structures remain central to mediating the social legitimacy of such land struggles, placing pressures for continual land redistribution.

The state faces pressure to expand agrarian support and market protection from various middle class and capitalist lobbies in farming, agro-industry and trade, including some formed specifically to access subsidies. Local farmers' associations aggregated their resources to tap public extension and inputs support and to negotiate markets (Moyo et al 2009; see Murisa 2010) and competed with middle-scale and larger-scale farmers within the bureaucracy to influence agrarian policy towards more subsidies. Recognising the input shortages facing peasants, the Inclusive Government gradually increased subsidies to them (MoF 2011), while negotiating new strategies to subsidise A2 farmers. Contract farming relations also drove the growth of local farming associations, reinforcing the influence of capital in agrarian markets with support from the bureaucracy. Meanwhile, many NGOs, which had stood aloof all along, now competed to mobilise small farmers into market-oriented input support schemes funded by donors in collaboration with older farmers' unions (see FAO 2011).

New agricultural commodity associations, mostly representing middle-scale farmers (e.g., the Sugar Cane Farmers Outgrowers' Association), also mobilised against former white landowners and estate managers over land, contract services and output prices, as well as for improved access to water for irrigation. The struggles for water now involved the private and public and other new black farmers upstream and downstream of the concerned waters. Agrarian politics are also being re-shaped by the changing local administrative and political power relations that resulted from replacing white farmers' control over land, territory and labour with local influence now being more broadly diffused. The landless remain the most vulnerable. Sparse local government authorities are ill-equipped to regulate the new, but ubiquitous, struggles over natural resource and mineral extraction in competition with agriculture, while hereditary chiefs demand more powers to oversee resource management (Moyo 2011d).

Despite this reconfiguration of agrarian politics, past discourses shaped by political party polarisation continued to cast a shadow over progressive egalitarian agrarian reforms. Such discourses have limited the prospects for further democratising the land administration system, regulating agrarian markets in favour of small producers and enhancing state agricultural support.

Conclusion

Zimbabwe's agrarian reform has reconstituted the structure and orientation of agricultural (and non-agricultural) production, mainly through expanding the numbers of small and middle-scale agricultural producers and reconfiguring rural labour relations. Large-scale farm holdings and plantations persist with disproportionately more land than is warranted, while agro-industrial estates were marginally restructured, by introducing more small-scale outgrowers and through the expansion of public estate production, increasingly in partnership with foreign capital. Agrarian labour relations are now dominated by self-employment in diverse farming and non-farm activities, with part-time wage-labour being more common, while prevailing agricultural wages and incomes remain repressed by low productivity, exploitative commodity markets and slow recovery of production in other economic sectors.

Overall, agricultural outputs had declined, immediately after the land transfers and as the policies failed to mobilise adequate agrarian finance, when capital had retreated. Outputs began to rise slowly and selectively from 2006, with a wider range of producers cultivating much more land, mainly for food, despite low land productivity due to limited access to farm input markets and subsidies. Many more producers are involved in producing diverse exports than before 2000 and these realise the greatest access to farming inputs. Nonetheless, there has been a general rise in self-generated income earnings from farming and accumulation of assets, as well as signs of a broad process of income redistribution.

While established agro-industrial conglomerates, merchants and banks had substantially retreated from supplying farming inputs and credit and the buying of commodities, by 2002, some were propped up by state subsidies. Their supplies were revamped when markets were re-liberalised from 2009. However, agricultural production is increasingly being organised through contract farming, focusing on export commodities, including new contractors from the East. Loans negotiated with China and Brazil supported the revival and reconfiguration of agrarian markets, luring back capital from the West. But such gains were circumscribed by increased export orientation and the ascendancy of agribusiness over popular markets, given the limited fiscal capacity of the state, reeling under sanctions.

Nonetheless, the restructuring of Zimbabwe's agrarian relations has the potential to deepen the autonomy of the peasantry and intensify productivity towards increasing supplies of more nutritious foods and raw materials for the home market. Class struggles over land, agrarian markets and labour now entail new and more broadly based forms of smaller producer organisations and protests, pitted against larger-scale farm producers and foreign monopoly capital, aligned to expanding multi-racial domestic capital, whose members traverse political party lines. Progressive agrarian policies should consolidate the trajectory of smallholder development by securing peasants' land and protecting their markets to enhance popular food sovereignty, productivity and cash-earnings from self-employment in farming and non-farm industry. This protection requires stronger producer and consumer cooperation against agri-business and contractors (from the West, East and South) that link input and credit supply to crop purchases at depressed prices and against trade policies which enhance the dumping of subsidised foods. Expanding the democratic space and regaining policy autonomy towards articulated national development will be critical to advancing policies in favour of the peasantry and industrial diversification. Redirecting the expanding rents from mining to support this agenda will be crucial. As elsewhere, agrarian change under contemporary imperialism is neither linear nor even.

Notes

1. The chapter relies on research undertaken through the African Institute for Agrarian Studies (AIAS) between 2002 and 2010 (see Chapter 1 on the methodology, including sources of data).
2. Eight of these estates, comprising 52,264 hectares, were leasehold lands belonging to Communal Areas.
3. Notable are the Openhiemer, Nicolle and Moxon families and the Charter Estates' families (Moyo 1998).
4. The A1 model targeted landless and poor families, providing land use permits on small plots for residence cropping and common grazing, while the A2 scheme targeted new 'commercial' farmers, providing larger individual plots on long-term leases to beneficiaries supposedly with farming skills and/or resources (including for hiring managers).

5. Tongaat Hulett Limited, a South African agro-processing group, owns 100% of Triangle Limited and in 2006 bought a 50.35% stake in Hippo Valley Estates.
6. This is distributed as 21,553 hectares by Triangle Limited, 19,917 hectares by Hippo Valley Estates and 442 hectares by Mkwasine Sugar Estates (Scoones et al 2010).
7. Zimbabwe Coffee Mills, 2010, Personal communication by email, October 2010.
8. Interview with Theo Khumalo of COLCOM, April 2011, Harare.
9. Zvimba Field Observation, 2008, observation by Prof. Sam Moyo during the field trip at Zvimba.
10. Mhondoro District Interview, 2009, interview of land beneficiaries by Prof. Sam Moyo during a field trip at Mhondoro, January 2009.
11. Triangle Ltd stopped producing ethanol for petrol blending in 1992, but in 2006, the GoZ's National Oil Company of Zimbabwe (NOCZIM) contracted Triangle Ltd to supply it with 20 million litres of ethanol (EU 2007).
12. The National Biodiesel Production Programme (GoZ 2007) promotes agro-fuel production from Jatropha for the remaining annual agro-fuel requirements on 120,000 hectares of small producers' land. 60,000 hectares had been planted in 2010 (interview with E. Mushaka, NOCZIM 2010), but this project stalled due to limited state financing.
13. Various interviews with central government officials between 2006 and 2008.

References

Abu-Basutu, K.N., 2010, 'Forestry and the land reform in Zimbabwe: challenges and the way forward', *SALARN Newsletter*, Vol. 1, No. 2, pp. 11.

AIAS, 2007, 'Inter-district household and whole farm survey database', *AIAS Baseline Survey*, Harare: African Institute for Agrarian Studies (AIAS).

Amin, S., 1972, *Neocolonialism in West Africa*, Harmondsworth: Penguin.

Anderson, J.A., 2007, 'How much did property rights matter? Understanding food insecurity in Zimbabwe: a critique of Richardson', *African Affairs*, Vol. 106, No. 425, pp. 681–90.

Arrighi, G., 1973, 'The Political Economy of Rhodesia', in G. Arrighi and J.S. Saul, eds., *Essays in the Political Economy of Africa*, New York and London: Monthly Review Press, pp. 336–77.

BBC, 2011, 'Are Zimbabwe's new farmers winning, 10 years on?'. (Http://www.bbc.co.uk/news/world-africa-15919538). 1 December 2011).

Bochwey, K., Collier, P., Gunning, J.W., and Hamada, K., 1998, *External Evaluation of ESAF: Report by a Group of Independent Experts. Part IV: Country Profile: Zimbabwe*, Washington DC: International Monetary Fund.

Bush, R. and Cliffe, L., 1984, 'Agrarian Policy in Labour Migrant Societies: Reform or Transformation in Zimbabwe?', *Review of African Political Economy*, Vol. 29, pp. 77-94.

Chambati, W. and Magaramombe, G., 2008, 'The Abandoned Question: Farm Workers', in S. Moyo, K. Helliker, T. Murisa, eds, *Contested terrain: land reform and civil society in contemporary Zimbabwe,* Pietermaritzburg: S&S Publishers, pp.207-238.

Chambati, W. and Moyo, S., 2009, T*he Political Economy of Farm Workers in New Resettlement Areas,* AIAS Monograph Series, Harare: African Institute for Agrarian Studies.

Chambati, W., 2011, 'Restructuring of agrarian labour relations after Fast Track Land Reform in Zimbabwe', *Journal of Peasant Studies*, Vol. 38, No. 5, pp. 1047-1068.

Chang, H., 2009, 'How to 'do' a developmental state: political, organisational and human resource requirements for the developmental state', in O. Edigheji, ed., *Constructing a democratic developmental state in South Africa: potential and challenges,* Cape Town: HSRC Press, pp. 82–96.

Cliffe, L., 2000, 'The Politics of Land Reform in Zimbabwe', in T.A.S. Bowyer-Bower, C. Stoneman, eds., *Land Reform in Zimbabwe: Constraints and Prospects,* Aldershot UK: Ashgate, pp. 35-46.

Cotton Ginners Association, 2010, 'Cotton pricing 2010', May, Harare.

Cousins, B. and Scoones, I., 2010, 'Contested paradigms of 'viability' in redistributive land reform: perspectives from southern Africa', *Journal of Peasant Studies,* Vol. 37, No. 1, pp. 31-66.

Cusworth, J. and Walker, J., 1988, *Land Resettlement in Zimbabwe: A Preliminary Evaluation*, Evaluation Report EV434, London: ODA.

CZI, 2010, *Back to Basics: Post-dollarization Manufacturing Recovery,* Confederation of Zimbabwe Industries (CZI) Manufacturing Sector Survey, 2010.

Edinger, H. and C. Burke, 2008, 'AERC Scoping Studies on China-Africa Relations: A Research Report on Zimbabwe', Centre for Chinese Studies, University of Stellenbosch, March 2008.

EU, 2007, 'Zimbabwean adaptation strategy to the European Union. Sugar regime reform: final report', February. Brussels: European Union.

EU, 2009, 'Annual Action Plan 2009 Zimbabwe: National Sugar Compact', Draft Final Report, EC Sugar Facility, March 2009.

FAO, 2008, 'Special Report: FAO/WFP Crop and Food Security Assessment Mission to Zimbabwe', 18 June 2008, Rome: Food and Agriculture Organization of the United Nations; Rome: World Food Programme.

FAO, 2009, 'Special Report: FAO/WFP Crop and Food Security Assessment Mission to Zimbabwe', 22 June 2009, Rome: Food and Agriculture Organisation of the United Nations; Rome: World Food Programme.

FAO, 2011, 'Conservation Agriculture (CA)', *Agriculture Coordination Working Group Journal,* Vol. 61, pp. 1-6.

GAIN Report, 2010, *Zimbabwe: Sugar Annual,* Global Agricultural Information Network (GAIN), USDA Foreign Agriculture Service.

GAIN Report, 2012, 'Zimbabwe: Sugar Annual', Global Agricultural Information Network (GAIN); USDA Foreign Agricultural Service.

Ghosh, J., 2008, 'The global oil price story'. (http://www.networkideas.org/jul2008/news28_oil_price.htm). 4 August 2008.

Gore, C.G., Katerere, Y., Moyo, S., Mhone, G.C.Z. 1992, 'The case for sustainable development in Zimbabwe: conceptual problems, conflicts and contradictions', Harare, Zimbabwe: Environment and Development Activities.

Government of Zimbabwe (GoZ), 2001, *Millennium Economic Recovery Programme (MERP).*

Government of Zimbabwe (GoZ), 2001, 'Statutory Instrument 235A of 16th July, 2001', Grain Marketing (Controlled Products Declaration) (Maize and Wheat) Notice, 2001, in terms of section 29 of the Grain Marketing Act (Chapter 18:14).

Government of Zimbabwe (GoZ), 2003, *National Economic Revival Programme: Measures to address the current challenges,* Ministry of Finance and Economic Development, February 2003.

Government of Zimbabwe (GoZ), 2008, 'Global Political Agreement' (GPA), Agreement between the Zimbabwe African National Union-Patriotic Front (ZANU-PF) and the two Movement for Democratic Change (MDC) formations, on resolving the challenges facing Zimbabwe, September 2008.

Government of Zimbabwe (GoZ), 2011, 'Indigenisation and Economic Empowerment Act', Act 14/2007.

GoZ, 1998, 'Land Reform and Resettlement Programme Phase II: A Policy Framework and Project Document', Harare: Government Printers.

Habib, A., 2011, 'Foreword: Reflections on the Prerequisites for a Sustainable Reconstruction in Zimbabwe', in H. Besida, ed, *Zimbabwe: Picking Up the Pieces*, New York: Palgrave Macmillan, pp. xi-xv.

IMF, 2005, *Zimbabwe: Selected Issues and Statistical Appendix. IMF Country Report No. 05/359,* October 2005, Washington, D.C.: International Monetary Fund.

Mafeje, A., 2003, 'The Agrarian Question, Access to Land and Peasant Responses in Sub-Saharan Africa', UNRISD, *Civil Society and Social Movements Programme Paper Number 6 May 2003*, Geneva: United Nations Research Institute for Social Development.

Matshe, I., 2004, 'The Overall Macroeconomic Environment and Agrarian Reforms', Harare: African Institute for Agrarian Studies Mimeo.

Ministry of Agriculture, Mechanisation and Irrigation Development (MAMID), 2012, 'Second Round Crop Assessment Report', Harare: Government Printers.

Ministry of Agricultural Engineering, Mechanisation and Irrigation (MAEMI), 2009, 'Agricultural Engineering Mechanization and Irrigation Strategy Framework: 2008 – 2058', Draft.

Ministry of Agriculture, Mechanisation and Irrigation Development (MAMID), 2010a, *Second Round Crop and Livestock Assessment Report,* 9 April 2010, Government of Zimbabwe, Harare: Government Printers.

Ministry of Agriculture, Mechanisation and Irrigation Development (MAMID), 2010b, *Agricultural Statistical Bulletin, 2010*, Harare: Government Printers.

Ministry of Agriculture, Mechanisation and Irrigation Development (MAMID), 2011, *Second Round Crop and Livestock Assessment Report, 14 April 2011*,Harare: Government Printers.

Mkandawire, T., 2010, 'On Tax Efforts and Colonial Heritage in Africa', *Journal of Development Studies,* Vol. 46, pp. 1647-1669.

MoF, 2010, 'Ministry of Finance Budget Speech 2010', Ministry of Finance of Zimbabwe. (http://www.zimtreasury.org/downloads/834.pdf). 22 December 2010.

MoF, 2011, 'The 2011 Mid-Year Fiscal Policy Review: 'Riding the Storm: Economics in the Time of Challenges', Presented to Parliament by Hon. T. Biti, MP, Minister of Finance, 26 July 2011.

Moyana, H.V., 2002, *The political economy of land in Zimbabwe*, Gweru: Mambo Press.

Moyo, S. and Yeros, P., 2005, 'Land occupations and land reform in Zimbabwe: towards the National Democratic Revolution', in S. Moyo and P. Yeros eds, *Reclaiming the Land: The Resurgence of Rural Movements in Africa, Asia and Latin America,* London: Zed Books, pp. 1-64.

Moyo, S. and P. Yeros, 2007, 'The Radicalised State: Zimbabwe's Interrupted Revolution', *Review of African Political Economy,* Vol. 111, pp. 103-121.

Moyo, S. and P. Yeros., 2009, 'Zimbabwe ten years on: results and prospects'. (http://www.pambazuka.org/en/category/features/54037). February 2009.

Moyo, S., 1998, 'The Land Acquisition Process in Zimbabwe (1997/8)', Harare: United Nations Development Programme (UNDP).

Moyo, S., 2000, *Land reform under structural adjustment in Zimbabwe: land use change in Mashonaland provinces, Uppsala: Nordic Africa Institute.*

Moyo, S., 1995, *The land question in Zimbabwe*, Harare: SAPES Books.

Moyo, S., 2001, 'The Land Occupation Movement and Democratization in Zimbabwe: Contradictions of Neoliberalism', *Millennium Journal of International Studies,* Vol. 30, No. 2, pp. 311-330.

Moyo, S., 2002, 'Peasant Organisations and Rural Civil Society in Africa: An Introduction', in S. Moyo and B. Romdhane, eds, Peasant Organisations and Democratisation in Africa, Dakar: CODESRIA Book Series, pp. 1-26.

Moyo, S., 2010, 'The Agrarian Question and the Developmental State in Southern Africa', in O. Edigheji, ed., *Constructing a democratic developmental state in South Africa: potential and challenges,* Cape Town: HSRC Press, pp. 285-314.

Moyo, S., 2011a, 'Three decades of agrarian reform in Zimbabwe', *Journal of Peasant Studies,* Vol. 38, No. 3, pp. 493-531.

Moyo, S., 2011b, 'Land concentration and accumulation after redistributive reform in post-settler Zimbabwe', *Review of African Political Economy,* Vol. 38, No. 128, pp. 257-276.

Moyo, S., 2011c, 'Agrarian reform and prospects for recovery', in H. Baseda, ed., *Zimbabwe: Picking Up the Pieces,* New York: Palgrave MacMillan, pp. 129–155.

Moyo, S., 2011d, 'Changing agrarian relations after redistributive land reform in Zimbabwe', *The Journal of Peasant Studies,* Vol. 38, No. 5, pp. 907–934.

Moyo, S., Chambati, W., Murisa, T., Siziba, D., Dangwa, C., Mujeyi K., and Nyoni, N., 2009, *Fast Track Land Reform Baseline Survey in Zimbabwe: Trends and Tendencies, 2005/06,* Harare: African Institute for Agrarian Studies (AIAS).

Murisa, T., 2010, 'Emerging forms of social organisation and agency in the newly resettled areas of Zimbabwe: the cases of Goromonzi and Zvimba districts', PhD. Thesis, Grahamstown: Rhodes University.

National Economic Recovery Council (NERC), 2006, 'Minutes of the 8th May 2006 Meeting', held in the Munhumutapa Building, Harare.

News24, 2003, 'Zimbabwe food price showdown', 18/07/2003, South Africa: News24 (http://www.news24.com/xArchive/Archive/Zim-food-price-showdown-20030718). 20 December 2010.

O'Laughlin, B., 2002, 'Proletarianisation, Agency and Changing Rural Livelihoods: Forced Labour and Resistance in Colonial Mozambique', *Journal of Southern African Studies,* Vol. 28, No. 3, pp. 511-530.

Patnaik, P., 2008. 'The accumulation process in the period of globalization'. (http://www.epw.in/perspectives/accumulation-process-period-globalisation.html).

Raftopoulos, B. and Mlambo,A., eds., 2009, *A History from the Pre-Colonial Period to 2008: Becoming Zimbabwe,* Harare: Weaver Press, pp. 201-232.

RBZ, 2006, 'Sunrise of Currency Reform, Monetary Policy Review Statement for First Half of 2006', Reserve Bank of Zimbabwe, Monday 31st July 2006.

RBZ, 2009, 'Exchange Control Directive Rk:39, issued in Terms of Section 35 (1) of the Exchange Control Regulations Statutory Instrument 109 of 1996', 23 February 2009.

RBZ, 2005, 'Monetary policy statement, issued in terms of the RBZ Act Chapter 22:15 Section 46', by Dr G. Gono, Governor of the Reserve Bank of Zimbabwe, 20 July 2005.

RBZ, 2007a, 'The Launch of Phase 1 of Agricultural Equipment Acquired under the RBZ Farm Mechanisation Programme', 11 June 2007, Reserve Bank of Zimbabwe.

RBZ, 2007b, 'The Impact of Sanctions against Zimbabwe, A supplement to the mid-term Monetary Policy Review Statement', October 2007, Reserve Bank of Zimbabwe.

RBZ, 2011, 'Monetary Policy Statement', January 2011, Reserve Bank of Zimbabwe.

Richardson, C., 2005, 'The loss of property rights and the collapse of Zimbabwe', *Cato Journal,* Vol. 25, No. 2, pp. 541–565.

Robertson, J., 2011, 'A Macroeconomic Policy Framework for Economic Stabilisation in Zimbabwe', in H. Besida, ed., *Zimbabwe: Picking Up the Pieces,* New York: Palgrave Macmillan, pp. 83-105.

Rukuni, M., Nyoni J., and Matondi, P., 2009, *Policy Options for Optimisation of the Use of Land for Agricultural Productivity in Zimbabwe,* Harare: World Bank.

Rukuni, M., Tawonenzi P., and Eicher, E. K., with Munyuki-Hungwe M., and Matondi, P., eds., 2006, *Zimbabwe's agricultural revolution revisited,* Harare: University of Zimbabwe Publishers.

Rutherford, B., 2003, 'Belonging to the Farm(er): Farm Workers, Farmers and the shifting politics of citizenship', in A. Hammar, B. Raftopoulos and S. Jensen, eds, *Zimbabwe's Unfinished Business: Rethinking Land, State and Nation in the Context of Crisis,* Harare: Weaver Press, pp. 191-241.

Sachikonye, L., 2003, 'The Situation of Commercial Farmers after Land Reform in Zimbabwe', report prepared for the Farm Community Trust of Zimbabwe.

Scoones, I., N. Marongwe, B. Mavedzenge, F. Murimbarimba, J. Mahenehene and C. Sukume, 2010, *Zimbabwe's Land Reform: Myths and Realities,* Harare: Weaver Press.

Seed Co., 2011, Maize seed supply data.

Sender, J. and Johnston, D., 2004, 'Searching for a Weapon of Mass Production in Rural Africa: Unconvincing Arguments for Land Reform', *Journal of Agrarian Change*, Vol. 4, Nos. 1 and 2, pp. 142-164.

Stoneman, C. and Bowyer-Bower, T.A.S., eds, 1988, L*and Reform in Zimbabwe: Constraints and Prospects,* Aldershot UK: Ashgate.

Sukume, C. and Moyo, S., 2003, 'Farm sizes, decongestion and land use: implications of the Fast Track Land Redistribution Programme in Zimbabwe', Harare: AIAS Mimeo.

The Herald, 2010, 'Chisumbanje: power with sweetness', 15 April 2010, Harare: Zimpapers.

The Herald, 2011b, 'Zimbabwe: Sugarcane Farmers, Tongaat Hulett Clash', 23 March 2011, Harare: Zimpapers.

The Herald, 2011c, 'Zimbabwe: Agriculture Stakeholders Hit Out at Commodity Exchange', 2 February 2011, Harare: Zimpapers.

The Standard, 2009, 'Government bungling stifles contract farming', 23 September 2009.

TIMB, 2010, *End of 2010 Tobacco marketing season final report,* Tobacco Industry Marketing Board, October 2010.

Timber Producers Federation, 2009. Zimbabwe timber industry statistics for the year ended 31 March 2009. Mutare: Timber Producers Federation.

Tripathy, D., C. Sukume, J. Nyoni and T. Dube, 2007, 'Grain Pricing and Marketing in Zimbabwe: Strategic Options', Report initiated by and submitted to the Ministry of Agriculture.

Tupy, M.L., 2007, 'A four-step recovery plan for Zimbabwe', CATO Institute, (http://www.cato.org/pub_display.php?pub_id¼8191). 9 October 2008.

UNCTAD, 2008, *UNCTAD Handbook of Statistics 2008,* New York and Geneva: United Nations.

UNDP (United Nations Development Programme), 2010, H*uman Development Report: The Real Wealth of Nations: Pathways to Human Development,* New York: UNDP.

UNDP, 2008, 'Comprehensive Economic Recovery in Zimbabwe: a Discussion Document'. Harare.

USAID, 2010, Zimbabwe Agricultural Sector Market Study, Weidemann Associates, Inc.

Utete, C., 2003, 'Report of the Presidential Land Review Committee, under the chairmanship of Dr Charles M.B. Utete, Vol. 1 and 2: Main report to His Excellency the President of the Republic of Zimbabwe', Presidential Land Review Committee (PLRC). Harare: Government Printers.

Weiner, D., 1988, 'Land and Agricultural Development', in Stoneman, ed., *Zimbabwe's Prospects: Issues of Race, Class and Capital in Southern Africa,* London: Mcmillan, pp. 63-89.

Wiggins, S., 2005, 'Southern Africa's food and humanitarian crisis of 2001-04: Causes and lessons', a discussion paper for the Agricultural Economic Society Annual Conference, Nottighham, 4-6 April 2005.

World Bank, 1991, Zimbabwe: Agriculture Sector Memorandum, Vol. 1 and Vol. 2, No. 9429-Zim, Washington DC: World Bank.

World Bank, 2006, *Agricultural Growth and Land Reform in Zimbabwe: Assessment and Recovery Options,* Report No. 31699-ZW, Harare: World Bank.

World Bank, 2012, 'Recovery and Growth of Zimbabwe Agriculture'. (www.networkideas.org/focus/may 2008/fo28_globalisation.htm).

Yeros, P., 2002, 'The political economy of civilisation: peasant-workers in Zimbabwe and the neo-colonial world', PhD. Thesis, University of London.

Zamchiya, P., 2011, 'A synopsis of land and agrarian change in Chipinge district, Zimbabwe', *Journal of Peasant Studies,* Vol. 38, No. 5, pp. 1093-1122.

Zimbabwe Independent, 2011, 'Agro bills likely undersubscribed', Thursday, 08 December 2011.

Zimbabwe Tea Growers Association, 2010, Tea and coffee production data.

7

Social Organisation in the Aftermath of 'Fast Track': An Analysis of Emerging Forms of Local Authority, Platforms of Mobilisation and Local Cooperation

Tendai Murisa

Introduction

The Fast Track Land Reform Programme (FTLRP) has led to significant social change, with approximately 160,000 families now settled in areas previously inhabited by approximately 4,000 large-scale farmers. As of 2005 most of this land was held through leases and permits issued by the state, as opposed to freehold land tenure (see Moyo, Chapter 2). Significant change was also evident in the manner in which the 'new' communities organised themselves to utilise land, with their land use preferences not necessarily conforming to the preferences of former large land owners (see Masuko, Chapter 4). In addition, the Fast Track resettlement process disrupted known ties of reciprocation developed within lineage and kinship ties in customary tenure areas.

This chapter tells the story of how the newly resettled land beneficiaries have been organising themselves. It discusses some of the factors that have shaped emerging social relations in the Newly Resettled Areas (NRAs) through an examination of the variety of arrangements which constitute rural social organisation, including the emerging social infrastructure, local institutions, customs and material and non-material relations. The chapter interrogates the forms of local authority, mobilisation and local cooperation that evolved among 'strangers' resettled together. It explores how the new land rights conferred on settlers through permits (for A1) and leases (for A2) influenced the relationship between land beneficiaries and local government structures and assesses how local government was being established within these areas.

It is important to note from the outset that, although there is an emerging consensus on the social and economic outcomes of land reform, the discussion on social organisation has been peripheral and the patterns remain contested. Matondi et al (forthcoming) have been dismissive of the process of social organisation and, instead, valorise the chaotic moment of the land redistribution, suggesting that the process has led to the emergence of conflict ridden communities and also negatively affected farm production. Moyo et al (2009) argue that the resettled are organising themselves, even though they are socially differentiated, based on their common needs to defend the newly found rights and also to improve their capacities for utilising land. Scoones et al similarly observe: 'social networks, replicating those found in communal areas have emerged in various forms… and these include work parties, funeral assistance and religious based interactions' (2010: 207). They argue that, 'religion and church affiliation have emerged as a vital component in the construction of social relations and networks on the new resettlements' (ibid: 71). Recent field-based studies (Murisa 2007, 2009; Masuko 2009) have found that, even though beneficiary selection did not prioritise resettling people of the same lineage and clan group on the same former large-scale farm, the newly- resettled beneficiaries have, on their own, begun a process of establishing networks of cooperation that include structured local farmers' groups in the few years of being settled together. Furthermore, earlier studies of the first phase of land reform (see, for instance, Barr 2004; Dekker 2004) found out that communities of Newly Resettled Areas were characterised by an accelerated process of investing in new social relations by entering into association within local organisations.

Field evidence from Goromonzi and Zvimba (Murisa 2009) concurs with Moyo et al (2009) and Scoones et al (2010). The Newly Resettled Areas have, since 2000, been characterised by the simultaneous emergence of local platforms of cooperation which include informal networks and structured groups on the one hand and the introduction of local structures of authority, especially traditional village heads and the mutation of local land occupations' Committees of Seven into Village Development Committees (VIDCOs) on the other. Furthermore, there has been an implicit reform of local government processes, which elevated the authority of traditional structures. Post-independence local governance reform effort in Communal Areas sought to limit the authority of traditional authority structures (Tshuma 1997), through the introduction of representative structures such as the Village Development Committee (VIDCO) and Rural District Council (RDC), although such elected structures were rarely effective in addressing the various grievances of rural communities. Instead, hereditary structures of power continued to influence the allocation of critical resources such as land since their authority was perceived, at times, to be superior to that of elected functionaries.

Post- independence rural social organization was shaped by four important developments: (i) slow movement on land reforms and economic development, (ii) local government reforms, (iii) restructuring of representative farmers' unions and (iv) the emergence and proliferation of non-state development agents in the form of NGOs and locally established voluntary Community Based Organisations (CBOs).

Evolution of forms of social organisation in rural Zimbabwe

The discussion in this section briefly examines the evolution of social organisation from the period of colonisation up to independence. Historically, social relations of production in Zimbabwe, like those in most of Africa, were structured around belonging within a defined lineage grouping that ensured access to land and related benefits (Adholla 1962: 23). Changes in the structures of the political economy from around 1903 accelerated the integration of the majority black indigenous population into the wage economy. The structural determinants of the proletarianisation process in the Rhodesian context were: diminishing access to land, increasing taxation and the inroads made by a government-protected and competitive white commercial agricultural sector (Van Onselen 1976). There was a gradual increase in the number of workers from 10,000 in 1909 to about 320,000 in 2000 (see Chambati and Moyo 2003; Chambati, Chapter 5). De Janvry (1981) observes that, under such conditions, rural labourers are prevented from getting access to land as freeholders and from capturing their opportunity cost on labour markets so as to make them dependent on both sectors of the economy.

The labour process came to be characterised by an enduring contradiction between proletarianisation and a politically engineered functional dualism, by which petty commodity production in the Communal Areas and unwaged labour (especially female and child labour) would subsidise the social reproduction of male labour power on mines and farms (Moyo and Yeros 2005).

This contradiction produced neither a settled industrial proletariat nor a viable peasantry, but a workforce in motion, straddling communal lands, white farms and industrial workplaces (Moyo and Yeros 2005). It sustains the infant capitalist system of production in two ways (Burawoy 1980): (i) the two processes take place in geographically different places and, at the level of institutions of reproduction, the institutions of maintenance are different from those of renewal; (ii) in the case of the family, geographical separation of the two processes is reflected in a corresponding division of labour and internal differentiation of the family unit. Under capitalism, the binding of production and reproduction is achieved through economic necessity: for the labouring population, work is necessary for survival. In colonial Zimbabwe, the system of migrant labour was reinforced through pass laws restricting urban or mine compound residence for workers only and also the low wages paid to mine and farm workers always necessitated the need to supplement through subsistence farming. In essence, therefore, the rural areas were seen as labour reserves for mining and settler agriculture.

Furthermore, the same household was subject to a dual form of authority in the form of the traditional authority (chief and his/her hierarchy of officials) and the domestic/'civil' (defined by the 'Master and Servant Ordinance' of 1901) in the Large-Scale Commercial Farm sector. The social organization that emerged in the Communal Areas was, therefore, a result of both traditional norms and colonial social engineering. Through a number of racially- motivated laws, from the creation of Tribal Trust Lands (TTLs) to the enactment of the Land Husbandry Act (LHA) in 1952, there was a remaking of traditional authority. The LHA (GoR 1951) was not only aimed at 'modernizing' the smallholder sector, but it also proved to be a watershed moment for the office of the chief as the colonial authorities strengthened the authority of the chief with the intention of using it to establish firmer control over the 'natives'.

The slow redistribution of land after independence ensured that the semi-proletarianisation continued with peasants remaining as generally petty commodity producers firmly located in the generalized system of commodity production that is capitalism (Yeros 2002). The grievances of the semi-proletariat were divided into two: the rural (family) and urban (workplace) grievances. In the majority of cases, the family farm shrank due to periodic sub-divisions necessitated by the results of reproduction (sons normally inherit from their father's portion), consequently negatively affecting farm yields. At the workplace, the wages were characteristically low, necessitating reproduction in the countryside and the conditions of employment were poor.

Local government reforms

The GoZ's position on local government has evolved from being suspicious and dismissive of the role of traditional authority in local government to a very difficult accommodation. From 1982 until 1984, the government was, through the Prime Minister's directive on local government engaged in attempts to introduce Village Development Committees (VIDCOs) and Ward Development Committees (WARDCOs) as part of a broader attempt to modernize local government and also to limit the functions of traditional structures. Further complications in terms of local government reform emerged when the Ministry of Land and Agriculture shifted the land reform discourse towards 'communal area re-organisation'. In 1992, a ministry official was quoted saying '...resettlement alone can never fully solve all the problems of the communal areas...the implementation has to be carried out in tandem with the programme of communal area re-organisation' (quoted in Von Blackenburg 1994: 37).

Local government reform in the decade prior to 'Fast Track' seemingly entailed undoing some of the democratisation gains that had been made in the 1980s in terms of institutionalising civil authority within the countryside, mainly due to policy recommendations from the Land Tenure Commission. The 1994 Commission on Land Tenure under the chairmanship of Prof. Mandivamba Rukuni recommended the abolition of the Village and Ward Development Committees and, in their place, recommended the enhancing of the administrative power of the Traditional Authorities, including over land matters. The association of Rural District Councils opposed the move, arguing that '...our traditional leaders are not known for accountability...a system that goes even beyond the dark ages (Chikate 1995: 43).

Whilst in the previous dispensation prior to the RDC Act (1996), the chiefs had been regarded as ex-officio members of the council, the new legislation did not make any reference to traditional leaders. The RDC Act (1996) was silent, not only on the relationship with the chiefs, but also on the council's role in terms of the communal lands. The Traditional Leaders Act (TLA 1999), on the other hand, created the impression that the two institutions of local government could easily work together. In terms of land, the TLA stated that the chief would:

> ensure that land is allocated in accordance with the Communal Land Act 1982 (20:041) and to prevent any unauthorised settlement or use of any land; and to notify the Rural District Council of any intended disposal of a homestead and the

> permanent departure of any inhabitant from his area and, acting on the advice of the headman, to approve the settlement of any new settler in his area (GoZ 1999).

The Communal Lands Act (CLA-1982, amended in 2002) ascribed land authority to the RDC. It stated that a person may occupy and use communal land for agricultural or residential purposes with the consent of the RDC established for the area concerned (GoZ 1982: 3). It went on to state that, when granting consent, the RDC should 'consult and cooperate with the chief appointed to preside over the community concerned in terms of the TLA (1999)'. It created the impression that the RDC was the initial point of contact in granting authority over land, whilst actual practice in the Communal Areas suggested otherwise. The concept of *kuombera,* which has now become embedded in customary practices, meant that outsiders seeking land approached the chief or lineage elder with a gift which symbolised the request to be considered as part of the lineage or clan – thus deserving land.

Terms such as 'consult', 'cooperate' and 'notify' within the TLA (1999) and the CLA (1982, amended in 2002) created an impression of a harmonious existence between the two institutions and failed to appreciate the real contestations and competition between RDCs and traditional authority. The Act stipulated that the RDCs should:

> grant consent only to persons who, according to the customary law of the community that has traditionally occupied and used land in the area concerned, are regarded as forming part of such community (CLA 1982, amended in 2002).

However, the act did not specify how the RDCs would verify this complex issue of belonging, especially considering the fact that lineage and clan affiliations were determined by the elders of those particular groups. The foregoing discussion suggests that there were real challenges associated with combining elected and hereditary structures of local government. Notably, the history of land alienation was closely associated with local government reform, which, during colonialism, marginalised traditional authority only to elevate such authority in the 1960s in response to escalating demands for land through the nationalist movement (see Nyambara 2001). The enduring popularity of the office of the chief among people derived from the fact that, despite the overt attempts at cooption by the state and the lack of explicit means of coercion, the former maintained a form of independence and autonomy in articulating the interests of the subject communities. Studies by Ranger (1999) in Matopos District and Alexander (2006) in Insiza District suggest that the relationship between traditional leaders and the state, whether colonial or post-colonial, have always been more nuanced and complex than is captured by the notion of cooption (Fontein 2009: 5). These nuances around the popularity of the office and the possibility of cooptation may explain the recent evolution of local government in Newly Resettled Areas.

Representative farmers' unions

This sub-section examines the origins and roles played by the apex level national farmers' unions. Towards the late 1980s, the GoZ pursued the 'one sector one union' policy, which envisaged the merger of all representative unions within the agricultural sector despite their different material interests. The unions, instead, formed a loose umbrella committee known as the Joint President Agricultural Committee (JPAC) comprising the presidents of the three unions, namely the Commercial Farmers' Union (CFU), the Zimbabwe National Farmers' Union (ZNFU) and the National Farmers' Association of Zimbabwe (NFAZ), as a forum to discuss marketing, pricing and related issues (Bratton 1994: 24). Eventually the GoZ used the bait of levying authority to coerce the NFAZ and ZNFU to merge into the Zimbabwe Farmers' Union (ZFU).[1] In 1992, some black commercial farmers formed the Indigenous Commercial Farmers' Union (ICFU), to represent them, independently of large white farmers and small black farmers, but it was only recognized by the GoZ in 1995.

Despite the proliferation and consolidation of these representative unions, demands for land reform remained poorly represented in formal pluralist policy lobbies (Moyo 1999: 15). The ZFU did not recognize the differentiation within the smallholder sector, where only 10 per cent could be described as capable, 50 per cent were poor (land-short, cattle-less and experiencing food deficit), whilst the remaining 40 per cent could barely break even from farming (ibid: 16). The ZFU leadership resisted identifying the different socioeconomic groups within their structures and potential membership, despite the fact that the union's structures of participation were designed according to landholding size: Small-Scale Commercial plot holders, indigenous large-scale and communal and resettlement area farmers (ZFU undated: 3). The leadership insisted that all its farmers had common interests with regard to agricultural issues (Bratton 1994: 27). However, the union advocated for land to be redistributed to the 'capable', instead of the 'needy'. Consequently, the demand for land during the 1990s could not be gauged on the basis of formal ZFU advocacy, but rather it was reflected in the informal land occupations and natural resource poaching (grazing, grass, wood and water), as well as localized demands for land redistribution through chiefs, parliamentarians, some NGOs and government staff (Moyo 1999: 19).

Development agents and rural social organisation

Rural social organisation in Zimbabwe has, in recent years, come under the modernising influence of external agents such as NGOs, church organisations and political parties, which seek to 'speak' for the rural poor (Moyo and Yeros 2005a: 41).[2] Although the NGOs are a relatively recent rural organisational form, compared to other social arrangements such as religious institutions, political movements, governments and transnational networks of various kinds (Bebbington et al 2008: 6), they have become a ubiquitous feature of development interventions in Africa (Moyo et al 2000a: ix; Helliker 2008: 240).

While Alexander (2006) has argued that it was not the chief running the countryside in the colonial era, but a wide range of state officials, one can also argue that, in the post-independence period, it was not only the chief and the newly established local structures in charge of the countryside, but also a variety of other non-state organisations such as farmers' unions, churches and NGOs engaged in various livelihood improving projects. In the process, NGOs facilitated the development of new social relations of production.

As of 2005, Zimbabwe had over 1,000 formal NGOs, which included local and national level organisations (Moyo 2005: 45) and which varied in size from those with over 100,000 members to smaller ones with 10 to 100 members or households. Some were membership-based and others positioned as vehicles of innovative interventions in rural development. In the 1990s, their presence through a variety of community development projects was pervasive in the Communal Areas. One community in Mhezi Ward (Chiduku District) was dealing with at least 15 NGOs operating in one ward alone (Moyo 1995: 43), while Makumbe (1996) notes that there were over 7 local and international NGOs in one ward (Makumbe 1996: 75). In this process, NGOs became intermediaries between donors and local communities, providing services such as project formulation, execution, training and consultancy. Historically, rural development NGOs in Zimbabwe have responded to four interrelated challenges affecting rural communities: (i) declining land quality as a result of continuous use and soil erosion, (ii) declining agricultural yields, (iii) inadequate farm-based incomes and (iv) inadequate social service provision (Helliker 2006; Murisa 2009). The NGOs designed a variety of interventions, some of which were influenced by Integrated Rural Development Programme (IRDP) philosophies. However, despite their increase in numbers and budgets, NGOs have had a very limited impact on rural development, mostly due to the fact that they have avoided the harder questions in development (such as skewed land ownership patterns) and so most of their interventions have not had a lasting impact beyond the lifetime of the project.

Nature and context of economic and social grievances since the FTLRP

Following the FTLRP, rural agency has been concerned with a variety of social and economic constraints, in addition to local democratic practise. Debates on the impacts of the FTLRP have tended to focus on the disruption of Zimbabwe's economy, including the decline in agricultural production, without noticing the recovery in the output of various crops (Moyo et al 2009; Jowah 2009; Chambati, Chapter 5). Rather than dwell on these productionist perspectives, this paper seeks to examine the challenges faced by the newly resettled farmers in the wider context of the broader effects of macro-economic declines on rural livelihoods.

The Zimbabwean economy faced a debilitating economic crisis from 1997, partly due to the negative effect of the economic reform programmes adopted in 1990 (Moyo 2001; Yeros 2002; Moyo and Yeros 2005; Murisa 2009), although others have argued that the crisis resulted from economic 'mismanagement' and the manner in which the Fast Track Land Reform was implemented (Richardson 2005). The economic decline worsened following the Land Reform Programme, weak macro-economy management frameworks, frequent droughts and an unfavourable external policy environment, including the impact of international isolation (World Bank 2006). Economic decline directly impacted rural livelihoods and potential for agricultural recovery, although it is accepted that land reforms generally lead to a transitory production decline (see Moyo and Yeros 2007a).

The social dimensions of the crisis were characterised by the decline of social service delivery in housing, health and education and the erosion of household incomes, as well as food insecurity and chronic vulnerability during drought years. The user fees at health centres introduced during the 1990s strained most rural households, while the capacity of health centres to effectively service communities was severely eroded, as evident in widespread shortages of essential medical supplies and qualified personnel (Human Rights Watch 2003), following the emigration of numerous health and education professionals.[3] Furthermore, in 2002 the Central Statistical Office (CSO) estimated that 70 per cent of the population was living below the poverty datum line, given the considerable increase in the price of food and other consumables due to inflation. The plight of the poor was exacerbated by a substantial shortfall in maize production. The GoZ estimated that in 2005 approximately 36 per cent (2.9 million) of the rural population would require food relief (Zimbabwe Vulnerability Assessment Committee [ZIMVAC][4]).

The cyclical linkages of support and cooperation that have traditionally existed between the rural and urban households that involved cash remittances from urban to rural and grains from the rural to the urban sector came under severe threat. Due to increased unemployment rates, urban to rural remittances for the purchase of essential inputs and working capital were reduced (Chigumira and Matshe 2004). A country that, in the late 1980s, was close to food self-sufficiency and closest to achieving the goal of health for all, now faced the reversal of these post-independence gains (World Bank 2006).

A variety of other social constraints affected land beneficiaries in the Newly Resettled Areas. The most apparent of which include the unavailability of suitable water for domestic use and lack of sanitation facilities, inadequate health and education facilities and general poor planning for any investment in social infrastructure. The health and schooling facilities prior to 'Fast Track' were barely sufficient to cover farm worker households: only 10.5 per cent of the households had access to clinics and only 12.9 per cent of children under the age of six were benefitting from early childhood education and care programmes (FCTZ 2001a & b). The increase in terms of population in formerly under-populated LSCF areas put a strain on the pre-existing social infrastructures and the situation was worsened by the vandalism that led to many resettled families using untreated and unsafe water from nearby rivers and dams (*The Standard,* 15 December 2002). Only 34.9 per cent of farm worker households had toilets of their own (ibid). The average distance to the nearest primary school was 14.3 kilometres. Table 7.1 below shows, in more detail, the nature of the social grievances affecting newly resettled households. Approximately 50 per cent of the respondents identified consultation fees at the local clinics as too high and felt that this discouraged the habit of seeking treatment at an early stage, whilst 16 per cent complained about the unavailability of essential drugs in the clinics. In terms of education, the respondents identified the challenges as unaffordable school fees (47.6%), unavailability of essential books (21.7%) and inadequate staffing levels (20.6%).

Table 7.1: Social grievances

Social grievances	Chipinge		Goromonzi		Zvimba		Mangwe		Kwekwe		Chiredzi		Total	
	No.	%	No.	%	No.	%		%	No.	%	No.	%	No.	%
Exorbitant consultation charges at clinic	17	39.5	28	51.9	43	55.1	4	57.1	10	41.7	18	42.9	120	48.4
Inadequate availability of drugs	6	13.6	9	16.4	16	21.1	2	14.3	6	22.2	3	6.5	42	16.0
Shortage of skilled personnel in the clinics	7	17.5	17	29.8	20	29.4	3	23.1	8	27.6	17	38.6	72	28.7
Exorbitant school fees	11	22.9	30	44.8	34	47.2	6	40.0	13	35.1	42	89.4	136	47.6
Lack of essential text books	4	8.3	19	30.6	3	4.5	2	13.3	19	51.4	13	27.7	60	21.7
Shortage of skilled personnel in the schools	2	4.2	17	26.6	9	13.4	6	40.0	13	35.1	10	21.7	57	20.6

The local state and competing local authorities in Newly Resettled Areas

Rural local government in Zimbabwe is composed of a combination of traditional and elected local authorities who do not necessarily complement each other, but are engaged in ongoing contestations for power and influence (see Murisa 2009). These conflicts have led to poor service delivery and have contributed to the proliferation of other platforms that organise local communities for improved social reproduction capacity, such as NGOs, churches and local groups. These non-state actors have gained legitimacy, even though they do not explicitly possess instruments of coercion, unlike the official agents of local government. The local state is thus a sum total of the different public spaces in which households interact with the aim of extracting specific social gains such as food security and improved health status.

The actual organisation of the process of land occupations was under the leadership of a new cadre of popular leadership. At the peak of the land occupations, the leadership, mostly war veterans, restricted traditional leaders to subordinate advisory roles and they invited them to identify ancestral lands and also to lead cleansing ceremonies after an occupation (Sadomba 2008a: 98). In some instances, however, such as in Svosve Communal Lands in the Marondera District neighbouring Goromonzi, traditional leaders were instrumental in mobilising subject communities to occupy lands on the basis of restitution claims (Moyo and Yeros 2005b: 187).[5]

The extension officer is the most visible local government agent on the ground, since the GoZ has resettled extension officers among the A1 land beneficiaries as part of efforts to increase farm production. They mostly reside in the former farm owner's house or the manager's quarters.[6]

The second most easily recognised actor of local government is the Ward Councillor, a locally- elected official of the RDC, responsible for establishing and chairing of the Ward Development Committee (WARDCO), which reviews development plans from the VIDCOs and for integrating them into a ward development plan for onward submission to the RDC (GoZ, Rural District Councils Act, 1996: 460). However, ward development activities have been limited by financial constraints. The ward councillors' popularity does not necessarily derive from their official roles as councillors, but rather from their roles as political functionaries at the forefront of mobilising the newly- resettled into political party activities.[7]

Competing sites of local government in the Newly Resettled Areas

Traditional authority at the village level

At the height of the land occupations, 'seven member committees' were established to facilitate the selection of land beneficiaries and to provide them with administrative support at the local level. Chaumba et al (2003: 10) describe these new village authority structures as 'a sudden emergence from seemingly nowhere', but in reality they were a slightly different version of the defunct VIDCOs that had been established in the first year of independence through the Prime Minister's directive of 1984. The seven member committees were an innovative, integrated, top-down system of governance in the new resettlements which were as 'striking as the dramatic physical transformation of the landscape' (Chaumba et al 2003: 11). They were characterised by a hierarchical committee-based structure with parallels to the decentralised local government development committee structures of the 1980s. Among other tasks, they were established at every occupied farm to ensure that land occupiers were not evicted by previous owners.

Various authors such as Moyo (2001), Moyo and Yeros (2005, 2007), Chaumba et al (2003) and Sadomba (2008a, 2008b) have disputed the chaos focused theory regarding the execution of the Fast Track Land Reform Programme and have (using field evidence) demonstrated that structures and procedures (albeit rudimentary) were established in land and beneficiary identification and allocation of plots. Smaller and localised administrative units capable of making decisions within a short amount of time were established in most of the locales (Moyo and Yeros 2007b). At the district level, the new land committees included local government ministry officials, traditional leaders, the ruling party, security organs and war veterans. Locally, the role and place of traditional leaders varied, in some instances, they were called in to lead the land identification process based on historical claims and, in others, they were asked to legitimise occupations and also to 'bless' the occupation (Chaumba et al 2003: 21). In certain instances, traditional leaders competed among themselves in defence of their territorial boundaries and also competed with local government structures and outsiders in order to settle their 'subjects' (Moyo and Yeros 2007b).

Furthermore, Chaumba et al (2003) noted the visible leadership role of war veterans and the replication of an 'army barrack' like form of organisation in which curfews were established and visitors had to report to the base commander. Initially, the Committees of Seven were dominated by war veterans, who, in many instances, occupied the post of Base Commander (Chairperson) and Head of Security (Chaumba et al 2003: 8; Masuko 2009: 5). Traditional authority functionaries such as chiefs and village heads were rarely part of the structures, but were consulted on some matters which included traditional cleansing ceremonies and beneficiary selection. The activities of the committee, in particular the pegging of plots, stand in stark contrast with the depiction of 'chaos' on the farms. Chaumba et al (2003: 17) state that, 'at the same time as they were riding roughshod over the rule of law, war veterans and other land occupiers employed the tools and practices of colonial land use planning to becoming visible and legitimate'.

The Committees of Seven were also responsible for ensuring that farm production commenced as soon as land beneficiaries had been allocated their individual plots (Sadomba 2008a: 115). However production on individual plots was not automatic: some well-endowed individuals managed to utilise the land in the first year of occupation, while others had to wait for state subsidies. Some of the Committees of Seven were instrumental in the institution of joint farm production when they demarcated plots for collective agriculture and provided farm inputs in the tradition of the *Zunde raMambo*[8] (Sadomba 2008a: 115). Beyond collective work on the *Zunde* fields, the occasion provided a moment of building solidarity among the land occupiers and an opportunity for the chief to legitimise the land occupations on the basis of a historical link with the occupied lands. Field observations in Glendale, Goromonzi and Zvimba indicate that the Committees of Seven are still thriving on many of the resettled farms, although their composition and names have changed. They are now mostly headed by village heads and are referred to as Village Development Committees.

In 2003, the GoZ issued a directive on local government which stated that, 'in terms of the Traditional Leaders Act (Chapter 20: 17), all resettlement areas shall be placed under the relevant traditional chiefs or headmen' (GoZ 2003: 4). The Traditional Leaders Act (TLA) empowered the chief to nominate village heads for appointment. The chiefs have, since the 2003 government directive on local government, been appointing village heads in the Newly Resettled Areas from among the land beneficiaries. The criteria and manner of appointment varies. Chief Rusike in Goromonzi has been combining arbitrary appointments with elections in areas where he is not familiar with the land beneficiaries and, in certain instances, appointments have been made after consultations with Ward Councillors.[9] The appointments without elections have been more common in areas contiguous to customary tenure areas, where the chief has appointed those related or belonging to a lineage group with a history of holding such office. In areas previously dominated by large-scale farms, such as in Bromley, the land beneficiaries have been asked to elect their village head, while Chief Matibiri of Zvimba mostly appointed land beneficiaries with whom he is familiar to positions of village head in consultation with extension officers. Most of the village heads appointed in the Banket area are from the neighbouring Zvimba customary lands.

The village heads chair the new Village Development Committees. Members of the VIDCO are directly elected into office by the members of the village. Within the new VIDCOs, a new post of war veteran representative was created and is reserved for one of the war veterans resettled on the farm. Other posts in the village council include officers responsible for village development, security, women's affairs, health and the youth.[10]

The village head has administrative oversight over the village and works with the VIDCO. The responsibilities of the new village authority include administrative functions such as developing mechanisms for sharing inherited infrastructure like dams, irrigation equipment and tobacco barns, conservation of natural resources and enforcing traditional norms and government directives on infrastructure use.[11] It seems, however, that most of the village heads are more focused on satisfying the demands of the local council, rather than those of the land beneficiaries. The four village heads interviewed listed their duties as ensuring that there is no unlawful cutting down of trees; ensuring that beneficiaries utilise received inputs instead of selling them; and, ensuring that there is no subletting of the plots. While the roles that were emphasised by the village heads are an essential part of local government, they are more oriented towards ensuring that government directives on agrarian reform are implemented at the local level. The village heads did not necessarily balance government's interest against the interests of the village members, which include a need for wood to build their homes and storehouses and being allowed to sublet some of their land in exchange for inputs and tillage support.[12] The new cadre of village leadership is different from the populist Committees of Seven which were more focused on the security of the land beneficiaries in the face of a real threat of counter eviction from the former owners. The village heads seem to be more content with policing the areas on behalf of the government.

Expansion of RDC frontiers and roles

The same GoZ directive on local government that empowered the chiefs to expand into Newly Resettled Areas also stated that, 'all resettlement areas shall fall under the jurisdiction of Rural District Councils and shall thus be incorporated into either existing wards, or new wards shall be created as necessary' (GoZ 2003). This policy vision was made official government policy in 2005 with the passing of the 17th Amendment to the Constitution. These local government reforms replicate the forms of local government in the Communal Areas, where elected ward councillors have to share political space with traditional authority functionaries within an unclear framework of local government. The clear delineation of roles, especially the responsibility over land allocation and adjudication between these two functionaries of local government in the customary areas, has been difficult in practice (Murisa 2009). Chiefs and lineage elites have historically disregarded the provisions of the Communal Land Act (1982, amended in 2002) in land allocation and in presiding over land conflicts. Anderson (1999) and Dzingirai (1994) have also shown how customary area dwellers continued to defer to chiefs, despite the envisioned consultations between the two institutions (RDCs and chiefs) within the legislation.

While traditional authority structures have been quickly introduced at the village level, the RDCs do not currently have similar levels of representation. Rather, the RDCs are dependent on village heads to enforce their policies. The current Village Development Committees, consisting of elected officials, operate as subordinate structures of the village head, unlike in customary areas where they were a parallel structure of local government made up of popular party functionaries elected by the villagers. Meetings of the VIDCOs in the Newly Resettled Areas are chaired by the village head, except for rare cases such as at Whynhill Farm in Zvimba, where they are chaired by the VIDCO chairperson (Murisa 2009).

Although the RDCs have slowly regained their authority over the informal structures established during the period of land occupations and are empowered to work with VIDCOs and Ward Development Committees (WARDCOs) to formulate area development plans, the coordination has not yet really taken off. According to the RDC Act (1996), WARDCOs are supposed to be chaired by the council's Ward Councillor, but, in reality, as in the customary areas, this structure does not exist. The councillors do not have any budgets to convene meetings and have made little progress in terms of establishing functioning WARDCOs. Traditional authority structures have not yet introduced the office of the *Sadhunu* (ward head) in these areas. In the customary areas, the *Sadhunu* works with a number of village heads operating within his *dhunu* (ward), but this office has been largely ineffective because of the arbitrariness of the establishment of wards, to the extent that in many circumstances the chiefs preferred to deal directly with the village head, also known as the Sabhuku.[13]

Local authorities such as the Goromonzi Rural District Council have developed a five-year strategic plan for infrastructure development without any significant input from the decentralised structures (VIDCOs). Village heads and members of the VIDCO in Goromonzi were not aware of the five year strategic plan that had been developed by the Goromonzi RDC. During interviews with officials of the Goromonzi Rural District Council, they revealed that the plan had been developed on the basis of perceived needs and in consultation with extension officers.[14] It is important to note that the district extension office is housed within the RDC's office, hence such consultation would have been more convenient for the council than getting their hands dirty by going to the farms. The RDC officials stated that they had not started implementing the plan because they were still waiting for the funds from central government. Although the lease document for the A2 scheme has a clause on land rentals, RDC personnel were not sure if they would be the ones to collect and use the rentals for local area development.[15] At the end of 2008, there were very few A2 plot holders (less than 6 per cent of the total number of beneficiaries) who had signed the lease agreements due to problems surrounding the legal registration of the lease (see Moyo 2007; Murisa 2009). Generally, local government authorities have been affected negatively by the hyperinflationary environment which characterised Zimbabwe from 2005 until the end of 2008, to an extent that a number of infrastructural development projects had to be suspended.[16]

The extension officer and local power relations in Fast Track areas

In the aftermath of the Fast Track reforms, the Ministry of Agriculture announced in 2004 that it would need 6,000 new extension officers in addition to the approximately 3,000 extension officers already in service (Mlambo 2006: 7). By the end of 2008, approximately 2,900 extension officers[17] had been recruited.[18] Besides the lack of personnel, the department faced numerous challenges, including, but not limited to, 'increasing budgetary constraints, poor remuneration and conditions of service and lack of transport and equipment and the fact that extension officers are expected to provide services over too wide an area' (Mlambo 2006: 8). As part of measures to address these challenges, the Ministry of Agriculture took the decision that extension officers involved in field demonstrations should be allocated A1 plots in the areas they cover and that they should olso be allocated houses on the former large-scale farms. However, the decision was made late and some of the farm-houses in A1 settlements had already been converted into social amenities such as schools or clinics or, in some instances, occupied by the Commanders of the Committee of Seven.[19] The majority of extension officers in Goromonzi are settled on A1 farms, but in Zvimba (especially in the Banket area) very few got accommodation on the farms.

During the period of Fast Track resettlement, extension officers worked with officials from the Ministry of Lands and the Surveyor General's office in the official demarcation of the new plots. The roles of the extension officers in the aftermath of land allocations included training on improved farming methods, assisting the newly resettled farmers in obtaining necessary farm inputs and monitoring the proper usage of received inputs on behalf of the government.[20] They were responsible for the relaying of information on crop prices and other changes to the marketing of crops and livestock.

Current extension support methods in the Newly Resettled Areas remain limited by the fact that there is not sufficient knowledge of the actual training needs and land use preferences of the newly resettled beneficiaries. The newly resettled households are made up of people from different socio-economic and professional backgrounds, literacy levels, skills and resource endowments.

The Department of Extension has not yet undertaken proper research on the specific needs of these communities. The majority of the extension officers still use top-down methods based on the transfer of knowledge, methods which have been challenged on the basis that they do not adequately consider indigenous knowledge (Mlambo 2006: 8). The methods preferred by extension officers, such as securing treated hybrid seeds and the use of inorganic fertilisers, increase the farmers' integration into and dependence on agricultural commodity supply markets.

Despite these shortcomings, locally-based extension officers have been at the forefront of introducing innovations in social organisation that are aimed at enhancing farm production such as the establishment of structured local farmer groups. In Wards 21 and 22 of Goromonzi (Bromley area), the extension officer has aggressively promoted the establishment of local farmers' groups (see Murisa 2009). The extension officers carry out their extension work within these groups and they facilitate the acquisition of farm inputs from the GoZ.

Changing land relations and the local state

Fast Track has not only transformed the physical agrarian landscape, but has also substantially changed agricultural land relations by extending state land ownership to the bulk of Zimbabwe's land and expanding new forms of landholding: leasehold and permissory forms of tenure (Moyo 2007: 8; Moyo, Chapter 2). Most discussions of FTLRP land tenure have been dominated by concerns about the use of land to access credit, rather than about the tenure security of the A1 permit and the broader implications on social organisation and farm production. Evidence from the field, however, shows that cases of eviction of resettled farmers have been very limited, with only 1.1 per cent cases of evictions reported (Moyo et al 2009).

According to officials from the Ministry of Lands (MoL) in Goromonzi and Zvimba, there have been cases of removals of resettled beneficiaries in both districts to suit the re-organisation of land reform models. The officials claim that most of the displaced were re-allocated land on other former large-scale farms within the area. However, Sadomba (2008a) argues that the onset of Fast Track was associated with a clandestine strategy to remove war veterans who had allocated themselves land during the earlier period of *Jambanja.* The actual figures of the number of war veterans who were affected are not readily available, but the practice seems to have been more prevalent in Mazowe, where more than four former large-scale farms that had initially been converted into A1 plots were later re-converted into A2 plots (Masuko 2009: 7).

However, there are nuanced differences between the land rights being bestowed in the Newly Resettled Areas and those existing in customary tenure areas. In the customary areas, land access and use rights are based on belonging and membership of a lineage group. Traditional institutions and authority are prominent in the distribution and administration of customary lands. Whereas traditional leaders have been at the forefront of land administration and allocation in customary areas, the A1 permit elevates the role of the state as the initial provider of land and is responsible for administrative oversight. The permit is silent on the role of traditional authority in the new landscape, despite the fact that chiefs were mobilised by the state to verify applicants for A1 farms.

Evidence from the field shows that village heads have appropriated for themselves the role of land allocation. The village head at Dunstan Farm in Goromonzi offered A1 plots to four households that had been excluded during the official demarcations.[21] At Whynhill Farm, the village head was removed from office by the chief and officials from the Ministry of Lands when he sold A1 plots that were vacant after the official beneficiaries had not taken the land.

It is important to note that land tenure is a social construct that is influenced by socialisation rather than legal contracts. While the legal contract through the permit seeks to promote a certain form of social relation based on individual rights and a direct relationship with the state, the new framework is yet to be adequately explained to the new beneficiaries who are mostly used to customary tenure. On the other hand, the introduction of traditional authority without the necessary rule book of what the village heads can and cannot do has contributed to misunderstandings concerning their powers in terms of governing the land. Their understanding of land tenure provisions is currently based on sketchy information from local leaders and government officials, to the extent that some of the beneficiaries interviewed believed that the chiefs have the authority to allocate land or to move them off the land. Some of the respondents indicated that they see no difference between the rights they have in the Newly Resettled Areas and those of their colleagues in the customary areas.[22] The situation is compounded by the fact that the majority of the beneficiaries are from customary areas, where a strong relationship between traditional structures and land ownership exists. The A1 beneficiaries in Goromonzi and Zvimba revealed that they expect traditional leaders to ensure that land and natural resources are being managed properly. Notably, most of the respondents did not see a problem with the expansion of traditional authority into the Fast Track resettlement areas.[23]

Despite the previous attempts to marginalise the office of the chief by both the colonial and post-colonial governments, it remains as one of the most enduring institutions in Zimbabwe. The authority of the chiefs derives from links with ancestors and certain beliefs in the protecting powers bestowed upon the chief. A similar connection between traditional authority and land was made by one of the chiefs when he said that 'the president does not own the land. The land belongs to the chiefs. The white settlers took the land from the chiefs and not the president' (Chief Charumbira quoted in *The Herald*, 3 December 2000).

Cooperation amongst resettled households

The manner in which individual households cooperate within their immediate neighbouring communities as part of a strategy to strengthen social reproduction capacities entails a bundle of strategies to overcome farm production and social constraints.

Maintenance of relationships with Communal Areas

The majority of the Fast Track Land Reform Programme beneficiaries are men, aged between 36 and 46 years, mostly from Communal Areas, with some form of secondary education (Moyo et al 2009). At independence, the oldest in this category of beneficiaries would have been in their early twenties and probably living with their parents. Given the dominant form of customary area organisation, some of the members of this demographic group could have left behind some small plots of land and even homes in the Communal Areas. Evidence from the field shows that social and economic interactions between the newly resettled land beneficiaries and their counterparts in the customary areas is ongoing, albeit at varying levels. There exist links between customary areas and land beneficiaries who used to live in customary tenure areas. Approximately 15 per cent still maintain homes there, for various reasons (see Table 7.2 below). The most (57.5%) commonly cited reason for the maintenance of a customary area homestead is because it is still home to other members of the extended family. There were very few instances of lineage groups being resettled together. During the survey, we only noted one A1 settlement on what used to be Dalkeith Farm in Zvimba District that was composed of land beneficiaries from the same clan. In the Shona and Ndebele cultural context, the concept of 'family' has a broad meaning, including what has generally been called the 'extended family'. In certain cases, these customary area homes are the location of gravesites for lineage members and, according to Shona custom, abandoning such homes would be seen as turning away from one's people (Bourdillon 1982: 3).

Table 7.2: Reasons for maintaining a Customary Area (CA) home

Reason for maintaining CA home	Chipinge		Chiredzi		Goro-monzi		Kwekwe		Mangwe		Zvimba		Total	
	No.	%	No.	%	No.	%	No.	%	No.	%	No.	%	No.	%
To boost production	2	4.3	4	13.8	27	22.3	2	5.1	1	5.9	15	27.3	51	16.6
To reduce risk of crop failure	-	-	1	3.4	4	3.3	-	-	1	5.9	1	1.8	7	2.3
In case of eviction	-	-	7	24.1	8	6.6	3	7.7	2	11.8	7	12.7	27	8.8
Sentimental values	5	10.6	7	24.1	9	7.4	9	23.1	5	29.4	6	10.9	41	13.3
Home part of the extended family	39	83.0	10	34.5	73	60.3	25	64.1	8	47.1	22	40.0	177	57.5
Children attending a local school	-	-	-	-	-	-	-	-	-	-	4	7.3	4	1.3
Business area	1	2.1	-	-	-	-	-	-	-	-	-	-	1	0.3
Total	**47**	**100**	**29**	**100**	**121**	**100**	**39**	**100**	**17**	**100**	**55**	**100**	**308**	**100**

Source: AIAS Household Baseline Survey, 2007, Household Questionnaire, N=2089

Furthermore, the movement of people from one locality to new areas that potentially provide more land for grazing and cropping, but still maintain the old Communal Area home for the rest of the family, has been common ever since the opening up of formerly largely wildlife areas, such as Gokwe in the 1970s and 1980s (Nyambara 2001: 773). Among those households maintaining customary area homes, 16.6 per cent were doing so as an attempt to boost production. There were more A1 farmers retaining use of Communal Area homes for production boosting purposes and this suggests that a category of A1 farmers have the capacity to utilise more land than they were allocated. Few land beneficiary households (1.29%) maintained a Communal Area home as a safety measure against eviction, providing a clue to the perception of security of tenure and suggesting that beneficiaries had an optimistic perception of the prevailing land relations, unlike the claims of insecurity of permit tenure.

These Newly Resettled Areas were not necessarily insular as they were made up of people who associated in various ways with the outside communities. The nature and form of association with neighbouring customary areas was defined by various factors, but the most important seemed to be physical proximity. The new communities associated with the customary areas to fulfil both economic and social needs. The activities ranged from utilising labour to establishing relations through marriage (see Table 7.3).

Historically, customary areas have served as a reservoir of cheap labour for the large-scale farms, mines and the urban formal sector. The tendency to recruit labour from customary areas is still prevalent, as 30.5 per cent of the beneficiaries were engaged in the practice as of 2006 (see Chambati 2009; Chambati, Chapter 5). Respondents revealed in 2007 that they usually go back to their area customary to recruit labour, especially during periods when a huge amount of manual labour is required. This practice suggests the initial signs of material differentiation between households within the NRAs and those in the customary areas.

Table 7.3: Cooperation between Customary (CA) and Newly Resettled Areas (NRAs)

Type of linkage	A1		A2		Total	
	No.	%	No.	%	No.	%
Farmers getting labour from CA	782	30.5	213	30.7	995	30.5
Farmers utilizing productive resources from CA	600	23.4	153	22.0	753	23.1
Farmers sourcing inputs from agro-dealer in CA	459	17.9	147	21.2	606	18.6
Students enrolled in schools in neighbouring CA	472	18.4	119	17.1	591	18.1
Access to health facilities in neighbouring CA	240	9.4	54	7.8	294	9.0
Farmers get seeds from CA farmers	11	0.4	3	0.4	14	0.4
Markets	1	-	2	0.3	3	0.1
Marriage	-	-	3	0.4	3	0.1
Total	**2565**	**100.0**	**694**	**100.0**	**3259**	**100.0**

Source: AIAS Household Baseline Survey, 2007, Household questionnaire, N=3 259

Approximately 33.1 per cent of the resettled households hired draught power and other productive resources from customary area farmers. Although the GoZ had been running a tillage programme through the District Development Fund (DDF) in most of the resettled areas, the service was stretched in terms of its capacity and was, at times, abused by the politically connected land beneficiaries within the A2.[24] There were slightly more A1 farmers (23.4%) in the six districts relying on productive assets from the customary areas than A2 farmers (22%).

These findings conform to the analysis of asset ownership among the newly resettled which indicated that the majority (95%) of A1 households only had full access to hand tools and that very few households (within less than 10%) had access to animal-drawn and power-driven implements such as planters, ridgers and trailers (Moyo et al 2009: 73). In many instances, customary area farmers who hired out their draught power also hired out their labour to operate the draught animals and the ploughs.[25]

There were also cases of land beneficiaries acquiring inputs and other productive resources from customary areas. Since the Fast Track reforms, agro-dealer activities declined and did not expand into the Newly Resettled Areas. While the former large-scale owners did not necessarily need local agro-dealers because they had capacity to move inputs from the neighbouring towns, the new breed of farm owners did not possess similar capacities. The GoZ revived the Grain Marketing Board (GMB) from being the buyer of last resort, to being the only supplier of inputs and buyer of controlled commodities such as maize and wheat (Govere 2006: 9). The price controls imposed on controlled commodities had a negative effect on the viability of agro-dealership networks that had contributed to the farm-input delivery. The GoZ's decision to supply A1 and A2 farmers with subsidised inputs dampened private sector participation in the inputs market. Approximately 18.6 per cent of the newly resettled have secured farm inputs from agro-dealer channels and informal markets within customary areas.

Due to the failure on the part of the government and other social service delivery agents, there is still a significant dependence on education and healthcare facilities in customary areas by the recently resettled A1 and A2 households. A total of 591 (18.1%) households have children attending schools in customary areas. There are 294 households (9%) utilising healthcare facilities based in customary areas. Nevertheless, investigations through key informant interviews revealed that resettled communities preferred to go to health centres in the satellite towns closer to their areas. In Goromonzi, they preferred to go to the district hospital at Goromonzi centre and some go to Harare.[26] In Zvimba, most of the resettled go to Chinhoyi, where there is a district hospital and a clinic. However, due to exorbitant commuter fares, they cannot afford to send their children to urban areas for schooling.

Networks of cooperation within the Newly Resettled Areas

Rural social organisation entails a complex array of social structures involving a number of interrelated associational forms of cooperation. Some of these are visible, whilst others are underground, within the context of wider local and central state authority structures and broader civil society structures (see Moyo 2002). The more underground forms tend to be more organically embedded in the every day practices of communities and mostly take on the form of solidarity in terms of defending livelihoods, territory, cultural spaces and strengthening one another in a crisis moment such as death. The motivation for cooperation ranges from various social to economic imperatives, largely at a very local level and rarely does cooperation and solidarity entail distant communities within the Newly Resettled or Customary Areas (Moyo et al 2009). Eight common areas of cooperation were observed (see Table 7.4).

Table 7.4: Forms of cooperation among farmers on newly resettled large-scale farms

Areas of cooperation	A1		A2		No. of Households (HH) in parentheses						
	No.	% of HH	No.	% of HH	Chipinge	Chiredzi	Goromonzi	Kwekwe	Mangwe	Zvimba	Total
sharing of productive infrastructure	887	53.7	205	46.8	168 (50.3)	141 (60.0)	349 (50.2)	234 (62.9)	104 (71.7)	96 (31.2)	1092 (52.3)
sharing of non-productive infrastructure	567	34.3	163	37.2	99 (29.6)	98 (41.7)	218 (31.4)	177 (47.6)	21 (14.5)	117 (38.0)	730 (34.9)
reciprocal hiring arrangements	568	34.4	164	37.4	166 (49.7)	43 (18.3)	252 (36.3)	110 (29.6)	42 (29.0)	119 (38.6)	732 (35.0)
reciprocal labour sharing	328	19.9	101	23.1	88 (26.3)	43 (18.3)	117 (16.8)	29 (7.8)	54 (37.2)	98 (31.8)	429 (20.5)
combined farming operations	144	8.7	28	6.4	17 (5.1)	8 (3.4)	63 (9.1)	5 (1.3)	2 (1.4)	77 (25.0)	172 (8.2)
membership in common agric and social groups	129	7.8	56	12.8	41 (12.3)	58 (24.7)	29 (4.2)	9 (2.4)	10 (6.9)	38 (12.3)	185 (8.9)
sharing of advice and information	419	25.4	152	34.7	162 (48.5)	93 (39.6)	193 (27.8)	28 (7.5)	45 (31.0)	50 (16.2)	571 (27.3)
sharing of seed and planting materials	101	6.1	28	6.4	44 (13.2)	14 (6.0)	43 (6.2)	4 (1.1)	7 (4.8)	17 (5.5)	129 (6.2)

Source: AIAS Baseline Survey, 2007

Joint use of productive infrastructure

The land redistribution process was implemented in such a way that it brought together strangers from different backgrounds, such as Communal Areas, urban areas and former Large-Scale Commercial Farms, to settle on previously large-scale farms that had been subdivided into smaller units for the land beneficiaries (Murisa 2009; Murisa 2011: 1146). Most of these previously large farms had immovable productive assets such as tobacco barns, dip-tanks, cattle- handling facilities and irrigation equipment and non-productive or social assets such as farmhouses and farm worker compounds (GoZ 2001: 3) that could not be utilised effectively by a single household resettled on 6 hectares of land. As part of its efforts to rationalise access to such equipment, the GoZ issued a directive that all A1 households should share the productive and social infrastructure left behind by the previous owner, but did not clarify how this would be done (GoZ 2001: 2). The farm divisions in A1 areas created common grazing lands which had to be utilised by the resettled beneficiaries on a particular former Large-Scale Commercial Farm, thereby suggesting another need for mechanisms of engagement and cooperation among the beneficiaries.

The Government of Zimbabwe policy on social infrastructure inherited from the previous owners in A1 areas was that it became state property to be used for state-specified public purposes, such as conversion of farm houses into schools or clinics, while productive facilities were to be used on a shared basis. This policy, in some cases, provided an impetus towards the establishment of social structures to coordinate the use and maintenance of these facilities (Murisa 2007: 39). Within the A2 scheme, the lease agreement provided the holder of the farm on which infrastructure was currently installed the authority to 'use and to sublet infrastructure to other farmers and obliges them to grant such subtenants right of access to the infrastructure' (World Bank 2006: 24).

Evidence gathered from the field showed that slightly more than half of the population (52.3%) of the total sample had entered into an arrangement of sharing productive infrastructure (Table 7.4), with the phenomenon of sharing being most common in Mangwe (71.7%) and Kwekwe (62.9%). The infrastructure that they shared included tobacco barns, irrigation equipment, farm compounds, dams, cattle-handling facilities and dip-tanks.[27] The initial stages of resettlement were characterised by vandalism and looting of productive assets from the farms, while equipment such as tobacco barns remained mostly unused in A1 areas, due to the changes in land use patterns. Soon after the Fast Track Land Reform in certain localities, the sharing of infrastructure was done through various arrangements, including the 'Committees of Seven', which contributed towards enhancing production capacities, although the holders of some plots with such infrastructure refused to share them with other land beneficiaries (Sunga and Moyo 2004: 7). In Goromonzi and Zvimba, the land beneficiaries also devised other mechanisms for managing the joint use of inherited infrastructure in order to ensure equitable access through the creation of sub-committees on equipment within the Committee of Seven (Murisa 2007, 2009, 2011), through the assistance of the local extension officer to coordinate the use of productive infrastructure in consultation with the village authority and the leadership of local farmers' groups.[28] A number of A1 settlements, including Dunstan Farm in Goromonzi District have come up with such an arrangement.

The GoZ's directives on cooperative access to and use of infrastructure on the farms also influenced cooperation in the sharing of other infrastructures not mentioned in the 2001 directive. For instance, at Whynhill Farm in Zvimba, a sub-committee of the Committee of Seven was established to coordinate the equal sharing and utilisation of inherited irrigation equipment and this sub-committee in 2008 mobilised funds from households utilising the equipment to purchase a new pump after the old one had been stolen and also to purchase new pipes in order to expand the area under irrigation.

Joint use of social infrastructure

Nearly 35 per cent of the respondents commonly used some of the inherited infrastructure for social purposes, such as health and sanitation facilities, water supply, classroom facilities and teachers' houses, farm worker compounds, service centres and recreational facilities. There was an equal distribution of households that jointly used inherited social infrastructure between the A1 (14.3%) and the A2 (14.8%) beneficiaries, despite the fact that the GoZ policy promoted autonomous use of inherited infrastructures on A2 individual plots, by asserting that it belongs to the beneficiary of the plot. This is one of the many areas in which local practice defies official policy.

The most shared assets among the A2 households were the farm worker compounds. At Warrendale Farm in Goromonzi, the land beneficiaries, both A1 and A2, agreed not to evict farm workers from their compounds and these workers provided labour on both temporary and permanent basis to those resettled on the former Large-Scale Farm. In A1 settlements, some of the farmhouses have been converted into schools, clinics, or houses for the extension officers; while, on some of the farms taken over during Jambanja, some of these properties were claimed by the base commander. At Dunstan Farm in Goromonzi, the former owner's double storey house was converted into a primary school: the top floor was subdivided into classrooms, while the ground floor was reserved as accommodation for the teachers.

Joint hiring of labour

There were reciprocal labour hiring arrangements in both districts. These arrangements included utilising one pool of semi-skilled workers, such as mechanics, tractor drivers, seedbed handlers and curing experts. Farmers then come up with an agreement as to when these workers' services would be required on each farm. In certain instances, these arrangements also applied to the hiring of general casual workers who would work as part of a labour collective on adjoining plots, especially during planting, weeding and harvesting periods.[29] Approximately 35 per cent of the resettled farmers engaged in such labour- hiring arrangements and the practice is evenly spread amongst both the A1 and the A2 farms.

There are cases of jointly carrying out farming operations such as land preparations, especially where the fields are adjacent to each other. Approximately 8 per cent of the respondents have been engaged in such combined farming operations since resettlement. The most common form of cooperation is the establishment of labour teams that work jointly on farms. This is a common labour supplementing practice in the customary areas which entails the grouping of available labour from nearby farms and then carrying out a specific task, such as harvesting on a single plot over an agreed amount of time before moving on to the next farm. The practice, known as *nhimbe*, was developed within a lineage framework of social organisation where members of the same lineage group would be organised into labour teams for ploughing, planting, weeding and other tasks. The practice of *nhimbe* is being exported into areas where the lineage framework is either non-existent or very weak. As the figures from 2007 suggest, the uptake on the practice is very low, probably suggesting the need to establish and strengthen a sense of belonging within a community amongst the beneficiaries before such practices can thrive.

Information and extension services

Although the GoZ has historically boasted of a robust policy on extension support, its effectiveness has been restricted by a number of logistical constraints. One of the ways in which land beneficiaries have responded to the non-availability of extension support has been through unofficial channels of technical advice. Some of the resettled farmers have experience in agriculture and they have been informally providing extension support to their colleagues. Approximately 27 per cent of the resettled households are engaged in providing local extension advice to others based on their previous experience. The issues that farmers advise each other on include where to buy inputs (seeds, fertilisers and chemicals), the dates on which to plant and responding to new market opportunities. The advice is normally reciprocal; instances where it has to be paid for are rare. Closely related to this practice is the sharing of seed and planting materials among resettled farmers. Only 6.2 per cent of the population is engaged in sharing seeds and planting materials. Although there has been an attempt since colonisation to commercialise all seeds for staple crops, there are certain crops, such as sorghum, cassava and sweet potatoes, for which seeds can still be obtained through various networks of cooperation (especially those from customary areas) and other local markets. However, the low numbers of households sharing seeds suggests the deepening integration of beneficiaries into the seed markets.

Structured multifaceted farmers' groups

There is a relationship between some of these unstructured networks and the emergence of more formal local farmers' groups, since many of the structured farmers' groups emerged from the former and membership was mobilised on the basis of previous membership in these networks.[30] Approximately 9 per cent revealed that they belonged to the same local groups as their neighbours.

There are many reasons for such association, including the previous socialisation of beneficiaries. Most beneficiaries came from customary tenure areas where NGOs (as discussed previously) had introduced various associational forms such as community-based organisations, farmers' associations, local clubs (for rotating savings and other income projects) and projects of cooperation. The second largest segment of beneficiaries came from the urban areas, where there are varied associational activities, ranging from religious activities to rotating savings and credit clubs among vegetable and other commodity vendors. The associational groups found in the NRAs were multi-focused: the most common activities included mobilising resources such as farm labour and productive assets into a common pool and sharing expert information, ensuring access to critical inputs such as fertilisers and seeds and mediation of farmers' grievances with regard to production.

Various layers of associational activity were in place by 2007. One layer consists of the representative associations established at the district, Intensive

Conservation Area (ICA)[31] and the ward level in many provinces. In Bromley (an ICA within Goromonzi District), for example, the land beneficiaries formed the Bromley Farmers Association (BFA) with approximately 250 active members drawn from the A1 and A2 farms. The association was formed in 2005 and seeks to address common grievances within the resettled community.[32] The association has been involved since its establishment in the bulk buying of inputs such as fertilisers and seeds for members. The executive committee of the association is almost entirely made up of A1 beneficiaries with just one A2 farmer as an ex-officio member. However, the association has been facing challenges since 2008 because it failed to secure inputs for its members and has not managed to come up with a constitution that clarifies its mandate and objectives.[33]

The Zvimba South Farmers' Association, to give another example, services half of the Zvimba District, which includes Banket and surrounding areas. The association has a pre-Fast Track resettlement history. It was created by local leaders (mostly politicians) to foster improved yields and nurture good agricultural practice among smallholder farmers, but was always hampered by low membership levels. In the aftermath of the Fast Track Programme, the association experienced a new lease on life. It was revived as a mobilising platform for those who had been offered land, but were struggling to obtain inputs.[34] The association represents all the newly resettled farmers and customary tenure area households. Since 2003, the association has been involved in securing inputs for its members through bulk buying or entering into contract farming arrangements. However, due to its broad-based membership, it has not adequately managed to satisfy the differentiated internal interests within the association and has been accused of prioritising the interests of the A2 farmers.

The second layer of associational activity is composed of loose networks of cooperation and structured local farmers' groups operating at the level of the A1 village. The village in the Newly Resettled Areas is composed of A1 households settled on what used to be one large-scale farm. In some cases, where there are more than 100 A1 beneficiaries, more than one village has been created within a previously single large farm. For instance, there are two villages at Dunstan Farm in Goromonzi, where there are a total of 115 beneficiaries and three villages at Chabwino Farm near the Juru business centre in Goromonzi. Approximately 40 per cent of the resettled households (both A1 and A2) belong to such farmers' groups (Moyo et al 2009; Murisa 2011). Membership levels vary from as low as ten (Goromonzi) to as high as 75 (Zvimba).

Table 7.5: Activity and number of groups involved

Benefit	A1		A2		Total	
	No. of HH	% of HH	No. of HH	% of HH	No. of HH	% of HH
Social support	21	16.2	4	10.5	25	14.9
Extension	67	51.5	17	44.7	84	50.0
Marketing support	5	3.8	6	15.8	11	6.5
Input procurement	29	22.3	10	26.3	39	23.2
Labour provision	1	0.8	0	0.0	1	0.6
Profit sharing	7	5.4	1	2.6	8	4.8
Total	130	100	38	100	168	100

Source: AIAS Baseline Survey (2007)

These groups serve a variety of purposes and at times duplicate the activities carried out within the unstructured networks discussed above. The most common activities carried out by the groups include mobilisation of resources such as savings, extension support, labour and asset pooling for production, input procurement and marketing of farm products (see Table 7.5).

The groups were formed as part of a bundle of strategies to improve the means by which inputs are obtained from government agencies. Some of the actors involved in these formations had prior experience of formalised associational activity and, in some cases, external agents such as lending institutions induced communities to establish one. Some of these groups were established in an opportunistic manner: for example, the availability of productive infrastructure such as irrigation equipment on the former large-scale farm has provided a justification for the establishment of groups.[35] Local farmers' group leaders and extension officers based on the resettled farms explained that some of these groups started off as informal networks of mutual cooperation, but eventually formalised themselves for a variety of reasons, which include an increase in membership, the need for increased accountability and transparency in the handling of resources, or to meet the lending conditions of financial institutions.

The local farmers' groups tended to be location- specific and most of them were not linked to the larger groups at district, provincial and national level. As of the end of 2009, there was no clear mechanism linking these fragmented organisations to the national unions, either to the ZFU or the CFU, or even to the new Zimbabwe Commercial Farmers' Union (ZCFU).

Nature of the emerging local state in Fast Track areas

When the GoZ regained the initiative in land redistribution in late 2001 through the Fast Track Programme, some observers (Sadomba 2008a&b) saw this process as aimed at marginalising the leadership of war veterans. The state sought to restore its planning authority (see Moyo 2005) using the Fast Track Land Reform legislation. According to one perspective, it focused on removing war veterans from the plots they had allocated themselves on the pretext of the need to re-zone the land either into A1 or A2 farms (Sadomba 2008a: 187). Although the programme was often perceived as 'chaotic' in execution, it was implemented through a centralised mechanism that controlled decentralised structures from the local (farm) level up to the central government level (Chaumba et al 2003: 9-10). In the process, the Fast Track approach gave 'a new impetus to local structures at a relatively low direct budgetary cost' (Moyo and Yeros 2007b: 108). The defunct Village Development Committees (VIDCOs) were revived and reconstituted in structure to be run by a seven member committee[36] (Chaumba et al 2003: 10). These were later subjected to the 'traditional' authority of village heads appointed by chiefs. At the district level there were District Land Committees (DLCs), which included the Rural District Council (RDC) Chairperson; the District Chairperson of the War Veterans Association; traditional leaders (chiefs and village heads); an officer from the President's Office (Intelligence), the Zimbabwe Republic Police (ZRP) and the Zimbabwe National Army (ZNA); and officials from the departments of Social Welfare, Health, Veterinary and Agricultural Research and Extension (AREX). The responsibilities of the DLC included identification of land for resettlement, beneficiary selection and attending to land disputes among the newly resettled farmers (GoZ 2001). The DLC reported to similarly constituted provincial land committees, coordinated by provincial governors, who in turn reported to the central government (Moyo and Yeros 2007b: 108).

Parallel to the reconfiguration of the local state, new power relations have emerged. While during the period of *Jambanja* the role of traditional leaders in beneficiary selection was overridden by war veterans, the former were elevated during the Fast Track Programme. The chiefs managed to weave their way into official structures and advocated for the expansion of their territories into neighbouring resettlement areas. While the pronounced role of traditional authority in beneficiary selection and the expansion of territorial control were logistically rational, especially in beneficiary selection, it reinforced customary authority as a whole, giving sustenance to possibilities of ethno-regional biases in land allocation (Moyo and Yeros 2007b: 111). The local leadership of war veterans, once dominant within the Committees of Seven at village level and within the DLCs, has increasingly been replaced by the various tiers of traditional authority, including chiefs (at district level), ward heads (at ward level) and village heads (at village level).

Government of Zimbabwe policy on local government in the Newly Resettled Areas seems intent on replicating the prevailing form of local government in customary areas despite the tenure differences between the two areas. In the Newly Resettled Areas, the permit creates a direct relationship between the land beneficiary and the state through civil courts, while in customary areas the traditional authority has considerable influence in land administration and allocation matters. While within the customary areas, allegiance to the chief and their structures is based on a historical claim to power and social relations that have been developed over a number of years, the Newly Resettled Areas are composed of a mixture of people from different social backgrounds who might not necessarily recognise the authority of the chief. The tensions that defined the relationships between war veterans and traditional authority leadership remain in a number of areas. In the selection of village heads, traditional leaders have used the criteria of belonging as the main qualification for one to operate in the office. Many of the war veterans, who were previously responsible for village administration on occupied farms, have been marginalised. The tension between settlers and the newly installed village heads is more pronounced in areas such as Goromonzi where land occupations were more dominant. At Dunstan Farm, one of the first farms to be captured by the land occupiers, resettled war veterans reiterated that they could not respect the newly installed village head because he did not participate in the land occupations.[37]

Conclusion

The Fast Track resettlement period and its aftermath offer contradicting opportunities. On the one hand, the physical and social changes to the agrarian terrain suggest the emergence of a pluralistic democratic form of social organisation with potential to nurture inclusive and participatory processes of local government that are amenable to a multiplicity of actors with conflicting agendas, but building towards more vibrant communities. At the same time, land reform has provided an opportunity for the expansion of traditional authority into areas that had previously been effectively dominated by the authority of large-scale farmers. Thus, while land beneficiaries are engaged in their own trajectory of forging relations of sociability that aim at enhancing farm production, state-based policy 'craftsmen' are attempting to replicate customary area forms of authority in the Newly Resettled Areas.

Land occupations initially contributed to the emergence of new forms of popular rural authority. However, since resettlement, these have gradually been replaced by the fusion of traditional and modern institutions which bring together customary and popular political functionaries to serve on the same platforms. In the post-Fast Track period, there has been a shift in the form of the local state towards re-establishing traditional structures in the Newly Resettled Areas. The manner in which the turn towards re-instituting traditional authority has taken place does not augur well for local democratic practice and is reminiscent of the manner in which the colonial state imposed traditional authority structures after the land alienations. The forms of social relations and action that have emerged in Newly Resettled Areas vary, but do not necessarily confront the turn towards the re-establishment of traditional authority. Rather, they respond to weak state delivery and limited market activity.

Although none of the existing Fast Track resettlement models provide for the creation of collective schemes, the resettled are combining individual and group action in response to different social reproduction constraints. Local networks of cooperation have been established to complement individual household efforts. A variety of local networks and associational forms have been formed recently on a number of former large-scale farms and these are established in various ways, including the very informal (with no structures) and at times invisible forms that are only activated during specific periods.

One of the notable outcomes of the Fast Track period has been the surge towards 'organic' association (in the sense that it has mainly been driven by the land beneficiaries with minimal input from external agents) and formalised associational activity. Approximately, 25 per cent belong to the more structured associational forms and there are more who belong to unstructured networks of cooperation.

Fast Track resettlement areas remain, not only isolated from the national smallholders' union, but also from global and national civil society comprising a complex web of networks involving local and international actors such as NGOs, unions and donors. The local farmers' groups that have emerged operate outside the parameters of this civil society. They sit uneasily in both the civil society and as subordinate agents of the state as they help their members to undertake productive and economic activities, a role associated with the state. They remain shunned and isolated by other civil society-based networks despite the state's attempts to civilise the Fast Track resettlement areas by ensuring that the land beneficiaries are legitimate property holders.

Notes

1. See Bratton 1994; Moyo 1999; Murisa 2009. The national leadership of the new union was elected at a national congress which the NFAZ leadership believed was a platform for discussing the process of merger, but it is reported that the Minister of Agriculture changed the agenda and called for elections. District- level and lower structures of the NFAZ complained, citing inadequate consultations prior to the merger.
2. Aggregate NGO budgets covering a wide array of social activities equal or are more than national budgets of some of the countries in which these organisations work (Bebbington *et al,* 2008:4).
3. It is estimated that more than 3.5 million people have left the country since 1997 *(Zimbabwe Independent*, 19 December 2004).
4. The Zimbabwe Vulnerability Assessment Committee (ZIMVAC) is a sub-committee of Poverty Eradication and Social Services Delivery Development Action Committee (PESSDDAC). This committee is chaired by the Food and Nutrition Council (FNC), which is part of the Scientific and Industrial Research and Development Centre (SIRDC), and is also composed of the UNWFP, FAO, UNICEF, OCHA, FEWSNET, SC (UK) and the University of Zimbabwe.
5. Interview with Chief Bushu, Nyanga, 2006.
6. Based on observations at Dunstan and Lion Kopje farms in Zvimba.
7. Based on Focus Group Discussions held in Goromonzi and Zvimba, June and September, 2006.
8. This is a long established tradition in rural Zimbabwe, where subjects of the *Mambo* (chief) work in the *Zunde* field. The harvest from the *Zunde* is then stored for times of crisis and various households rely on these reserves only when their food stocks run out (Sadomba 2008a: 114).
9. Interviews with Extension Officer in Wards 21 and 22 of Goromonzi, September 2008.
10. Based on interviews with headmen based at Dunstan and Buena Vista farms, September 2008.
11. Based on interviews held with the village headmen at Dunstan Farm and Lot 3 of Buena Vista Farm, September 2008.
12. Based on focus group discussion with local farmers' group members, held at Dunstan and Lot 3 of Buena Vista, September 2008.
13. Based on interviews carried out with Chief Bushu, June 2006.
14 Interviews with Goromonzi RDC, Chief Executive Officer, September 2006.
15. Interview with Goromonzi RDC, Chief Executive Officer, September 2007.
16. Interview with Zvimba RDC, Chief Executive Officer, June 2007.

17. Consisting of graduates from the University of Zimbabwe, Bindura University and other agriculture colleges.
18. In fact, the GoZ led drive for extension officers was so great that it had to lower the entry level requirements for this post. Whereas previously one had to have at least two 'A' level passes at and a Diploma from a recognised tertiary college in the period just after Fast Track resettlement, the GoZ recruited into the position of extension officer even those with only five 'O' level subjects, as long as one of these was Agriculture.
19. Based on interviews held with Zvimba and Goromonzi District extension officers, September 2008.
20. Interview with Acting Zvimba District Extension Officer, September 2008.
21 Interviews with Dunstan Farm Village Head, September 2008.
22. Focus group discussions in Zvimba, September 2008.
23. Focus group discussions, Goromonzi and Zvimba, September 2008.
24. Interviews with Goromonzi District Extension Officer, September 2008.
25. Goromonzi district focus group discussions, 2008.
26. Interview with Goromonzi Rural District Council official, June 2008.
27. Interviews with AREX officers in Goromonzi and Zvimba, September 2005 and 2006.
28. Interviews with Goromonzi District Extension officer, September 2009.
29. Interview with Goromonzi AREX extension officer, September 2008.
30. Interview with local farmers' group leader in Goromonzi, September 2008.
31. An ICA comprises 4-5 administrative wards and an average of 6 ICAs make up the district.
32. Interviews with BFA members, September 2008.
33. Interview with AREX officer, September 2008.
34. Interview with Executive Committee member of the Association, August 2006.
35. See Murisa (2009) and Murisa (2011) for a more detailed discussion on the emergence of local farmers' groups in A1 settlements.
36. The seven member committee comprised a Chairperson (usually a war veteran), Vice Chairperson, Secretary, Vice Secretary, Treasurer, Security Officer (usually a war veteran) and one ordinary member.
37. Goromonzi, key informant interviews, September 2008.

References

Adholla, M.S.E., 1962, 'Traditional Society and Cooperatives', in C.G. Widstrand, ed., *Cooperatives and Rural Development in East Africa,* Uppsala: Scandinavian Institute of African Studies.

AIAS Household Baseline Survey, 2007, Household Questionnaire.

Alexander, J., 2006, *The Unsettled Land: State-Making and the Politics of Land in Zimbabwe 1893-2003*, Oxford: James Currey Publishers.

Anderson, J.A., 1999, 'The Politics of Land Scarcity: Land Disputes in Save Communal Land', *Journal of Southern African Studies*, Vol. 25, No. 4, pp. 553-578.

Barr, A., 2004, 'Forging Effective New Communities: The Evolution of Civil Society in Zimbabwean Resettlement Villages', *World Development*, Vol. 32, No. 10, pp. 1753-1766.

Bebbington, A., Hickey S. and Mitlin, D., 2008, *Can NGOs Make a Difference? The Challenge of Development Alternatives*, London: Zed Books.

Bourdillon, M.F.C., 1982, *The Shona Peoples*, 2nd edition, Gweru: Mambo Press.

Bratton, M., 1994, 'Micro-Democracy? The Merger of Farmer Unions in Zimbabwe', *African Studies Review*, Vol. 37, No. 1, pp. 9-37.

Burawoy, M., 1980, 'Migrant labour in South Africa and the United States', in T. Nicholas, ed., *Capital and Labour*, London: Fontana.

Chambati, W., 2009, 'Land Reform and Changing Agrarian Labour Processes', MM Thesis, University of Witwatersrand.

Chambati, W. and S. Moyo, 2003, *Land Reform and the Political Economy of Agricultural Labour*, AIAS Monograph Series, Harare: AIAS.

Chaumba, J., Scoones aI. nd Wolmer, W., 2003, 'New Politics, New Livelihoods: Changes in the Zimbabwean Lowveld since the Farm Occupations of 2000', *Sustainable Livelihoods in Southern Africa Research Paper 3*, Brighton: Institute of Development Studies.

Chigumira, G. and I. Matshe, 2004, *Agricultural Productivity and Efficient Management of Resources Study: The Overall Macroeconomic Environment and Agrarian Reforms*, AIAS Monograph Series, Harare: AIAS.

Chikate, S., 1995, 'More Revenue to Local Authorities', Harare: The Association of Rural District Councils.

CSO, 2004, *Zimbabwe National* Census 2002, Harare: Government Printers.

de Janvry, A., 1981, The Agrarian Question and Reformism in The Latin America, Baltimore, MD and London: John Hopkins University Press.

Dekker, M., 2004, 'Risk, Resettlement and Relations: Social Security in Rural Zimbabwe', Tinbergen Institute Research Series No. 331, Amsterdam: Thela Thesis and Tinbergen Institute.

Dzingirai, V., 1994, 'Politics and Ideology in Human Settlement: Getting Settled in the Sikomena Area of Chief Dobola', *Zambezia*, Vol. XXI, No. 1, pp. 167-176.

FCTZ, 2001a, 'Farm Community Trust of Zimbabwe Annual Report 2000-2001'.

FCTZ, 2001b, 'The impact of land reform on commercial farm workers' livelihoods', 9 May.

Fontein, J., 2009, 'We Want to Belong to Our Roots and We Want to be Modern People: New Farmers, Old Claims Around Lake Mutirikwi, Southern Zimbabwe', *African Studies Quarterly*, Vol. 10, No. 4. (http://www.africa.ufl.edu/asq/current.htm).

Govere, I., 2006, 'Should the Government Distribute Agricultural Inputs to Farmers?', Paper presented at AIAS-ZIMPRO Land and Resource Rights Dialogue, 16 November, Harare.

Government of Rhodesia (GoR), 1951, *National Land Husbandry Act*, Salisbury (now Harare): Government Printers.

Government of Zimbabwe (GoZ), 1982, *Communal Lands Act,* Harare: Government Printers.

Government of Zimbabwe (GoZ), 1996, *Rural District Councils Act,* Harare: Government Printers.

Government of Zimbabwe (GoZ), 1999, *Traditional Leaders Act*, Harare, Government Printers.

Government of Zimbabwe (GoZ), 2001, 'Land Reform and Resettlement Programme – Revised Phase II, Ministry of Lands, Agriculture and Rural Resettlement', unpublished.

Government of Zimbabwe (GoZ), 2003, 'Ministry of Local Government's Directive on Local Government Procedures in Newly Resettled Areas', unpublished.

Government of Zimbabwe (GoZ), 2005, Zimbabwe Vulnerability Assessment Committee.

Helliker, K.D., 2006, 'A Sociological Analysis of Intermediary Non-Governmental Organisations and Land Reform in Contemporary Zimbabwe', DPhil. Thesis, Grahamstown, SA: Rhodes University.

Helliker, K.D., 2008, 'Dancing on the Same Spot: NGOs', in S. Moyo, K.D. Helliker and T. Murisa, eds., *Contested Terrain: Land Reform and Civil Society in Contemporary Zimbabwe,* Pietermaritzburg: S&S Publishers.

Human Rights Watch (HRW), 2003, 'Not Eligible: The Politicisation of Food in Zimbabwe'. (www.hrw.org/en/reports/2003/10/23/not-eligible). 18 April.

Jowah, E., 2009, 'Rural livelihoods and food security in the aftermath of Fast Track Land Reform in Zimbabwe, MA thesis, Grahamstown, SA: Rhodes University.

Makumbe, J., ed., 1996, *Participatory Development: The Case of Zimba*bwe, Harare: University of Zimbabwe Publications.

Masuko, L., 2009, 'Land Occupation Movement: Nyambira-Mazowe War Veterans Association as a Case Study', *AIAS Occasional Paper Series* (unpublished, available at the AIAS repository).

Matondi, P., ed., Forthcoming, *Zimbabwe's Fast Track Land Reform,* London: Zed Books.

Mlambo, B., 2006, 'The Fast Track Land Reform Programme in Zimbabwe: A Critical Analysis of Changes in Land Tenure Regimes, Land Use and Agricultural Production Trends', AIAS monograph (unpublished, available at the AIAS repository).

Moyo, S., 1995, *The Land Question in Zimbabwe,* Harare: SAPES Books.

Moyo, S., 1999, 'Land and Democracy in Zimbabwe', Monograph Series No. 7, Harare: SAPES Trust.

Moyo, S., 2001, 'The Land Occupation Movement and Democratization in Zimbabwe: Contradictions of Neoliberalism', *Millennium Journal of International Studies,* Vol. 30, No. 2, pp. 311-330.

Moyo, S., 2002, 'Peasant Organisations and Rural Civil Society: An Introduction', in M. Romdhane and S. Moyo, Eds., *Peasant Organisations and the Democratisation Process in Africa,* Dakar: CODESRIA.

Moyo, S., 2005, Land Policy, Poverty Reduction and Public Action in Zimbabwe, ISS/ UNDP Land, *Poverty and Public Action Policy Paper No. 11,* The Hague: ISS.

Moyo, S., 2007, Emerging Land Tenure Issues in Zimbabwe, AIAS Monograph Series, 1/2007, Harare: AIAS.

Moyo, S. *et al,* 2000, *Land Reform under Structural Adjustment in Zimbabwe: Land Use Change in the Mashonaland Provinces,* Uppsala: Nordic Africa Institute.

Moyo, S., Chambati, W., Murisa, T., Sibiza, D., Dangwa, C. Mujeyi K. and Nyoni, N. 2009, *Fast Track Land Reform Baseline Survey in Zimbabwe: Trends and Tendencies,* 2005/06, Harare: AIAS Monograph.

Moyo, S. and Yeros, P., 2005, 'The Resurgence of Rural Movements under Neoliberalism', in S. Moyo and P. Yeros, eds., *Reclaiming the Land: The Resurgence of Rural Movements in Africa, Asia and Latin America,* London: ZED Books.

Moyo, S. and Yeros, P., eds., 2005a, *Reclaiming the Land: The Resurgence of Rural Movements in Africa, Asia and Latin America,* London: ZED Books.

Moyo, S. and Yeros, P., 2005b, 'Land Occupations and Land Reform in Zimbabwe: Towards the National Democratic Revolution', in S. Moyo and P. Yeros, eds., *Reclaiming the Land: The Resurgence of Rural Movements in Africa, Asia and Latin America,* London: ZED Books.

Moyo, S. and Yeros, P., 2007a, 'Intervention, the Zimbabwe Question and the Two Lefts', *Historical Materialism,* Vol. 15, No. 3, pp. 171-204.

Moyo, S. and Yeros, P., 2007b, 'The Radicalised State: Zimbabwe's Interrupted Revolution', *Review of African Political Economy,* Vol. 34, No. 111, pp. 103-121.

Murisa, T., 2007, 'Social Organisation and Agency in the Newly Resettled Areas of Zimbabwe: The Case of Zvimba District', AIAS Monograph Series, Harare: AIAS.

Murisa, T., 2009, 'An Analysis of Emerging Forms of Social Organisation and Agency in the Newly Resettled Areas of Zimbabwe, the Case of Goromonzi and Zvimba Districts', PhD Thesis, Grahamstown, SA: Rhodes University.

Murisa, T., 2011, 'Local Farmer Groups and Collective Action within Fast Track Land Reform in Zimbabwe', *Journal of Peasant Studies,* Vol. 38, No. 5, December 2011, pp. 1145-1166.

Nyambara, P., 2001, 'Immigrants, "Traditional" Leaders and the Rhodesian State: The Power of "Communal" Land Tenure and the Politics of Land Acquisition in Gokwe, Zimbabwe, 1963-1979', *Journal of South African Studies,* Vol. 27, No. 4, pp. 771-791.

Ranger, T.O., 1999, *Voices from the Rocks: Nature, Culture and History in the Matopos Hills of Zimbabwe,* Indiana: Indiana Press.

Richardson, C., 2005, 'The Loss of Property Rights and the Collapse of Zimbabwe', *Cato Journal*, Vol. 25, No. 3, pp.541-565.

Sadomba, W.Z., 2008a, 'War Veterans in Zimbabwe's Land Occupations: Complexities of a Liberation Movement in an African Post-Colonial Settler Society', PhD Thesis, Wageningen Agricultural University.

Sadomba, W.Z., 2008b, 'Movements within a Movement: Complexities within the Land Occupations', in S. Moyo, Helliker, K.D., and Murisa, T., eds, *Contested Terrain: Land Reform and Civil Society in Contemporary Zimbabwe*, Pietermaritzburg: S&S Publishers.

Scoones, I., Marongwe, N., Mavedzenge, B., Mahenehene, J., Murimbarimba F., and Sukume, C.v 2010, *Zimbabwe's land reform: Myths and realities*, Suffolk: James Currey.

Sunga, I. and Moyo, S., 2004, *Productive and Social Infrastructure in Models A1 and A2 Resettlement Schemes,* AIAS Monograph Series 8/2004, Harare: AIAS.

The Herald (Zimbabwe), 3 December 2000.

The Standard (Zimbabwe), 15 December 2002.

Tshuma, L., 1997, *A Matter of (in) Justice: Law, State and the Agrarian Question in Zimbabwe*, Harare: SAPES Books.

Van Onselen, C., 1976 (rpt. 1980), *Chibaro: African Mine Labour in Southern Rhodesia 1900-1933,* London: Pluto Press.

Von Blackenburg, P., 1994, 'Land Reform in Southern Africa: The Case of Zimbabwe', in *Land Reform; Land Settlement and Cooperatives,* Rome: FAO.

World Bank, 2006, *Agricultural Growth and Land Reform in Zimbabwe: Assessment and Recovery Options,* Report No. 31699-ZW, Washington DC: World Bank.

Yeros, P., 2002, 'The Political Economy of Civilisation: Peasant-Workers in Zimbabwe and the Neo-Colonial World', PhD Thesis, University of London.

Zimbabwe Independent, 'Zimbabwe', 19 December 2004.

ZIMVAC, 2005, 'Zimbabwe's Rural Food Security and Vulnerability Assessments', June Report No. 5. (www.zimrelief.info/files/attachments/doclib/003.pdf).

ZFU, undated (c.1994), 'An Introduction to the Zimbabwe Farmers' Union', Harare: ZFU.

8

Media Framing of Land Reform in Zimbabwe

Tendai Chari

Introduction

That the land issue has been the epicentre of Zimbabwe's socio-political and economic struggles since colonial times is hardly disputable. The extensive coverage of the country's land revolution in the local and global media, particularly after the launch of the Fast Track Land Reform Programme (FTLRP) in the year 2000, attests to the potency of the mass media in public opinion formation. The news media determine which issues members of the public think and talk about. Through various discursive practices and interpretative fram eworks, the media direct the public's attention to certain issues and formulate certain mental pictures and perceptions in readers (McCombs 2002; Lipman 1922). This chapter employs framing analysis to examine the representation of land and agrarian issues in the Zimbabwean and international media, in order to better understand the role played by the media in moulding public opinion and perceptions of land and agrarian issues in the post-FTLRP period. The key question posed is: how were the various social and political actors presented by these media and how were their perspectives on land and agrarian issues represented? A purposive sample of news articles published in Zimbabwean publicly-owned newspapers (mainly *The Herald* and the *Sunday Mail*), privately-owned newspapers (mainly *The Daily News, The Zimbabwe Independent, The Standard a*nd *The Financial Gazette)* and selected international news organisations, as published between January 2000 and November 2007, were subjected to content analysis. In addition, interviews were conducted with key informants and documents were analysed, in order to complement data from content analysis. Newspaper articles lend themselves to content analysis because they are retrievable from archives. Also, content analysis enables the researcher to make replicable and valid inferences from data, in order to provide 'knowledge… insights and representation of facts and a practical guide to action' (Krippendorff 1980: 21). The study is grounded in framing theory and social construction theory, which accentuate the role of the media in the construction of social reality (Tuchman 1978). According to Entman:

> To frame is to select aspects of perceived reality and make them more salient in communicating text in such a way as to promote a particular problem definition, causal interpretation, moral education and/or treatment recommendation for the item prescribed (1993: 52).

For Melkote, framing involves ignoring certain aspects of an issue, creating an artificial balance, exaggeration, lack of analysis of events and the use of a narrow selection of experts (2009: 549). Parenti argues that:

> The most effective propaganda term is that which relies on framing rather than on falsehood. By bending the truth rather than breaking it, using emphasis, nuance, innuendo and peripheral embellishments, communicators create a desired impression without resorting to explicit advocacy and without departing too far from appearance of objectivity (1993: 200).

Hence, framing influences how people think about issues by invoking certain interpretations of information. The way in which the news is packaged, the amount of exposure or placement given to an issue and the overall accompanying headlines and visual effects, engender certain ways of interpreting reality. The extensive coverage of land reform and agrarian issues in the local and international media could have encouraged certain interpretations of these issues. Unpacking these perspectives is a core objective of this chapter. Although land and agrarian issues have been perennial issues on Zimbabwe's media agenda since colonial times, there is a general agreement that media attention increased significantly after the year 2000, when government instituted a constitutional amendment that empowered government to expropriate land without paying compensation.[1] The FTLRP initiated in July 2000 radically transformed the agrarian sector in a manner that had far-reaching socio-political ramifications (see Moyo and Yeros 2008). Its execution and implementation invited diverse responses from both domestic and international media. Some critics accuse both the local and international media of various shortcomings in their reportage of the land reform.

While some accuse the media of propagating reform 'distortions' and 'misconceptions' about the land reform programme (Stone 2007; Taylor 2007; Chari, 2010; Elich 2011), others charge the media of 'less comprehensible' coverage, resulting in the propagation of 'myths' rather than reality (Scoones and Mavedzenge 2010). As a consequence, media representation of the land reform in Zimbabwe has been a terrain for the contestation of hegemonic and counter-hegemonic discourses. The challenges faced by the media in representing the multi-layered conflicts and complex elements of this issue has been alluded to by various scholars who acknowledge that the centrality of the land question is intricately linked with the race question (see Mamdani 2008; Muzondidya 2011; Gowans 2008; Elich 2011).

Mkodzongi succinctly comments on the shortcomings in analysis of the land reform thus:

> An analysis of the arguments against radical land reform reveals a chronic failure by both journalists and academics to provide a balanced view of the Zimbabwean land issue; the causal factors of landlessness steeped in the country's history are often ignored. There is a tendency to confuse the land issue with Mugabe's political expediency and in the process the baby is thrown away with the bath water. The genuine need for land, which is reflected in many rural areas across the country, is simply dismissed as Mugabe's political posturing. What is often forgotten is that not very long ago millions of Africans were deliberately disenfranchised by a system of state managed repression, segregation and violence (2010: 2).

There is merit in Mkodzongi's argument in the sense that both academics and journalists have exhibited a tendency to engage in emotive debates that centre on personalities rather than issues, thereby missing opportunities to critically evaluate Zimbabwe's radical land reform programme.

Media coverage of the land issue during the period under examination reveals competing versions of reality epitomised by vested group interests in the context of a bifurcated political economy of the media. While reportage is largely event-based, scholarly literature on the subject tends to be highly opinionated, selective, emotional and personalised (see Curtin 2008; Blair 2002; Meredith 2002; Bond and Manyanya 2002).

An academic inquiry on how the perspectives of different actors have been articulated in the media is still missing. Nor have the implications of media reporting of land and agrarian issues on policy matters and public opinion been adequately interrogated. The ideological assumptions underpinning the content and its possible impact on readers and the socio-political climate in which the content was produced are examined in this chapter.

Perspectives on Zimbabwe's land question

The land issue in Zimbabwe is both a consequence and a cause of the struggle for liberation and has always been at the core of the country's political, economic and social struggles, beginning with the first Chimurenga (*Imfazwe*) in 1896. That the land issue threatened to derail the 1979 Lancaster House negotiations for independence between liberation movements and Ian Smith's regime, demonstrates the emotive nature of the land issue. After striking a compromise, the Patriotic Front later announced that:

> We have now obtained assurances that Britain, the United States of America and other countries will participate in a multinational donor effort to assist in land, agricultural and economic development programmes. These assurances go a long way in allaying the great concern we have over the whole land question arising from the great need our people have for land and our commitment to satisfy that need when in government (Utete 2003: 16).

Even though the pledge by the British and the Americans to fund land reform was not inscribed in the constitution, the Patriotic Front was persuaded to accept the compromise after being put under pressure by the Front line States who had been their benefactors. In addition, the Declaration of Rights (Section 16 of the Zimbabwe Constitution) circumscribed the compulsory acquisition of any property, including land, for a period of ten years after the date of independence. Any constitutional amendments during this ten year period needed a 100 per cent majority, something that was impossible given the fact that whites had 20 seats reserved for them under the same constitution for the next seven years.

Realising the duplicity of the deal, the then president of Tanzania, Julius Nyerere commented that it would be impossible to:

> ...tax Zimbabweans in order to compensate people who took it away from them through the gun. Really the British cannot have it both ways. They made this an issue and they are now making vague remarks mixing rural development with the question of land compensation. The two are separate... The British paid money to Kenya. That the future government of Zimbabwe must pay compensation is a British demand and the British must promise in London to make the money available (Utete 2003: 17).

As a result, white farmers who were reluctant to relinquish their land sold land that was mostly in poor ecological regions through the 'willing seller-willing buyer' arrangement, resulting in land reform moving at a very slow pace during the first few years after independence. The situation was compounded by the fact that the Government of Zimbabwe did not have enough funds to procure land for resettlement programmes intended to decongest rural areas (Stoneman 1988). A formal announcement by the new Labour government in Britain in 1997, that it had no obligation to fund land reform in Zimbabwe, marked the turning point in relations between Zimbabwe and its former colonial master. This change in position was articulated by the then British Secretary for International Development, Claire Short, who wrote a letter to the Zimbabwean government stating thus:

> I should make it clear that we do not accept that Britain has a special responsibility to meet the costs of land purchase in Zimbabwe. We are a new government from diverse backgrounds without links to former colonial interests. My own origins are Irish and as you know we were colonized and not colonizers (Utete 2003: 15).

After the rejection of a government- sponsored draft constitution in the February 2000 referendum, the government amended the constitution by retrieving a clause from the rejected draft constitution to give effect to Constitutional Amendment Act Number 16 (Act 5/2000), which empowered the government to compulsorily acquire land without compensation. The 'historic' nature of this amendment was dramatised in the state daily, *The Herald* of 7 April 2000, thus:

> Zimbabwe yesterday took a giant leap towards correcting the historical imbalances in land ownership when Parliament passed a Bill which gives Government the power to compulsorily acquire land for resettlement without paying compensation. The MPs [who voted in favour of the law], who included Vice-President Muzenda and Msika, immediately broke into the liberation war song "Zimbabwe Ndeyeropa" [Zimbabwe's independence was won through bloodshed] soon after the bill was passed as British High Commission Officials trooped out of the Speaker's Gallery. Some MPs could not contain their joy and swayed to the rhythm of the song, while others clapped and banged benches in ecstasy (cited in Willems, 2004: 167).

The period following the amendment witnessed an intensification of occupations of white- owned commercial farms. The farm occupations and the subsequent FTLRP in July 2000 became major talking points in the local and international media, soliciting varied interpretations in relation to their causes, objective and impacts on the economy and social relations.

The media situation during the colonial period

The media in Zimbabwe traces its history to the colonial era, with the establishment of the Argus Printing and Publishing Company, which formed The Herald and The Bulawayo Chronicle in 1892. Editorially, the two newspapers sought to protect the economic and political interests of the minority whites, who had taken the land from the indigenous blacks by violent means. In its maiden issue, The Herald stated that:

> The aims they (the publishers) will keep steadily in view will be to advance to the fullest of their powers the mining and agricultural interests, to discuss and criticise moderately, but without fear or favour the topics of the day or hour and to promote fellowship and unity amongst all classes of sections of the white community (Gale 1962: 19).

This introduction shows that these newspapers had the fundamental objective of underpinning the economic and commercial interests of the ruling elite. For instance, *The Rhodesian Herald* (now *The Herald*) of July 1893 justified land grab from the indigenous people by stating that blacks were not using a 'large portion of their rich and fertile country and the indemnity for expenses incurred could be paid without hardships to the natives in farms and mining gold' (Utete 2003: 10). The newspaper denigrated blacks, in order to justify land dispossession. In 1895, the newspaper continued in the same vein, stating, for example, that:

> For Rhodesians it was absurd to take the untutored savage, accustomed as he is from time immemorial to superstitious and primitive ideas of law and justice and suddenly to try and govern him by the same code of laws that govern a people with many centuries of experience and enlightenment (*The Herald,* April 1895, as quoted in Gale 1962).

The Herald was also used to celebrate the work of white farmers and to project them as the messiah of the blacks. White commercial farmers were lionised for their farming prowess as a way of justifying colonialism. A case in point is *The Rhodesian Herald* of 22 December 1893, which featured a letter complementing white farmers in the country. The story was headlined 'Go Ahead Farmer' and read thus:

> A gentleman recently from a trip in the country writes to us-"I have done a good deal of travelling in this and other districts. I am inquisitive wherever I go and always have thought farming the real mainstay of any healthy country. For these reasons I bring to your notice the good work being done by a farmer (Mr. Tapsell) about 14 miles east of Salisbury. Real progress characterises his farm. He has nearly completed a water furrow 10, 000 feet long, which will bring 500 acres of excellent wheat land under cultivation. Half of this ground is already ploughed and ready for seed. Mr. Tapsell has also prepared a site for a flour mill and will order machinery for three pairs of French Burr stones with dressers and cleaning

> apparatus driven by turbine. On the farm there is a piggery of 50 grunters-most excellent porkers they are too. Energy of this sort is the highest compliment that can be paid to the fertile nature of Mashonaland and if there were many more like Mr. Tapsell, breadstuff and bacon would soon be at reasonable rate (*The Rhodesian Herald,* 22 December 1893).

This article shows that the colonial press was keen to portray white farmers as hardworking, self less and patriotic citizens who were determined to see their nation prosper. This portrayal was an open endorsement of the status quo, which was characterised by skewed racial ownership patterns of land ownership. This clearly demonstrates that, contrary to the myth that the media seeks to report issues objectively and impartially, their primary aim is to serve the hegemonic interests and aspirations of those who own them. The story about Mr Tapsell cited above shows that the colonial press was no exception to the rule.

As has been observed above, the colonial media developed a representation of indigenous blacks as 'untutored savages' who did not know how to fully use the land and of white settlers as energetic, knowledgeable gentlemen who understood how to make the fertile lands produce.

The media context after independence

At independence in 1980, Zimbabwe inherited a relatively diversified media terrain, including the blossoming of the private press during the political and economic transitions dominated by white capital. The private sector grew rapidly between 1980 and 1990 and the number of privately- owned publications tripled in the first six years of independence (Saunders 1991: 3).

The government acquired the Rhodesia Printing and Publishing Company from the South African- based Argus Newspaper Group using a grant provided by the Nigerian government. This acquisition made the government the major shareholder in the newly- created Zimbabwe Newspapers (1980) Ltd (popularly known as Zimpapers), owned by the Zimbabwe Mass Media Trust (ZMMT), which also owned the Zimbabwe Inter-Africa News Agency (ZIANA), the Zimbabwe Information Service (ZIS) and the Community Newspapers Group (CNG).

During the first decade of independence, the media enjoyed relative independence and freedom to publish or practice, the only requirement being to register with the Post Office. The only other legal restrictions related to pornographic material and public decency, racist material and information deemed to threaten state security such as the disclosure of military secrets.

The next decade would be remembered as 'the golden age' of the press in Zimbabwe as it saw unprecedented growth of new privately-owned publications. A factor in this growth was the quest for alternative political voices generated by the formation of a new political party, the Zimbabwe Unity of Movement (ZUM) and the changes arising from the Economic Structural Adjustment Programme (ESAP). Some publications folded, among them the first privately- owned daily newspaper, *The Daily Gazette* and its sister publication, The *Sunday Gazette*. The latter folded due to 'undercapitalisation, lack of feasibility studies prior to launching, competition from the Zimpapers titles, high interest rates and lack of advertiser support' (Kupe 1997: 27).

The launch of *The Daily News* in 1999 by Associated Newspapers of Zimbabwe (ANZ), a group backed by a 'consortium of institutional and private investors from Zimbabwe and abroad' (Waldahl 2004; Ronning and Kupe 2000) marked the beginning of a new era in Zimbabwean media history. *The Daily News* grew rapidly to threaten the dominance of the state-controlled daily, *The Herald. The Daily News* and its allies in the private press were stridently critical of government policies. The private press subscribed to an independent watchdog role, seeking to expose the corrupt practices of government officials (Waldahl 2004). Much of the private press initially offered unqualified support to the MDC and the government came to label *The Daily News* 'an opposition mouthpiece' (Chikowore 2000). While *The Daily News* and other privately-owned publications saw themselves as independent, balanced and impartial, the conflict of views with the state-controlled media drove the two newspapers beyond the boundaries of professional and ethical journalism (see Chari 2007, 2009, 2010). As a result, their readers found it impossible to maintain a critical distance on national issues and chose to become captive to the passions of the political protagonists. Media reportage of the land reform, therefore, took place in a context of intense political and media polarization, which epitomised a fractured society. This polarization was dramatised by a newspaper columnist thus:

> The polarization in our society today is best depicted in the press. Basically, the press is either pro-government or anti-government. Sometimes objectivity is sacrificed on the altar of expedience in order to be true to their chosen position. If you buy newspapers from one divide, you will get half the story (MMPZ 2002: 82).

It is against this backdrop of political polarisation that various positions about the land reform were articulated by the different media. An appreciation of this broader context will, therefore, help one understand the dynamics that shaped their framing of the land reform in Zimbabwe.

Approaches to land and agrarian reporting since 2000

This section examines literature on media representation of land and agrarian issues after the introduction of the FTLRP in 2000. Although the primary focus of this chapter is media representation of land and agrarian issues, it is imperative to acknowledge the outpouring of literature on land reform and agrarian issues from other standpoints, particularly after the FTLRP (see Alexander 2006; Moyo 2001, 2007; Moyo and Yeros 2005, 2007a, 2007b; Sachikonye 2003; Scoones et al 2010). Sachikonye (2003), for example, examines the impact of FTLRP on farm workers, noting that less than 5 per cent of the total number of farm workers in the country benefited from the FTLRP and less than 20 per cent of women in the country got land under the programme. This study gives a historical account of the land issue since 1980. Government efforts to address the issue are acknowledged and obstacles therein are highlighted. The study notes that during the period 1980-2000, farm workers were marginalised in the land reform and agrarian discourse. After 2000, the *Jambanja*[2] period, the discourse on farm workers became more polarized between those who supported *Jambanja* and those who opposed it. A limitation of this study is its failure to critique the 'schizophrenic' tendencies of farm workers who could not identify themselves either with the peasant land movements or the urban working class that was clamouring for land. As a consequence, farm workers became pawns on the political chessboard as the state was keen to use them as evidence that beneficiaries of the land reform programme came from all classes, while white commercial farmers were keen to use them as 'human shields' for stalling the land reform.[3] Although Sachikonye acknowledges that his study was provisional, as the full ramifications of the land reform on farm workers and the broader economy would take longer (Sachikonye 2003: 25) he does not acknowledge government efforts to cushion farm workers from the vagaries of the FTLRP at the time and also downplays how some commercial farmers were not willing to adhere to the legal instruments put in place by the government to cushion farm workers from the negative effects of the Fast Track Land Reform Programme.[4]

Alexander (2006) gives a useful historical account of the land problem in Zimbabwe, although the drawback of her study is over-reliance on journalistic sources and analyses and also its failure to transcend the polarization characterising the Zimbabwean society at the time.

Moyo locates land occupations within a global context, pointing out that events of the late 1990s in Zimbabwe are a manifestation of a 'larger phenomenon underway in the South' (2001: 3110). The popular view that land occupations in Zimbabwe were stage-managed by the state, in order to retain power is interrogated and a more nuanced approach demonstrates that land occupations were linked to unresolved grievances associated with the failure of the developmental prescriptions of the North on the South. He notes that attempts to settle the land question in Zimbabwe using market instruments had failed and that the urban- based civil society in Zimbabwe had never prioritised land reform, resulting in it being alienated from the 'rural civil society' (Moyo 2001: 313).

Moyo also demonstrates that land occupations in the country have been an ongoing phenomenon in the rural areas, both before and after independence. He argues that land occupations represent 'an unofficial or underground social pressure' to force land redistribution and further argues that the '2000-2001 land occupations mark the climax of a longer, less public and dispersed struggle over land, under adverse economic conditions that have been exacerbated by the onset of economic and political reform' (Moyo 2001: 314). This argument debunks popular perceptions that land occupations were a 'new phenomenon' and that they were necessarily sanctioned by the state all the time.

Moyo and Yeros (2007a) identify the state as the locus of the land reform in Zimbabwe, arguing that 'peripheral capitalism' has been unable to resolve the national and agrarian question over the years, resulting in these problems recurring as social and political crises with a potential to escalate to revolutionary situations. They note that the Zimbabwean state has, from time to time, shown tendencies of 'radicalisation' which reached a climax between 2000 and 2003. This radicalisation had begun with government interventions in the economy in 1997, the suspension of the ESAP and the listing of 1471 white owned farms for expropriation.

Moyo and Yeros (2005) also explore the ideological shifts within academia from the late 1990s and how these shifts have shaped the framing of debates on land reform. They argue that assumptions about concepts such as neoliberalism, sovereignty and self-determination tend to be emptied of their content in the euphoric discourses on 'democratisation, human rights and good governance', such that the neoliberal frames of analyses have resulted in the demotion of fundamental human rights such as the right to 'self-determination' which can only be fully realized through land redistribution.

In spite of the existence of a significant body of literature on land and agrarian issues, there is, however, a troubling paucity of literature that examines media discourses on land reform and agrarian issues per se. Considering the role played by the media in mediating the land reform programme in Zimbabwe, this is baffling. This section, therefore, examines the extant literature on media representation of land and agrarian issues in Zimbabwe. Such literature is thus far fragmented, as it is scattered in a few journals, opinion pieces and reports by Non-Governmental Organisations (see MMPZ 2000, 2002; Willems 2004; Harvey 2000). The bulk of writings about Zimbabwe's land issue are journalistic exposés in local and international newspapers and online websites of major news organisations such as the BBC, CNN, AFP and others. These journalistic exposés are largely authored by undercover correspondents[5] (because of the country's restrictive media environment), who have very little knowledge of the country's socio- political context (see MISA 2009).

As a result, some scholars have criticised western journalists for 'distorting' the truth about the land reform in Zimbabwe. Elich, for example, writes:

> For years, Western journalists have castigated Zimbabwe's land reform program. From afar, they pronounced land reform a failure for having brought about the total collapse of agriculture and plunging the nation into chronic food insecurity. Redistributed land, we are continually told, went to cronies with political connections, while ordinary people were almost entirely excluded from the process. Farmland went to ruin because of the incompetence of the new owners. These were simple messages, drilled into the minds of the Western public through repetition. For Western reporters, certain that they owned the truth, emotion substituted for evidence (2011).

Apart from the journalistic writings, the Media Monitoring Project Zimbabwe (MMPZ)[6], a Non-Governmental Organisation and 'media watchdog', has published reports on local media coverage of various issues in the country. However, these reports tend to be quantitative since they focus on how much broadcast time or space in newspapers is devoted to particular issues. As a result, these reports lack qualitative analytical depth since they are primarily concerned with the question of media bias. Holsti (cited in Riffe et al 2005) argues that quantitative analysis trivialises issues and fails to show their significance, since it is preoccupied with frequency of stories, rather than the social, political and economic conditions in which those stories are produced. Apart from methodological limitations, the MMPZ itself tends to uncritically applaud the private media, while routinely criticising the state- owned media, giving an impression that the MMPZ itself is biased and, therefore, unable to maintain critical distance.

Willems (2004) uses qualitative content analysis to examine the coverage of Zimbabwe's land reform by *The Herald* and *The Daily News*. Her study only covers the period immediately after the 2000 referendum and is limited to two newspapers, thus limiting the range of discourses on the land issue to the two newspapers. She does, however, bring out useful insights on the shortfalls of the two newspapers in their coverage of the land issue. In 'Remnants of Empire: British media reporting of Zimbabwe', Willems (2005) argues that the manner in which the British media cover the land issue in Zimbabwe reflects the capitalist interests and colonial legacy of Britain in that country.

Harvey employs Chomsky and Herman's propaganda model to examine the coverage of the land issue in Zimbabwe by western news agencies. He notes how British, United States of America and New Zealand media 'manufactured consent' through slanting and spinning stories on the land issue in Zimbabwe (2000: 1). News selection was used to disguise the colonial links that Britain has with Zimbabwe and Britain's moral 'indebtedness' to that country. Farm occupations, the plight of white farmers and their families, were given more priority in the news. These were covered with what Harvey refers to as 'heavy sentimental rhetoric, in order to wring the sympathy of the international audience' (Harvey 2000: 9). Harvey argues that the British media sought to divert attention from Britain's 'blameworthiness' over Zimbabwe by focusing on the personal plight of white commercial farmers whose farms were being occupied. The British media presented white commercial farmers as 'worthy victims' by featuring them 'prominently and dramatically' in the news so as to evoke the sympathy of the international audience who are dominantly white. On the one hand, because America does not have any colonial links with Zimbabwe, the American media did not 'excessively humanise' the white farmers. On the other hand, New Zealand, a former colony of Britain, had its media parroting the British media because of the colonial ties between the two countries. A significant number of commercial farmers (who lost their land during the violent take-over of white owned farms period) immigrated to New Zealand, meaning that New Zealand was 'bearing the brunt' of the land reform. Also, by virtue of New Zealand being a member of the Commonwealth, the New Zealand government had an interest in the issue. Through 'mass media sourcing', the New Zealand media managed to saturate the news with its government's voice as government officials were given unlimited opportunities on air (ibid.: 10). Thus, Harvey provides important insights on the dynamics of the land issue in Zimbabwe, even though his study covers only a brief period.

Thus far, scholars who have analysed the international media's representation of the land reform in Zimbabwe have focused on the shortfalls of media coverage of the land reform. Key points are that:

- Media coverage of land reform in Zimbabwe is highly contested;
- Media coverage of Zimbabwe's land reform programme reflects ideological bias and distortions and lacks objectivity;
- Media representation of land reform is heavily influenced by the colonial links between Zimbabwe and Britain; and
- The global capitalist interests of the West also influence the way the foreign media represent land reform in Zimbabwe.

Public media framing of land and agrarian issues

The public media, namely newspapers under the Zimpapers stable and the Zimbabwe Broadcasting Corporation, generally echoed the government's position on issues such as farm occupations, violence in the farms, conflicts over land, the food security situation, sanctions, productivity, effects of farm occupations and other events.

This point of view can be seen with regard to many issues. Land occupations were judiciously defended by these media and they were described as 'demonstrations by land hungry peasants'. For example, *The Herald* of 4 April 2000 described land occupations as spontaneous uprisings by the masses, claiming that:

> ...land hungry war veterans poured into at least 30 commercial farms countrywide last month after the rejection of the draft constitution and to exert pressure on the government to speed up the resettlement programme.

The rejection of the draft constitution in February 2000 was described as a 'temporary set-back on the revolution'. The killing of white commercial farmers on farms was interpreted as 'unfortunate' and, in some instances, the murders were blamed on the victims who were accused of fanning violence. The rejection of the draft constitution was framed as 'a vote against land reform'. Similarly, *The Herald* of 7 April 2000 quoted President Mugabe, who said:

> To us as government, what the war veterans have done is a clear demonstration that the government has delayed in redistributing land. This is a clear peaceful demonstration and there is no problem with that...We warned the farm owners not to resist, fight or take up arms. And should they do that, we shall not be responsible for the consequences. Those who have tried to fight have created problems for themselves. It is difficult for us to protect them should they trigger violence.

The public media, therefore, took a cue from pronouncements by ruling party officials in its coverage of the land reform, thus showing that ownership and control played a crucial role in shaping discourses about land reform.

A perception was created that the violence on white commercial farms had the tacit approval of the government. As a result, the private media published a flurry of news stories alleging that certain arms of the government endorsed the land occupations. The public media downplayed the killings of white commercial farmers by maintaining silence about the news. In some cases, the killers remained nameless (e.g. unknown assailants, 'alleged war veterans', or just 'two gunmen'), (see *The Herald,* 9 May 2000; *The Sunday Mail,* 2 July 2000; *The Herald,* 8 August 2000; *The Herald,* 16 May 2001). Where a black farm worker was killed, the death would be given prominence, while the death of whites was downplayed. Typical headlines that presented white commercial farmers as villains and not victims include: 'White commercial farmers perfect economic terrorism' (*The Herald,* 24 September 2001), 'Nyamandlovu farmer dies in shootout with war veterans' (*The Herald,* 19 April 2000), 'Beatrice farmer murdered' (*The Herald,* 9 May 2000), 'Farmers on warpath' (*The Herald,* 8 August 2000), 'Odzi farmer kills resettled farmer: resettled man ran over, dragged for 20m' (*The Herald,* 16 May 2001) and 'Farmers organize attacks on war vets, police issue stern warning' (*The Sunday Mail,* 2 July 2000).

The overall impression created through these headlines was that white commercial farmers were the aggressors, rather than victims. Where victims of violence were black villagers or settlers, they were given extensive and prominent coverage by the public media. A case in point was when a new settler, one Mapenzauswa, was allegedly killed by a white commercial farmer, Bezuidenhout, in Mutare, in 2001. Mapenzauswa's death was widely covered by both the print and electronic public media, while the private media gave it very little attention, often describing the deceased as 'an invader', thus creating the impression that his death was deserved.

State-owned newspapers published opinion pieces historicising the land issue, primarily suggesting that it was an unfinished historical item on the decolonisation project. Examples include an opinion piece by Kenneth Kaunda headlined 'Western Countries Wrong' (*The Herald,* 14 June 2007) and Neil Thomas's 'Zim's suffering externally driven' (*The Herald,* 14 January 2007). The public media, therefore, harked back on history to show that the ZANU-PF leadership was being unfairly blamed for the situation unfolding in the country, when, in fact, the West, particularly Zimbabwe's former colonial master, Britain was to blame for the crisis.

International condemnation of the land reform programme and domestic resistance was buttressed by a well orchestrated propaganda machine led by news organisations such as the BBC, CNN, Sky News and Reuters. This international propaganda machine forced the public media within Zimbabwe to adopt a defensive stance, particularly in the initial stages of the FTLRP. As a consequence of the international condemnation of the land reform, the public media became defensive and started publishing news articles that gave even the slightest endorsement of the land reform programme, particularly by people from outside the establishment. Support from the SADC region and beyond was particularly viewed positively (See Box 8.1 below for examples).

Box 8.1: Public media land headlines on regional responses

- 'German delegation implores state to intensify land reform process' (*The Herald*, 10 March 2000)
- 'Over 50 per cent of South Africans support ex-combatants occupation of farms' (*The Herald*, 04 May 2000)
- 'Anglican Church backs Land Reform Programme' (*The Herald*, 13 April 2000)
- 'We support Zanu (PF) on Land Issue: Former ZIPRA members' (*The Herald*, 25 April 2000)
- 'Methodist Church welcomes land talks' (The Herald, 02 May 2000)
- 'ZCTU calls for speedy land redistribution' (*The Herald*, 02 May 2000)
- 'Communal farmers support farm invasions' (*The Herald*, 09 June 2000)
- 'SA demonstrations support Zimbabwe war veterans' (*The Herald*, 27 May 2000)
- 'Zambia backs land reform programme', (*The Herald*, 09 June 2000)

The public media accentuated the view that land redistribution was necessary to address imbalances created by many years of British colonialism. The land issue was often projected as a bilateral dispute between Zimbabwe and its former colonial master. The former was criticised for attempting to internationalise what was 'clearly a bilateral issue'. Britain was often criticised for showing sympathy for its 'kith and kin' (meaning white commercial farmers whose land had been expropriated). For instance, the then Minister of Information and Publicity, Professor Jonathan Moyo was quoted by New Zealand TV1 (07 May 2000) saying:

> ...when black people die like one million people die in Rwanda, the whole world, (Western World) is not worried. When two whites originally from Britain (sic) die in Zimbabwe, the whole world press is descending on us, just for two whites who have died (Harvey 2000: 3).

This statement shows that the Zimbabwean government perceived Britain as 'ethinicising' the land issue. Some scholars attribute the fallout between Zimbabwe and Western governments to the land reform (Chingono 2010; Chigora and Dewa 2009). Chingono argues that it is because of the land issue that Zimbabwe was put under sanctions after 'a fatal politicization and tragic internationalization of the land issue'. These views show that the land reform in Zimbabwe is complex and has many facets to it. By concentrating on the historical aspects, the public media failed to critique the impact of violence on society and on the economy as a whole. In addition, they also failed to expose other dimensions of the land issue such as skewed gender and class relations in the land reform discourse.

Some of the problems noted in the various land audits (e.g., the Flora Buka Report of early 2003 and the Utete Report also of 2003) were suppressed by the public media (see Chabarika 2003; Mphisa 2009). Problems that were suppressed included multiple farm ownership, low uptake of farms and the manner in which powerful elites took advantage of the land redistribution exercise at the expense of the poor peasants.

In addition, irregularities such as the expropriation of farms protected by Bilateral Investments Protection Agreements (BIPA) and those protected through the Zimbabwe Investment Centre and Export Processing Zones were ignored. Legal impediments facing the FTLRP, haphazard allocations of land and selection of beneficiaries, as well as problems of insecure tenure and collateral security were overlooked in spite of the fact that these were acknowledged government authorities (Utete 2003: 21).

If the public media had exposed some of the irregularities in the implementation of the land reform, the necessary remedies could have been taken and some of the negative effects of the land reform programme would have been rectified. By presenting the land reform through 'rose tinted lenses', the public media abdicated their responsibility to inform and educate citizens.

It is also worth noting that, while some social groups such as war veterans were projected as taking a leading role in the FTLRP, their voices were eclipsed by those of ruling party elites. War veterans were only depicted as active agents during violent confrontations with white commercial farmers. In most instances, they were lumped together with 'land hungry peasants' or masses.

During the 'land invasions', war veterans were largely projected as the vanguard of the 'Third Chimurenga'[7] and the public media often described them as 'peaceful demonstrators'. It was often argued that war veterans' patience had run out. It was said that the war veterans were merely showing their displeasure at the slow pace at which the land distribution exercise was moving (e.g., 'War veterans vow not to leave farms', The Herald, 3 April 2000).

Whereas the private media projected war veterans as violent or rogue elements, the public media portrayed them as peace- loving citizens. In order to reinforce this notion, they were reported as cordially co-existing with white commercial farmers. A case in point is when war veterans agreed on a peace deal (brokered by Father Fidelis Mukonori of the Catholic Church) with the Commercial Farmers Union (CFU) in April 2000 (*The Herald,* 20 March 2000). War veterans were described as 'peace-loving' people whose quest for justice was not being appreciated by the detractors of land reform. The refusal by some war veterans to vacate white- owned farms was justified on the grounds that war veterans had long historical ties with the land (ibid.: 1). A statement by Andrew Ndlovu, one of their leaders, is instructive in this regard:

> the people want their land now. They do not want any obstruction. If we moved off the farms now, then we would have waived our right to land because that will be a violation of our rights as citizens of this country. Our historical background speaks for itself. Moving off the farms will be tantamount to disregarding the sacrifices of the people who fought for this country. Law is law and politics is politics. How do you marry the two? (*The Herald,* 20 March 2000).

On the one hand, veterans were projected by the public media as positive agents of change with a history of fighting for justice. On the other hand, white commercial farmers were largely presented by the public media as anti-land reform, racist and selfish (e.g., 'Anglo-Saxon racism at war with Zim over land' (*The Sunday Mail,* 20 April 2008). The CFU was projected as uncooperative and was blamed for the violence in occupied farms, with the CFU portrayed as preoccupied with selfish interests (e.g., 'CFU in new bid to remove occupiers', *The Chronicle,* 11 April 2000). The view that white farmers were not against the land reform per se, but the method used to acquire it, was dismissed by the public media.

The public media also gave the impression that the land reform programme enjoyed support that cut across the social and political divide. For example, support from civil society organisations[8] was given prominence in the news headlines. Hence, headlines such as 'Anglican Church backs land reform programme' (*The Herald,* 13 April 2000), 'Methodist Church welcomes land talks' (*The Herald*, 02 May 2000), 'Churches urge state to speed up land reform' (*The Herald*, 18 December 2001), 'ZCTU calls for speedy land redistribution' (*The Herald*, 02 May 2000) were sometimes found in the public media since they endorsed the land reform. These headlines show that the government was concerned with moral issues in the implementation of the FTLRP, particularly the issue of violence on the farms.

In the same vein, when the International Socialist Organisation (ISO), through its leader, former opposition Movement for Democratic Change (MDC) Member of Parliament for Highfield, Munyaradzi Gwisai, endorsed the FTLRP in 2002, his position document entitled 'MDC Go Back to the People' was quoted extensively in the public press to legitimate the Fast Track Land Reform (for example: 'Gwisai dismissal from MDC unjustified', *The Herald*, 8 December 2002; 'Revolt against MDC leaders Gwisai urges supporters', *The Herald*, 5 December 2002). Gwisai was one of the very few personalities from the opposition who openly supported the land reform. As a result, he was quoted saying:

> ...if you look at what has happened in the world, what is clear is that when land is taken from people, it is not a tea party. Thousands of our people were killed and massacred by the colonialists in order for them to get the land. Tens and thousands of people were murdered, were robbed and were raped in the 1890s (SW Radio 2005).

The co-optation of 'unfamiliar sources' such as Gwisai by the public media shows that the government was keen to make the land reform as all encompassing and inclusive as possible. The sidelining of 'civil society' groups perceived to be anti-land reform resulted in a very narrow perspective of the land reform in the public media.

Apart from civil society groups, farm workers were opportunistically represented in the public media. While their voices were also scarce, they were conveniently used to magnify the cruelty of their white bosses. Farm workers were thus portrayed as victims of exploitation by commercial farmers. Stories focused more on how white farmers were coercing their farm workers to vote for MDC. For instance, *The Herald* reported that:

> Zimbabwe's farm workers normally treated with contempt by their 'baases' suddenly have new importance thrust upon them as potential voters for the Movement for Democratic Change as the farmers desperately try to keep the status quo on land in place. Vote for Zanu (PF) and you are out, they are threatened by the farmers, who hope to keep their stranglehold on vast tracts of fertile, idle land with a possible change in government (25 April 2000).

It does not come as a surprise that farm workers benefited the least from the FTLRP because their voices were marginalised in the public media. The public media could not positively influence policy decisions in relation to the plight of farm workers. Clarke notes that farm workers were:

> ...seldom interviewed in the media or by other branches of the media. Their high rate of illiteracy imposes a severe disability upon them in a word... These workers have no collective voice at a national level...It is not surprising then that the debate and discussions on farm labour policy proceeds in a way which excludes the subjects of the discussion, as if by some stroke of magic the very people most concerned about were not even there, except as objects of manipulation in varying degrees of benevolence (cited in Sachikonye 2003: 23).

This shows that media representation of land and agrarian issues could have been broader so as to provide an outlet for marginalised voices. By so doing, the multiple dimensions that characterised land reform could have been unravelled.

Private media framing of land and agrarian issues

With the exception of *The Daily Mirror* and *The Sunday Mirro*r[9], private media generally gave the FTLRP negative coverage. They focused on negative elements of the FTLRP such as the violence on farms and the negative impact of farm occupations on food security and the environment. Examples include: '$75b farm equipment vandalized, stolen', *The Zimbabwe* Independent, 15 August 2003; 'Land crops set to decline by 60%', *The Zimbabwe Independent*, 30 May 2003; and 'War vets illegally auction farm equipment', *The Zimbabwe Independent*, 8 February 2002.

Unlike the public media, the privately owned media did not make much effort to historicise the land issue. This a-historical approach resulted in the private media labelling the Fast Track Land Reform a 'political gimmick' by the ruling ZANU-PF party. For example, headlines such as, 'Land Reform: a Revolutionary Move or Political Gimmick?' (*The Daily News*, 3 March 2003), 'What is Mugabe's real motive on the land issue?' (*The Daily News*, 24 April 2001) and 'Corrupting the law' (*The Standard*, 22 August 2002) illustrate the point. These headlines show that the private media placed emphasis on the property rights of the white commercial farmers, while ignoring the unlawful way in which blacks were disposed of their land during the colonial era. The cynical tone of these headlines betrays the private media's ahistorical approach as it creates the impression that the land reform programme was merely a propaganda tool meant to divert the nation's attention rather than to correct colonial injustices.

Thus the 'No' vote in the referendum was described as 'a victory for democracy' and a vote of no confidence in President Mugabe and his government. This representation also demonstrated the personalisation of national issues. A senior journalist with *The Daily News* wrote that:

> The referendum turned into a trial of Mugabe's rule as well as a crucial assessment of his government's legitimacy. While Mugabe's drubbing gave the presidential court a scare it also shook the foundation of the political establishment which had confidently assumed that the draft would be nodded through. Most here believe the 'No' vote in the referendum augurs well for the country's democracy (Thondlana 2000).

Reducing the 'No Vote' to a defeat for Mugabe was one of the most serious shortcomings of the private media in the sense that the opportunity to broaden the scope of debates on key national questions of historical significance such as the land question was squandered as reason gave way to emotion.

Land occupations were characterised as 'primitive', 'barbaric', 'land grab', while the FTLRP was described as 'chaotic', 'violent', 'anarchic'. At the height of the Fast Track Land Reform, the private media carried more stories that sought to de-legitimise the FTLRP. Typical headlines are shown in Box 8.2 below:

Box 8.2: Typical private media headlines

- 'Avoid primitive solutions to land' (*The Daily News*, 25 March 2000)
- 'EU supports calls for rule of law' (*The Daily News*, 11 April 2000)
- 'Government deploys army to direct farm invasions' (*The Zimbabwe Independent*, 20 April 2000)
- 'Mugabe gets tough with land grabbers' (*The Financial Gazette*, 22 July 2004)
- 'Government's chaotic land reform vexes planners' (*The Standard*, 28 November 2004)

The private media suggested that the economy and not land was the top priority of the nation in contradistinction to the ruling party's election campaign theme, 'Land is the Economy and the Economy is Land'. In order to buttress this view, *The Daily News* published an opinion survey in which it concluded that land was not the most pressing issue in Zimbabwe. Instead, top priorities were identified as rising prices of commodities, unemployment, poverty, corruption and the falling value of the local currency. An opinion piece in *The Daily News* of 1 October 2002 stated that:

> To say land issue is the issue in Zimbabwe is a paralysis of analysis. Land is a smokescreen and Mugabe knew he could exploit the mistrust and differences in the global village to cobble up an excuse for lawlessness, the dictator's haven. That it took 20 years for Mugabe to act on the land question seems lost to the solidarity bloc who feels he is righting colonial wrongs (Guma 2002: 2).

Thus the view that the land reform was a diversionary tactic by an opportunistic leader and a government whose popularity was waning was bolstered by such headlines as: 'What has Zanu PF been doing for 20 years?' (*The Zimbabwe Independent*, 21 December 2001) and 'Zim's land allocation deeply flawed' (*The Zimbabwe Independent*, 20 May 2005). *The Daily News* of 18 April 2000 published an article by one David Mills which sought to prove that land was not a priority issue in Zimbabwe, but that the economy was.

> There is need to understand and appreciate that our attention is being diverted from the real issue and threat confronting Zimbabwe. We are being drawn into debates on the land issue and the rule of anarchy, when the most serious and pressing problem that we should be debating and concentrating on is the rapidly declining state of the economy and how will the economy be restored to a more viable one (Mills, *The Daily News*, 18 April 2000).

Attempting to separate land and the economy was flawed in the sense that Zimbabwe's economy is agro-based and the fact that the country's secondary and tertiary industries were intricately connected with the agricultural economy. It is therefore inconceivable how the economy would industrialise if its major resource, the land, remained in the hands of a few white commercial farmers.

The private media were keen to prove that the Zimbabwean government was wrong in embarking on the Fast Track Land Reform and did not listen to wise counsel. As the land reform was singled out for the country's multifarious problems, the impact of droughts, the decelerated foreign direct investment and the deleterious effects of sanctions imposed on the county by Western countries were completely ignored by the private media. This myopia was in spite of the fact that even the most strident critics of the ZANU-PF government, such as the Media Monitoring Project Zimbabwe (MMPZ), acknowledged the negative effects of sanctions. The MMPZ noted that:

> ...none of the private newspapers offered informed analysis on the effect that freezing international financial assistance would have on the economy and therefore the generality of Zimbabwe (MMPZ 2002: 69).

Also, *The Financial Times* (Britain) acknowledged the negative effects of economic sanctions on the country when it noted that:

> Mr. Mugabe and his regime have been remarkably resilient. The country is enduring de-facto sanctions; the IMF and World Bank have frozen loans, aid is limited to humanitarian needs and foreign investment has dried up (*The Financial Times*, 18/08/01, cited by Fahim Ahmed 2002: 5).

Moyo and Yeros demonstrate how the Zimbabwean economy was subjected to economic sanctions since 1998, when Britain imposed a military embargo on the country, the IMF and World Bank suspended lending to the country in 1999 and all donor development assistance was frozen after the year 2000 (Moyo and Yeros 2007: 14-15). Development assistance contracted from $562 million in 1994 to $190 million by 2000 (ibid.: 15). Thereafter, donor assistance (except humanitarian aid) ceased after the enactment of the Zimbabwe Democracy and Recovery Act (2001) by the United States of America government (ibid.: 15).

Most recently, the Minister of Finance in the coalition government, Tendai Biti, admitted that sanctions, primarily the Zimbabwe Democracy Act (ZIDERA), were hurting the economy. He noted that:

> ... if you consider for instance the World Bank right now has billions and billions of dollars that we have to access but we can't access unless we have dealt with and normalised our relations with IMF. We cannot normalise our relations with IMF because of the voting power. It's a veto of America and people who represent America on that board (who) cannot vote differently because of ZIDERA, so it is critical (NewZimnbabwe.com, 3 May 2009; *The Herald*, 2 May 2009).

This goes to show that the privately-owned media failed to present a broader perspective of the causes of Zimbabwe's economic dislocation by singling out land reform as the sole source of the country's economic problems while ignoring or minimising the impact of sanctions imposed on the country by Western countries.

This selective memory on the causes of the country's economic collapse exposed the uncritical journalism on the part of the privately-owned media and punctures the myth that the private media in Zimbabwe is the beacon of journalistic independence and excellence. The private media accused the government of resting on its laurels for many years by not instituting land reform with vigour after 1980, only to wake up when its grip on power was threatened. The plethora of legal and political obstacles that stood in the way of land reform, such as the property safeguards in the Lancaster House Constitution, the fact that the British and American governments reneged on their financial obligations and the resistance by white farmers were ignored by these media.

Ahmed (2002), however, notes that, although Britain and the United States of America pledged $2 billion to compensate white farmers, they failed to pay this money. Resistance by white farmers against land reform technically disabled the government, rendering it a sitting duck for the greater part of the post-independence period. Ahmed notes that:

> ...the commercial Farmers Union of white farmers blocked many initiatives for rural relocation. They controlled 90% of all the agricultural production, paid one-third of the country's goods. The continuing colonial mentality of the land lords was evident from the fact that they carried on voting for the former party of apartheid, the Rhodesian Front, until recently that is when they struck on a more sophisticated weapon; the Movement for Democratic Change (MDC) (Ahmed 2002: 2).

Again, the private media's stance that war veterans and peasants lacked a genuine desire for land and that they were merely political tools of the government was an inadequate analysis of the situation on the ground. On the one hand, private media discourse portrayed war veterans as 'thugs', 'terrorists', 'murderers' and 'henchmen'. On the other hand, newly resettled farmers were caricatured as 'cell phone farmers', 'Mugabe cronies', 'idle party hacks', or people with no desire for farming. Such representations gave the impression that the only beneficiaries of the land reform were the elites. Findings of recent studies on the land reform contradict this view (see Moyo and Yeros 2009; Scoones et al 2010; Scoones and Mavedzenge 2010; Elich 2011; Winter 2010; Mataire 2010). For example, a study conducted by Ian Scoones and colleagues at the Institute of Development Studies at the University of Sussex (UK) (*Zimbabwe's Land Reform, Myths and Realities*) dispels the myth that the major beneficiaries of the land reform were elites. In addition, the study also debunks the myths that:

- Land reform has been a total failure;
- There is no investment on the resettled land;
- Agriculture is in complete ruins, creating chronic food insecurity; and
- The rural economy has collapsed (Scoones and Mavedzenge 2010).

Scoones notes that: 'What we have observed on the ground does not represent the political and media stereotypes of abject failure; but nor indeed are we observing universal, roaring success' (cited by Winter 2010: 1). Since no attempts were made to consider 'positives' of the land reform programme, implying, for instance, that all new farmers were lazy and lacked the desire to farm, the private media fell prey to partisan interests the same way the public media did. Their uncritical condemnation of newly resettled farmers gave an impression that they endorsed the status quo.

An example of their stereotypical lazy farmer is found in an article published in *The Zimbabwe Independent* which states that:

> Here in lies the biggest problem with our new farmers and government is playing right into the hands of greedy sharks out to make a quick buck. Some of the people who got huge tracks of land not only lack the skills and interest in farming, they also have no culture of long term investment and sacrifice. While the white farmers who were removed from the land had spent painstaking years borrowing and investing in infrastructure, from dams to irrigation equipment, the new guys want everything on a silver plate, so they can become instant millionaires (*The Zimbabwe Independent*, 06 May 2005).

The image of the new farmer as a pathologically lazy person in the privately-owned newspaper is a sharp contrast to the white commercial farmer who is portrayed as a 'jolly good fellow' who has fallen victim to the whims and caprices of a 'deranged regime'. In the private media, as much as in the foreign media, the white farmer is a 'messiah' who can extricate his nation from the jaws of the worst famine ever in history by producing unlimited quantities of food. Hence, some stories focused on the 'brutal' murder of white farmers, providing lurid details of these 'gruesome' acts, in order to invoke the sympathy of readers and to project the government as a devil. Examples are: 'Commercial farmer under siege at farm' (*The Standard*, 11 July 2004), 'Ex-fighters hold farmer hostage for three days' (*The Daily News*, 9 July 2001) and 'Another white farmer killed' (*The Daily News*, 12 December 2000).

'Victims of violence' were given a long leash to empty their souls. The Commercial Farmers' Union (CFU) and Justice for Agriculture (JAG) bodies, which represented white farmers, particularly enjoyed the generosity of the private media as sources of news. For example, in *The Daily News* (27 March 2002), the CFU spokesperson Jenni Williams was given a long leash to make allegations about 'suspected' ZANU-PF supporters who allegedly had descended on some farms in Marondera, harassing commercial farmers in the area. Alleged perpetrators of the violence were not given an opportunity to rebut these allegations in spite of their seriousness.

It has been argued that anti-land reform lobbyists used the issue of 'wrong methodology' in land expropriation as a smokescreen. Ahmed, for instance, notes that 'In 1990, parliament passed the Land Reform Act, which proved popular among the majority of the workers and peasants, but evoked fierce resistance from wealthy whites' (2002: 2). Goncalves (1993) chronicles government efforts to speed up land reform after the expiration of the ten year period imposed by the Lancaster House Constitution. These include the 1990 Constitutional Amendment and the 1992 Land Designation Act (LDA), both of which were attempts to create leg room in order to tackle the land problem (Goncalves 1993: 6). In 1997, the government earmarked 1470 white owned farms for compulsory acquisition, but did not succeed due to legal impediments mounted by the farmers. Evidence provided by the scholars cited above testifies that, while an orderly land redistribution exercise was desirable, legal impediments and the intransigence of white commercial farmers made it impossible for government to expedite the land redistribution process. The 'orderly' process between 1980 and 1999 had not achieved much and, perhaps, a more radical approach was inevitable in order to pacify the restive peasant population which was clamouring for land. Thus the charge that the government only became serious about land reform after the rejection of the February 2000 referendum is a mis-analysis of the Zimbabwean state. Tendencies of 'radicalisation' were present even before the 2000 referendum (Moyo and Yeros 2007a). After the Land Designation Act (1992), some farmers challenged its constitutionality and took the government to court, resulting in President Robert Mugabe vowing that he would disregard any court decision that would stand in the way of land reform. He told a ruling ZANU-PF party central committee meeting that:

> I, Robert Mugabe, cannot be dragged to court by a settler ... if white settlers took the land from us without paying for it, we can in a similar way just take it from them, without paying for it, or entertaining any ideas of legality and constitutionality (Goncalves 1993: 7).

This also shows that claims by some sections of the privately-owned media that Mugabe used the land reform programme to retain his waning power need to be moderated as much as it is a truncation of history. Thus the a-historical stance taken by the private media in their coverage of land reform prevented them from properly explaining the complex nature of the land issue, the Zimbabwean state and President Robert Mugabe as a person.

Unlike the white farmers who enjoyed generous coverage, farm workers were hardly interviewed in the private media. They were more spoken about than they spoke. When they were spoken about, the objective was to portray

Mugabe and his government as cruel and uncaring. Examples of headlines conveying such a message include: 'Displaced farm workers now destitute' (*The Standard*, 25 April 2004), 'Women farm workers bear the brunt of land seizures' (*The Standard*, 9 January 2005) and 'Disaster strikes-Ethnic Cleansing-mass displacements' (*The Zimbabwe Independent*, 31 August 2001). The emphasis on the 'victimhood" of farm workers in the private media gave the impression that the condition of farm workers in the new dispensation was worse than during the era of white commercial farmers (see Thornycroft 2009). A report commissioned by the General Agricultural and Plantation Workers Union of Zimbabwe (GAPWUZ) echoed similar sentiments. A farm worker who was interviewed by researchers who compiled the study is reported to have said:

> I would like Murungu[10] to come back because these issues of trauma were not there during Varungu's time. You just knew that you would go to work and get paid at the end of the month and if you did not perform well you would deal with the foreman and be sent back home (GAPWUZ 2009: 55).

As a consequence of them being projected as passive victims of the land reform, farm workers were not adequately given voice. Emphasis on their victimhood meant that they appeared more as pawns on the chessboard of political machinations, rather than as active agents. Like their public media counterparts, the privately owned media became hostages of political forces, thus failing to represent the land reform from a much broader perspective. However, unlike the public media which attempted to historicise the land issue, the privately owned media accentuated the property rights of the white commercial farmers, democracy, human rights and rule of law. Partisanship and their a-historical approach gave way to selective coverage of issues related to land reform, distortions and an oversimplification of complex issues around the land reform, thereby abdicating their responsibility to inform and educate the Zimbabwean public and the global community about the situation in Zimbabwe.

Framing of land and agrarian issues in the international media

Representation of land and agrarian reforms in the international media since 1997 was shaped by the foreign policies of the various western countries towards Zimbabwe. Following the election of the Labour Party in Britain in 1997, relations between Zimbabwe and Britain deteriorated (Chigora 2006: 61). The FTLRP and President Mugabe's refusal to renew a second round of the economic structural adjustment programmes or what was described as a 'conflict of values' has been cited as the harbinger of the fallout between the West and the

ZANU-PF government (Chigora 2006). Reportage on land reform inevitably foregrounded the 'breakdown of rule of law', 'good governance' and 'democracy' in Zimbabwe, thus mirroring the contours of Western foreign policy.

The land redistribution was often described as 'land grab', 'land seizures', 'theft', 'violent' and 'barbaric'. For example, *The Telegraphy* (16 July 2000) reported that land grab chaos was looming in Zimbabwe. This report was after the then Vice President, the late Joseph Msika, promised war veterans that the government would accelerate the land redistribution exercise. Other news headlines which conveyed similar messages included: 'court backs land seizures' (CNN.com, 4 December 2001), 'Zimbabwe presses its seizures of farms' (*The New York Times*, 12 November 2000) and 'Zimbabwe's large farms face squeeze' (BBC News.com, 6 January 2000). These headlines were consistent with the views of most Western countries which saw land reform as a violation of property rights.

Ankomah (2000) contends that, when covering Africa, the western media are guided by four codes, namely, their country's national interest; their government's lead; government leaning; and advertisers and readers. By projecting human rights issues, the international media sought to divert attention from the indebtedness of countries like Britain to Zimbabwe since Britain had reneged on its pledge to fund land reform. Instead, the blame was laid on President Mugabe's 'corrupt' government. Some critics argue that the West has unjustifiably placed primacy on issues of democracy, good governance and rule of law at the expense of Zimbabwe's national question, which is the land issue. *The Scrutator*, for example, argues that in:

> ...the absence of political economy context and theoretical framework, much of our writings on human rights, rule of law, constitution etc, uncritically reiterate or assume neoliberal precepts. Human rights is not a theoretical tool of understanding social and political relations. At best, it can only be a means of exposing a form of oppression and, therefore, perhaps, an ideology of resistance (*The Scrutator*, cited in Raftopoulos 2005: 2)

Moyo and Yeros (2008: 2) concur that the issue of 'democracy is intrinsic to both the agrarian and the national questions'. They add that in Zimbabwe, democracy was a result of the overthrow of colonialism, but this democracy fell far short of addressing historical imbalances.

In terms of human rights (in this case the property rights of whites), President Mugabe's persona became synonymous with the 'Zimbabwean crisis'. Examples of headlines which suggest this personalisation of the land issue include: 'Is Mugabe's strategy working?' (The BBC, 13 April 2000), 'Mugabe defends land seizures' (CNN, 13 April 2000), 'Stay Cool on Zimbabwe Crisis, Mugabe says' (Reuters, 13 April 2000), 'Embattled Mugabe confronted by rule of law' (*The Guardian*, 13 April 2000), 'Mugabe warns boers to leave' (BBC News, 8 April 2000) and 'Mugabe "will not negotiate" over white land' (*The Times*, 8 April 2000). Personalisation resulted in the oversimplification of the issue, as important dimensions of the land question were marginalised. The excessive focus on the deaths of white commercial farmers, while downplaying the plight of black victims of the violence amounted to 'ethnicisation' of the land issue (Willems 2005). Ethnicisation meant that more sympathy was shown towards white victims of the land occupations, while black victims were ignored. While more space was devoted to recounting the ordeals of white commercial farmers, black victims were conspicuous by their absence. Numerous stories were devoted to the deaths of white farmers such as David Stephens and Martin Olds.[11] Examples of such headlines include: 'White farmer killed in Zimbabwe' (*BBC News*, 18 March 2002), 'White farmer killed by Zimbabwean war veterans' (*The Guardian*, 8 August 2001), 'White farmers in Zimbabwe struggle against increasing violence' (*The Telegraphy*, 11 June 2010), 'Mugabe warns 'Boers' to leave' (*The Observer,* 8 April 2000) and 'Seventh white farmer killed in Zimbabwe' (*The Independent* (UK), 13 December 2000).

On 8th March 2002, the BBC (Online) published a story about the death of Terry Ford, who was allegedly shot dead on his farm near Norton. Terry was found 'propped against a tree outside his homestead'. In the same story, we are told about the death a black security guard, who had been beaten to death at a farm outside the town of Marondera. While a lot of detail is furnished about the white farmer (the method used to kill him, the place of killing and the tragic manner he died, as well as the fact that he was the tenth white farmer to be killed under similar circumstances), very little information is supplied about the black security guard. This shows that western media sought to racialise the violence associated with the farm occupations. Doing so diverted attention from the legitimacy of land redistribution in the country.

Harvey notes how the BBC and *The London Times* gave reports that were 'saturated with humanistic rhetoric that supported the tremendously good white farmers and their families' (2000: 5). According to Harvey, an excessive amount of detail was devoted to 'irrelevant information' in order to incite the sympathy of readers. Information – on how white farmers were attacked, their life styles, names and other minute details – not linked to the story was often too detailed for a news article.

Harvey argues that the key ideological positions projected in these reports are imperialism and the humanist values that uphold it. This representation of white farmers sharply contrasted with that of farm workers who were largely marginalised in the land reform discourse. Taylor argues that:

> Western interest and media coverage are inseparable when it comes to Zimbabwe; here we witness the enforcement of the myth that whiteness is power. In particular, a western assumption about the worth of white life over black life is so clearly exposed in recent western media reporting in Zimbabwe (2007: 3).

Thus, white stories of 'victimhood' continue to trump black stories of racial injustice. 'And yet on what grounds can it be argued that white suffering is more important than the past and present suffering of black Zimbabweans?' (ibid: 3).

Whereas in the initial phases of the FTLRP, attention was mainly focused on human rights violations, in the post FTLRP phase, the international media sought to vindicate themselves by focusing on the 'negative consequences' of the FTLRP foretold at the onset of the programme. During this phase, stories on social and political calamities befalling Zimbabwe, such as those occasioned by food shortages, shortage of basic commodities, 'plummeting production' levels and 'drying' of foreign currency reserves, were the staple diet of the international media (See Box 8.3 for examples).

Box 8.3: Typical headlines on the 'crisis' outcome

- 'White land grab policy has failed, Mugabe confesses' (Reuters, 03 March 2005)
- 'Food Crisis in Zimbabwe Worsens' UN (AFP, 03 October 2003)
- 'Zimbabwe's food crisis: What went wrong' (Reuters, 01 August 2002)
- 'Mugabe Blockading Food Relief– Zimbabweans Starve' (*The Mail and Guardian*, 17 November 2002)

The number of Zimbabweans 'facing starvation' as a result of the land reform was a common feature in the news. Figures varied from publication to publication. For example, *The Mail and Guardian* reported that 'Six million Zimbabweans face starvation' (17 November 2002), while *The Times* (London) put the figure at 'five million' (14 October 2008). The emphasis on food security and the insinuation that only white farmers could save Zimbabwe from starvation was an implicit endorsement of the skewed racial ownership of land that existed before the FTLRP.

Another aspect which was emphasised by the international media was the so-called 'contagion effect' of the Zimbabwe land reform. There were concerns about Zimbabwean- style farm invasions spilling into neighbouring countries where the land issue had not been resolved, primarily, South Africa and Namibia. If what was happening in Zimbabwe went unchecked, these countries would sooner or later catch 'the Zimbabwean disease', so went the reasoning. In the news, South Africa and Namibia were 'warned' not to emulate the Zimbabwe- style of land reform. Such headlines include: 'Regional concern over land crisis' (BBC News, 16 May 2000), 'SA land reforms walk uneasy path' (Reuters, 28 October 2004) and 'Reform to be according to Law', Nujoma (AFP, 23 April 2004).

The entrenched commercial interests of the West influenced the Western media to avoid looking at the land issue in Zimbabwe in a dispassionate manner, resulting in numerous stories that were left yearning for attention. Thus, the entrenched positions resulted in crucial facts such as those exposed by Moyo et al (2009) and Scoones et al (2010), studies (see above) either obfuscated, distorted, convoluted, or completely omitted from the discourse altogether (see Box 8.4 below for more missing stories about the land reform).

Box 8.4: Missing stories in the domestic and international media

- Facts about current distribution of land
- Actual impact of the Fast Track Land Reform
- Impact of droughts on the economy
- Land policy relating to tenure and farm size
- Actual number of people needing food assistance
- The food security situation in the country
- Production levels
- The main beneficiaries of the land reform

Moyo also notes that social facts on the ground show that land redistribution has redressed the imbalanced racial legacy, but has at the 'same time spawned new inequalities which are less sharp, while challenges to the outcome by former land owners remain' (Moyo 2007: 1). Distortions and omissions in the media resulted in a paucity of information on a number of issues and the public are ill-informed, confused, or completely ignorant about certain important issues. Representation of land and agrarian issues in the international media demonstrates that institutional, ideological, political and other facts can impose limitations on the media, thereby preventing them from properly executing their democratic mandate of informing the citizens.

Conclusion

This chapter has discussed framing of land and agrarian issues by the local media (private and public) and the international media between 2000 and 2007.

A review of headlines in the post-2000 era indicates that private media accentuated property rights, democracy and rule of law in their framing of FTLRP, while the public media projected the necessity of correcting historical imbalances and social justice and the international media focused more on humanitarian aspects, violence and human rights issues. During this time period, both private media and international media replicated the dichotomy defined by colonial media: These representations set up a contrast between the well-meaning, responsible, skilled white farmer and the incompetent, greedy and dangerous black interloper. Zimbabwe's state media, post-2000, has positioned itself in opposition, justifying black ownership and the credibility of indigenous Zimbabweans as farmers. These extreme positions reflect the political polarization of the media at the time.

Representation of land and agrarian issues reflects the existence of conflicting ideological values. On the one hand, the state media foregrounded the necessity of land reform in order to correct historical imbalances. On the other hand, the local-privately owned media and the international media accentuated neoliberal democratic values such as 'property rights', 'rule of law' and 'democracy and good governance'. The contention in this chapter is that media framing of Zimbabwe's land reform programme, particularly by the privately-owned and the corporate- funded Western media reflected a simplistic and dichotomised view of the land reform, whereby the Zimbabwean state under the leadership of President Mugabe has been constructed as both another African dream that has become a nightmare (Akpabio 2008) and a titan 'at the forefront of the battleground against Western imperialism' (Rutherford 2005). These generalisations have resulted in numerous blind spots in the land and agrarian discourse, what one can call a 'crisis of framing' land and agrarian issues, epitomised by selective voicing of social and political discourses on land reform, generalisations and self-serving evidence of failure or success of the land reform programme. Rutherford rightly points out that: 'these competing generalisations neglect some of the complexities associated with the current Zimbabwean conflict. What they overlook are the overlapping 'territorializing projects', 'the varied political attempts being made to control and influence the Zimbabwean people and their relations' (p. 103). Findings from this study bring under the spotlight neoliberal conceptualisations of the media as 'public spheres' or 'watchdogs', particularly in societies plagued with socio-economic conflicts. It is therefore imperative to re-think these classical formulations of the media in order to locate the proper function of the mass media in transitional societies.

Notes

1. After the rejection of the draft constitution in February 2000, the ZANU-PF government passed Constitution of Zimbabwe Amendment Number 16 (2000), which empowered it to expropriate white- owned land.
2. Loosely translated, '*jambanja*' means violence in Shona. Shona is the main vernacular language spoken in Zimbabwe.
3. There were reports in the public media that farm workers were being mobilized by their employers to vote against the government-sponsored draft constitution as an incentive to protect their jobs.
4. For instance, the government put in place a statutory instrument in terms of Section 17 of the Labour Relations Act, which prescribed the requirement and mode of compensation to all farm workers affected by the land reform. This law entitled all affected farm workers to receive severance packages calculated on the basis of their current salaries and their period of service.
5. For instance, BBC and CNN, which were banned by the Zimbabwean government because of their hostile reporting, were allowed back into the country in July 2009 after the formation of the coalition government.
6. According to the Report by the Commonwealth Observer Group (2000), the MMPZ is funded by the Norwegian International Development Agency (NORAD) and the Open Society Initiative for Africa (OSISA). The MMPZ is biased in favour of the private media. Its weekly reports routinely criticize the public media, while lauding the private media.
7. The fast-track land reform was officially known as *The Third Chimurenga.*
8. Mainstream civil society organizations in the country do not regard the War Veterans Association of Zimbabwe as civil society, presumably because of its alliance with the state.
9. The two newspapers pursued a middle of the road approach characterized by neutrality on land reform, for example: 'More resources- for land survey '(*The Sunday Mirror*, 12 November 2006) and 'The hidden hand in Zim politics' (*The Sunday Mirror*, 9 July 2006).
10. "Murungu" is singular for 'white man' and "Varungu" is the plural.
11. These were the first two white farmers to be killed after the land occupations started.

References

Ahmed, A., 2002, 'Stop Imperialist Intervention in Zimbabwe, Global Analysis Zimbabwe'. (http://www.agitprop.au.nowar).15 April 2008.

Akpabio, E., 2008, 'Framing of the Zimbabwean Crisis in Botswana Print Media', A paper presented at The International Association of Media and Communication Conference, University of Stokholm, Sweden, 20-25 July 2008. (http://www.imcr.org/virtual-paper-room-secjjournandeducation-370). 20 September 2009.

Alexander, J., 2006, *Unsettled Land: State Making and the Politics of Land in Zimbabwe*, Oxford: James Currey.

Ankomah, B., 2000, 'In the Name of the National Interest', *New African Magazine*, July.

Blair, D., 2002, *Degrees in Violence: Robert Mugabe and the Struggle for Power in Zimbabwe*, London: Continuum Books.

Bond, P. and S. Manyanya, 2002, *Zimbabwe's Plunge: Exhausted Nationalism, Neoliberalism and the Search for Social Change*, Pietermaritzburg: University of Natal Press.

Buka, F., 2003, 'A Preliminary Audit Report of Land Reform Programme', Harare: Government of Zimbabwe.

Chabarika, S., 2003, 'Land Reform: A Revolutionary Move or Political Gimmick?' *The Daily News*, 3 March.

Chari, T., 2007, 'Rethinking Ethical Issues in African Media', *African Identities*, Vol. 5, No. 1, pp. 39-60.

Chari, T., 2009, 'Ethical challenges in the context of the media in Zimbabwe', *Global Media Journal*, Vol. 3, No. 1, pp. 46-79.

Chari, T., 2010, 'Salience and Silence: representation of the Zimbabwean crisis in the local press', *African Identities*, pp. 131-150.

Chigora, P. and Dewa, D., 2009, 'Surviving a hostile Environment: An analysis of Zimbabwe's Foreign Relations in 21st Century International Relations', *African Journal of Political Science and International Relations*, Vol. 3, No. 3, pp. 92-98.

Chigora, P., 2006, 'On Crossroads: Reflections on Zimbabwe's Reactions with Britain at the New Millennium Alternatives', *Turkish Journal of International Relations*, Vol. 5, No. 3, pp. 61-76.

Chikowore, T., 2000, 'Business Newsgathering and Media Independence', MA Dissertation, Department of English and Media Studies, University of Zimbabwe.

Chingono, H., 2010, 'Zimbabwe sanctions: An analysis of the "Lingo" guiding the perceptions of the sanctioners and the sanctionees', *African Journal of Political Science and International Relations*, Vol. 4, No. 2, pp. 66-74.

Curtin, T., 2008, 'The economic history of land tenure in Zimbabwe', Paper Presented at the Conference of The African Studies Association of Australasia and the Pacific, held at the John Curtin School Of Medical Research, Australian National University, 31st January 2008.

Elich, G., 2011, 'Challenging Western Distortions about Zimbabwe's Land Reform', *Black Agenda Report.* (http://www.blackreport.com/content/challenging-western-distortions-about-zimba). 20 May.

Entman, R.M., 1993, 'Framing: Toward Clarification of a fractured paradigm', *Journal of Communication*, Vol. 43, No. 4, pp. 51-58.

Gale, W.D., 1962, *History of the Rhodesian Printing and Publishing Company*, Harare: Mardon Press.

GAPWUZ, 2009, If something is wrong…The Invisible suffering of commercial/farm workers and their families due to "Land Reform", Harare: GAPWUZ.

Goncalves, F., 1993, 'The Land Issue in Zimbabwe: Mugabe stands firm on defiant white farmers', *SAPES Southern Africa and Economic Monthly*, Vol. 17, No. 1, pp. 6-11.

Gowans, S., 2008, 'Cynicism as a substitute for Scholarship', Global Research. (http://www.globalresearch.ca/index.php?context-Gowans). 20 May 2008.

Guma, L., 2002, 'Zimbabwean crisis: paralysis of analysis', *The Daily News*, 1 October.

Harvey, J., 2000, 'Media Coverage of the Zimbabwe "land crisis"', *Deep South*, Vol. 6, No. 3, Spring 2000, pp.1-4.

Krippendorff, K., 1980, *Content Analysis: An Introduction to its Methodology*, London: Sage Publications.

Kupe, T., 1997, 'Voice of the Voiceless; Popular magazines in a Changing Zimbabwe, 1990-1997', PhD. Thesis, University of Oslo.

Lipman, W., 1946, *Public Opinion*, New York: Penguin.

Mamndani, M., 2008, 'Lessons of Zimbabwe', *Pambazuka News*, 3 December 2008. (http://www.pambazuka.org/en/category/features/52407). 15 June 2009.

Mataire, L.R., 2010, 'Zimbabwe land reform critics eat a humble pie', *The Herald*, 9 December.

McCombs, M., 2002, *The agenda-setting function of the mass media.* (http://www.infoamerica.org/documents-pdf/mccombs01.pdf). 8 September 2009, 11 November 2010.

Melkote, S.R., 2009, 'News Framing during a time of war: An examination of coverage in *The New York Times*', *International Communication Gazette*, Vol. 71, No. 7, pp. 547-559.

Meredith, M., 2002, *Our votes, Our Guns: Robert Mugabe and the Tragedy of Zimbabwe*, New York: Public Affairs.

Mills, D., 2000, 'Government Side-stepping on real issues', *The Daily News*, 18 April 2000.

MISA-Zimbabwe, 2009, 'Ban on BBC and CNN lifted'. (http://www.ifex.org/zimbabwe/2009/07/30/bbc_cnn_ban_lifted). 13 April 2011.

Mkodzongi, G., 2010, 'Zimbabwe's land Reform is Common Sense', *Pambazuka News*, 3 November 2010. (http://www.pambazuka.org/en/category/features/62917). 26 May 2011.

MMPZ, 2000, *Election 2000: The Media War*, Harare: Media Monitoring Project Zimbabwe.

MMPZ, 2002, *Media Under Siege: Report on Media Coverage of the 2002 Presidential and Mayoral Elections in Zimbabwe*, Harare: Media Monitoring Project Zimbabwe.

Moyo, S., 2007, 'Zimbabwe Land Reform and the Way Forward', *The Mail & Guardian,* 05 April 2007.

Moyo, S. and Yeros, P., 2007a, 'The Radicalised State: Zimbabwe's Interrupted Revolution', *Review of African Political Economy*, Vol. 34, No. 111, pp. 103-121.

Moyo, S. and Yeros, P., 2007b, 'The Zimbabwe Question and the Two Lefts', *Historical Materialism*, Vol. 15, No. 4, pp. 171-204.

Moyo, S. and Yeros, P., 2009, 'Zimbabwe ten years on: Results and prospects. *Pambazuka* issue No. 419. (http://www.pambazuka.org/en/category/courses/54037). 15 July 2010.

Moyo, S., 2001, 'The Land Occupation Movement and Democratisation in Zimbabwe: Contradictions on Neo-Liberal Millennium', *Journal of International Studies*, Vol. 30, No. 2, pp. 311-330.

Moyo, S. and Yeros, P., 2005, 'Land Occupation and Land Reform in Zimbabwe', in S. Moyo and P. Yeros, eds., *Reclaiming the Land: The Resurgence of Rural Movements in Africa, Asia and Latin America*, London: Zed Books.

Mphisa, A., 2009, 'Land Reform Source of Future Conflict', *The Zimbabwean Times*, 4 December 2009. (http://www.thezimbabweantimes.com/p=25599). 9 May 2011.

Muzondidya, J., 2011, 'The Elephant in the Room: A critical reflection on race in Zimbabwe's protracted crisis'. (http://www.solidaritypeacetrust.org/834/the-elephant-in-the-room-a-critical-reflection-on). 12 May 2011.

NewZimbabwe.com Interview with Finance Minister Biti, 3 May 2009. (http://www.newzimbabwe.com/pages/interview87.19761.html). 20 February 2010.

Raftopoulos, B., 2005, 'The Zimbabwe crisis and the challenges of the left', Paper presented at the University of Kwazulu-Natal, 23 June 2005. (http://www.kubatana.net.docs/opin/050623_left_and_zimb_crisis.pdf). 21 September 2009.

Parenti, M., 1993, *Inventing Reality: The Politics of News Media*, New York: St Martins Press.

Riffe, D., Lacy S., and Fico, F., 2005, *Analysing Media Messages Using Quantitative Content Analysis in Research*, London: Lawrence Erlbaum.

Ronning, H. and Kupe, T., 1997, 'The dual legacy of democracy and authoritarianism: the media and the state in Zimbabwe', in J. Curran, M. Park, eds., *De-Westernizing Media Studies*, London: Routledge.

Rutherford, B., 2005, 'The Rough Contours of Land in Zimbabwe', *Fletcher Forum of World Affairs*, Vol. 29, No. 2, Summer 2005, pp. 103-115.

Sachikonye, L., 2003, 'The Situation of Commercial Farm Workers after Land Reform in Zimbabwe: A Report Prepared for the Farm Community Trust of Zimbabwe', March 2003, Harare: Farm Community Trust of Zimbabwe.

Saunders, R., 1991, 'Information in the Interregnum: The Press State and Civil Society in Struggles for Hegemony, Zimbabwe 1980-1990', PhD Thesis, Carleton University.

Scoones, I. and B. Mavedzenge, 2010, 'Don't condemn Zimbabwe: Despite the global outrage, Robert Mugabe's land reforms have had some successes and are boosting trade', *The Guardian*, 8 November 2010.

Scoones, I., Marongwe, N., Mavedzenge, B., Murimbarimba, F., Mahenehene, J., and Sukume, C. 2010, *Zimbabwe's Land Reform: Myths and Realities*, Harare: Weaver Press.

Stone, B., 2007, 'Zimbabwe's Different path and Penalty Incurred', *Global Research*, 22 May 2007.

Stoneman, J., 1988, 1988, *Zimbabwe's Revolution*, Cambridge: Cambridge University Press.

SW Radio, 2005: Behind the headlines talks to Munyaradzi Gwisai, Interview with Lance Guma, 4 May 2005. (http://maravi.blogspot.com/2009/09/sticky-kubatana-behind-headlines-talks.html). 14 December 2010. (http://www.newzimbabwe.com/pages/interview87.19761.html). 14 May 2009.

Thornycroft, P., 2009, 'Union Says Zimbabwe Farm Workers Worst Abused Sector in Past 10 Years'. (http://www.voanews.com). 10 May 2011.

Taylor, Y., 2007, 'The Construction of White Victims in Zimbabwe', Paper presented at the workshop on Cultural Construction of Zimbabwe, 20 -22November 2007, Harare.

Thondlana, B., 2000, 'Zimbabwe: Beginning of the end of Mugabe', *African Business*, April.

Tuchman, G., 1978, *Newsmaking: A Study in the Construction of Reality,* New York: Free Press.

Utete, C., 2003, *Report on the Presidential Land Review Committee on the Implementation of the Fast Track Land Reform Programme, 2000 – 2001,* Volume One, Harare: Government of Zimbabwe.

Waldahl, R., 2004, *Politics and Persuasion: Media Coverage of Zimbabwe's 2000 Election,* Harare: Weaver Press.

Willems, W., 2004, 'Peaceful demonstrators, Violent invaders: Representation of the Land questions in the Zimbabwean Press', *World Development*, Vol. 32, No. 10, pp. 1767-83.

Willems, W., 2005, 'Remnants of Empire? British Media Reporting on Zimbabwe', *Communication and Culture, Special Issue on Zimbabwe,* October: pp. 91-108.

Winter, J., 2010, *Zimbabwe land reform 'not a failure'*, BBC. (http://www.bbc.co.uk/news/world-africa-117640042M). 26 May 2011.

International Media

'White farmers in Zimbabwe struggle against increasing violence', *The Telegraph*, 11 June 2010.

The Times (London), 14 October 2008.

'More resources for land survey', *The Sunday Mirror*, 12 November 2006.

'The hidden hand in Zim politics', *The Sunday Mirror*, 9 July 2006.

'White land grab policy has failed, Mugabe confesses', Reuters, 03 March 2005.

'SA land reforms walk uneasy path', Reuters, 28 October 2004.

'Reform to be according to Law', Nujoma, AFP, 23 April 2004.

'Food Crisis in Zimbabwe Worsens' UN, AFP, 03 October 2003.

'Mugabe Blockading Food Relief – Zimbabweans Starve', T*he Mail and Guardian*, 17 November 2002.

'Zimbabwe's food crisis: What went wrong', Reuters, 01 August 2002.

'White farmer killed in Zimbabwe', BBC News, 18 March 2002.

'court backs land seizures', *CNN.com*, 4 December 2001.

'White farmer killed by Zimbabwean war veterans', *The Guardian*, 8 August 2001.

'Seventh white farmer killed in Zimbabwe', *The Independent* (UK), 13 December 2000.

'Zimbabwe presses its seizures of farms', *The New York Times*, 12 November 2000.

The Telegraph, 16 July 2000.

'Is Mugabe's strategy working?', *The BBC*, 13 April 2000.

'Mugabe defends land seizures', *CNN*, 13 April 2000.

'Stay Cool on Zimbabwe Crisis, Mugabe says', Reuters, 13 April 2000.

'Embattled Mugabe confronted by rule of law', *The Guardian*, 13 April 2000.

'CFU in new bid to remove occupiers', *The Chronicle*, 11 April 2000.

'Mugabe warns boers to leave', *BBC News*, 8 April 2000.

'Mugabe warns 'Boers' to leave', *The Observer*, 8 April 2000.

'Mugabe "will not negotiate" over white land', *The Times*, 8 April 2000.

'Zimbabwe's large farms face squeeze', *BBC News*.com, 6 January 2000.

'Regional concern over land crisis', *BBC News*, 16 May 2000.

Private Media

'Zim's land allocation deeply flawed', *The Zimbabwe Independent*, 20 May 2005.

The Zimbabwe Independent, 06 May 2005.

'Women farm workers bear the brunt of land seizures', T*he Standard*, 9 January 2005.

'Mugabe gets tough with land grabbers', *The Financial Gazette*, 22 July 2004.

'Commercial farmer under siege at farm', *The Standard*, 11 July 2004.

'Government's chaotic land reform vexes planners', *The Standard*, 28 November 2004.
'Displaced farm workers now destitute', *The Standard*, 25 April 2004.
'$75b farm equipment vandalized, stolen', *The Zimbabwe Independent*, 15 August 2003.
'Land crops set to decline by 60%', *The Zimbabwe Independent*, 30 May 2003.
'Land Reform: a Revolutionary Move or Political Gimmick?', The Daily News, 3 March 2003.
'Corrupting the law', *The Standard*, 22 August 2002.
'War vets illegally auction farm equipment', *The Zimbabwe Independent*, 8 February 2002.
'What has Zanu PF been doing for 20 years?', *The Zimbabwe Independent*, 21 December 2001.
'Disaster strikes-Ethnic Cleansing-mass displacements', *The Zimbabwe Independent*, 31 August 2001.
'Ex-fighters hold farmer hostage for three days', *The Daily News*, 9 July 2001.
'What is Mugabe's real motive on the land issue?', *The Daily News*, 24 April 2001.
'Another white farmer killed', *The Daily News*, 12 December 2000.
'Government deploys army to direct farm invasions', *The Zimbabwe Independent*, 20 April 2000.
'EU supports calls for rule of law', *The Daily News*, 11 April 2000.
'Avoid primitive solutions to land', *The Daily News*, 25 March 2000.

Public Media

'Repeal Sanctions Act, Biti urges US', *The Herald*, 2 May 2009.
'Anglo-Saxon racism at war with Zim over land', *The Sunday Mail*, 20 April 2008.
Kenneth Kaunda headlined 'Western Countries Wrong', *The Herald*, 14 June 2007.
Neil Thomas's 'Zim's suffering externally driven', *The Herald*, 14 January 2007.
'Gwisai dismissal from MDC unjustified', *The Herald*, 8 December 2002.
'Revolt against MDC leaders Gwisai urges supporters', *The Herald*, 5 December 2002.
'White commercial farmers perfect economic terrorism', *The Herald*, 24 September 2001.
'Odzi farmer kills resettled farmer: resettled man ran over, dragged for 20m', *The Herald*, 16 May 2001.
'Farmers on warpath', *The Herald*, 8 August 2000.
'Farmers organize attacks on war vets, police issue stern warning', *The Sunday Mail*, 2 July 2000.
'Communal farmers support farm invasions', *The Herald*, 09 June 2000.

'SA demonstrations support Zimbabwe war veterans', *The Herald*, 27 May 2000.
'Beatrice farmer murdered', *The Herald*, 9 May 2000.
'Over 50 per cent of South Africans support ex-combatants occupation of farms', *The Herald*, 04 May 2000.
'Methodist Church welcomes land talks', *The Herald*, 02 May 2000.
'ZCTU calls for speedy land redistribution', *The Herald*, 02 May 2000.
The Herald, 25 April 2000.
'We support Zanu (PF) on Land Issue: Former ZIPRA members', *The Herald*, 25 April 2000.
'Nyamandlovu farmer dies in shootout with war veterans', *The Herald*, 19 April 2000.
'Anglican Church backs Land Reform Programme', *The Herald*, 13 April 2000.
'War veterans vow not to leave farms', *The Herald*, 3 April, 2000.
'German delegation implores state to intensify land reform process', *The Herald*, 10 March 2000.

9

The Zimbabwe Model: Radicalisation, Reform and Resistance

Sam Moyo and Paris Yeros

Introduction

The world system has entered a period of prolonged crisis which is already producing a new generation of radicalisms. If we were to judge from previous periods of systemic transition, the current one is likely to evolve through a series of revolutionary situations and eventually yield a handful of revolutionary ruptures, which will unleash tidal waves throughout the system. But every radicalisation and revolution will obtain distinct characteristics, in accordance with local conditions, with some being more innovative than others in confronting universal challenges. This is the case of Zimbabwe's radicalisation.

Although it occurred a decade before the 'Arab Spring', Zimbabwe's radicalisation has not aroused as much intellectual interest. The propaganda war by corporate media has a large share of responsibility for this indifference (see Chari, Chapter 8), but also the larger process of 'intellectual structural adjustment' that has been underway since the 1980s (Moyo and Chambati, Chapter 1). Thus, a genuine confrontation with imperialism has been roundly dismissed as a case of African 'despotism', requiring 'regime change' (Moyo and Yeros 2005b). It is, indeed, an irony that on the eve of systemic transition, 'regime change' is being promoted as the only relevant historical category: not only has Zimbabwe's rebellion been condemned for not meeting historical criteria, North Africa's rebellions must now be made to conform to them, by force. Yet, both are cases of robust revolutionary situations under contemporary imperialism and comparable to other cases in Latin America and Asia (Moyo and Yeros 2005a, 2011a).

The researchers involved in this book have been engaged in rigorous debate over the last decade and find that a distinct and advanced case of radicalisation, structural reform and resistance to imperialism has been in progress in Zimbabwe. In fact, we may speak of a 'model': not a model in the Weberian sense of 'ideal types'; nor a model which deserves uncritical emulation; nor a model of revolution, for it did not result in one (Moyo and Yeros 2005b, 2007a, 2011b). But it is a model of radicalisation which stubbornly escalated through most of the contradictions of a contemporary revolutionary situation and offered a number of lessons along the way. Zimbabwe has undergone a multi-class, rural-urban political mobilisation; suffered international sanctions, political destabilisation and militarisation; and experimented with a new economic structure with a diversified set of external economic relationships.

This concluding chapter elaborates on six points that make the recent Zimbabwean experience distinct and innovative. They include: (i) the character of the land movement, which has been multi-class, decentralised and anti-bureaucratic, but also united by radical nationalism; (ii) its capacity to articulate grievances across the rural-urban divide; (iii) the radicalisation of its petty-bourgeois components; (iv) the resulting creation of a tri-modal agrarian structure as a matter of state policy; (v) experimentation with state dirigisme, developmentalism and an emerging popular cooperativism; and (vi) a new non-alignment policy termed 'Look East'.

But before we delve into these issues, it is important briefly to interrogate the historical context of Zimbabwe's decolonisation, as a counterpoint to a revisionist historiography which has emerged in the course of radicalisation.

The decolonisation of Zimbabwe: why history matters

This is not the first time that Zimbabwe has been seen as a model. In the 1980s, it was actively promoted as a model of political transition in the settler societies of Southern Africa, whereby majority rule was to be conditioned on property guarantees. It also became a pilot project for market-led land reform, which later flowered into general World Bank policy. What is different now is that Zimbabwe has proposed new ways of deepening the transition to majority rule, by means of radical land reform and, as happened elsewhere in Africa after decolonisation, through an 'indigenisation and empowerment' programme.

After the stalling of decolonisation in Southern Africa in the 1960s, the process was re-launched by a combination of armed and political struggle, leading to military victories in Mozambique and Angola against Portugal and negotiated transitions in Zimbabwe, Namibia and South Africa. This was an integrated, thirty-year regional conflict, at a time of wider imperialist crisis. Crucially, the post-independence invasion and destabilisation of Angola and Mozambique by the apartheid regime in South Africa was used by imperialism as a lever of negotiation, until the region as a whole succumbed to a generalised pact in the 1990s: peace, independence and majority rule, in return for property guarantees, plus economic opening to monopoly and finance capital. Unlike in the rest of the South, decolonisation and neoliberalism in Southern Africa coincided, the one being conditional on the other. But the pact was unstable from the beginning.

In the case of Zimbabwe (Moyo and Yeros 2011b), the pact included those in the nationalist movement, led by the Patriotic Front parties (ZANU-PF and PF-ZAPU), who viewed the pact as a strategic objective, seeking piecemeal reforms and eventually the growth of a black middle class; as well as those in the movement who saw the pact as a tactical move, intended to consolidate political gains and prepare for the next the economic phase of the struggle. For imperialism, it was a tactical retreat, aiming to cut its losses and rely on economic statecraft to maintain its monopoly position. A previous, watered-down plan for 'independence', by which white political privileges would have been retained indefinitely, was negotiated with a colonial proxy advocating peace with Rhodesia (Abel Muzorewa's UANC), but was defeated. Nonetheless, the above 'trifurcated' contestation, born of the Lancaster House negotiations in 1979, was never laid to rest, even as, in the closing years of the Cold War, domestic political forces were temporarily co-opted into accepting the pact as final. The adoption of structural adjustment in 1990 and, thereafter, the generalisation of the pact to the region, raised false hopes for a peace and development dividend all around.

To understand the subsequent radicalisation in Zimbabwe, the character of the decolonisation pact must be clarified. This character remains important, given the emergence of a revisionist historiography, which claims to be more peaceful and democratic than the violent 'patriotic history' of the nationalist leadership (Ranger 2004: Raftopoulos and Mlambo 2009). This revisionism is essentially the reincarnation of a liberal form of settler-colonial political compromise. Despite the phasing out of white political privileges, Zimbabwe remained a racially divided society, in which the defence of 'human rights' served mainly to protect white property and race-based privilege. Neocolonialism in Zimbabwe, not only relegated the majority population to a permanent process of semi-proletarianisation and super-exploitation, it also excluded the possibility of the emergence of a black middle class with roots of its own in the economy.

The structural violence inherent in this 'post-white settler' type of neocolonialism (Mandaza 1985) was never to be pacified by piecemeal reforms. As the country entered structural adjustment in the 1990s, even the visible social gains of the prior decade were reversed.

Another peculiar political dynamic was also in place. Like in Angola, Mozambique and Namibia, where liberation was obtained by armed struggle, the security apparatus of the new state was rapidly taken over by guerrilla commanders. Other branches of the state were 'Africanised' in due course. But from early on, control over the security apparatus became a political resource for petty-bourgeois struggles. Since there were diverse elements among the security forces with varying inclinations vis-à-vis the independence 'pact', any perceived sign of contestation among the parties over the military apparatus, the electoral dispensation and worse, suspicion of South African involvement, at a time when apartheid destabilisation was rife, tended to seriously unsettle the balance. This political dynamic degenerated into a fratricidal conflict in Matabeleland from 1983 to 1987. In effect, petty accumulation impulses, instead of challenging racial inequalities and defending against de-stabilisation, were channelled into a violent, 'ethnic' competition over exclusive control of the state apparatus. But from a longer historical perspective, it is also clear that the accumulation needs of the petty-bourgeoisie could not be realised. Under the different neoliberal conditions of the 1990s, marked by ongoing obstacles to accumulation and social differentiation, the petty-bourgeoisie was forced back into a popular, inter-class black alliance against the status quo dominated by settler and foreign capital.

Various aspects of this history and the subsequent process of radicalisation have been analysed in detail elsewhere (Moyo and Yeros 2005b, 2007a, 2011b; Sadomba 2008). The irony is that, by the time Zimbabwe entered the process of re-radicalisation, intellectual discourse had already suffered a historic reversal, to such a degree that a settler-inspired revisionist history, based on an awkward confluence of liberalism, Weberianism, post-structuralism and pseudo-Gramscianism, could now pass as 'progressive' and even dominate publication outlets with a 'radical' tradition (Moyo and Yeros 2007b; Moyo and Chambati, Chapter 1).

Radicalisation and its mode of mobilisation

The land movement: decentralised, anti-bureaucratic agency

The land movement was initiated by popular rural and urban mobilisation against the immediate policy of the ruling party and the state, under the leadership of liberation war veterans. This point is affirmed by Wilbert

Sadomba and Luis Masuko (Chapters 3 and 4) and is contrary to the 'land-grab orchestration' scenarios (Hammar et al 2003), or to supposedly 'agnostic' assessments as to the 'impossibility of generalisations' (Scoones et al 2010). The nationalist leadership dragged its feet until 1997, stepping in only when it risked losing its most critical social bases, the peasantry and the war veterans, the latter permeating the security forces of the state apparatus. The purpose of the nationalist leadership was to control and co-opt the land movement, as well as to open a political space for the expression of pent up land demands among layers of the population, some of which were not directly organised by war veterans. Most crucially, it did so to accommodate the interests of the aspiring black bourgeoisie, through a bifurcated land redistribution programme, providing for both peasant and small-scale capitalist farming (see Moyo, Chapter 2). It also spared from redistribution certain farms owned by foreign capital, the state and public trusts, ostensibly to maintain some critical food supplies and agro-industrial capacity.

Streamlining the land movement was critical to the state, by creating Land Committees at district and provincial levels, as well as Committees of Seven on the farms, while diminishing the powers of local war veterans who were the vanguard of the land occupations (see Sadomba, Masuko and Murisa in this volume). In their place, civil servants, chiefs and other war veterans, not connected directly to local struggles, were installed, thereby broadening and diluting the representation and class character of the land movement. Over the following years, gaining firm control over the movement was, however, made difficult by the war veterans' decentralised and anti-bureaucratic character. This form of agency was enabled by historic and organic roots of social mobilisation developed during the armed struggle, as well as by the pre-existence of localised land movements.

This decentralised and anti-bureaucratic nature of the land movement is its first distinctive characteristic, essential for understanding the success of this mass mobilisation. Formally constituted and bureaucratised organs of political representation, such as political parties, farmers' unions, trade unions and NGOs lacked either the interest or the organic roots to mobilise a radical land movement (see Moyo 2001; Yeros 2002; Moyo and Yeros 2005b). The formally constituted war veterans' association, the Zimbabwe National Liberation War Veterans Association (ZNLWVA), was also lacking in this regard (Sadomba, Chapter 3). This characteristic may be seen as having parallels with the recent North African mass mobilisations, with the exception that in Zimbabwe this decentralised nature has been based on a unifying principle of radical nationalism.

Bridging the rural-urban divide against occupational corporatism

The second distinctive characteristic of Zimbabwe's radicalisation was the extensive rural-urban spread of the land movement, in terms of active membership and physical participation in the land occupations. The leadership of the land movement included local peasant leaders, local war veterans, spiritual leaders, some chiefs and various working class activists, intellectuals and political party leaders, in a cross-class alliance. But war veterans and various local leaders played a vanguard role in galvanising the mobilisation of long standing grievances over land and racial inequality.

If political parties, farmers' unions, trade unions and NGOs have lacked sufficient interest or organic roots in the land question, they have also been structurally incapable of bridging the rural-urban gap in the interest of mass mobilisation. The land movement did manage to bridge this gap, by both incorporating urban elements into rural land reform and promoting land occupations in urban areas for residential purposes. Thus, the land movement overcame the occupational corporatism of trade unions and farmers' unions and the often divisive strategies of political parties. This form of mobilisation is a rare phenomenon, which has some parallels in the contemporary world, namely in Bolivia and Nepal and, to a lesser degree, in Venezuela, but apparently also in Egypt (see Moyo and Yeros 2011a).

It remains important to emphasise the bureaucratic sclerosis and the sources of political polarisation that have accentuated the rural-urban divide. By the mid-1990s, trade unions, led by the Zimbabwe Congress of Trade Unions (ZCTU) had completely abandoned land reform as a political project (Yeros 2002). Previously, the labour centre had maintained in its analysis and political discourse, however superficially, an acknowledgement of the importance of the land question. Yet, as the ZCTU pried itself away from the control of the ruling party and the state in the late 1980s and also articulated a critique of structural adjustment in the early 1990s, it gravitated increasingly to a political project of 'good governance' and 'regime change', promoted by foreign donors and international trade unions. In so doing, it joined forces with a broad array of liberal, urban-based, middle-class, donor-dependent NGOs, including the National Constitutional Assembly (NCA). By the time the ZCTU founded the Movement of Democratic Change (MDC) in 1999, all the 'pro-democracy' forces had been completely overwhelmed by white-settler interests and foreign donors (see also Gwisai 2002).

Farmers' unions representing the peasantry had also, in the 1990s, distanced themselves from the land reform agenda (Skalnes 1995) as petty-bourgeois interests prevailed among their ranks, to focus mainly on access to state services and subsidies. Although they did not expressly oppose the land reform, they were both uninterested and unable to mobilise a constituency in the interest of repossessing land. On the other hand, the white-settler Commercial Farmers' Union (CFU), in alliance with GAPWUZ, the farm workers' trade union, mobilised both its membership and international public opinion against the land occupations.

It has been claimed that such 'pro-democracy' alliances have been the vanguard of 'progressive' politics in Zimbabwe and the reason why the ruling party opted for a radical position on the land question (Raftopoulos 2009), as if there were no real political or historical basis for such a position. With the mounting evidence of an extensive land redistribution, there has now been a veiled acknowledgment of the vanguard role of the land movement. However, this role is rendered as a mere component, together with the MDC alliance, of a broader 'passive revolution', as per Gramsci, that has 'remained largely under the control of the state' and that has 'largely politically marginalised the majority of the population' (Raftopoulos 2010: 707). Such an interpretation serves only to obscure the distinctive features of a rare mass mobilisation which confronted the white agrarian monopoly and the imperialist alliance as a whole, to the effect of liquidating the settler element and broadening the social base of the economy.

Petty-bourgeois radicalism: an unexpected factor?

There are outstanding issues regarding the relationship of the land movement to the nationalist leadership. The difficulty of interpretation lies in the fact that the ruling party, having succumbed to structural adjustment, changed course in the late 1990s to enter a process of radicalisation, even as it sought to streamline and control the land movement. Most analyses have adopted a 'neopatrimonial' conceptual framework, for which the only relationship that exists in society is between rapacious black capitalists and their ethnicised client networks. Even the so-called Gramscians have replicated this imagery, seeing in the above 'passive revolution' a 'destructive party accumulation project' (Raftopoulos 2010: 706), not a radicalisation of an array of forces, which included the semi-proletariat and aspiring black capital, all against monopoly capital. Others, despite their keener interest in class analysis (see Masuko and Sadomba, in this volume), have inclined in a similar direction, arguing that black capital never really broke ranks with monopoly capital, acting solely on the latter's behalf to control the land movement. We have argued elsewhere that the process of radicalisation integrated diverse class interests, including the petty-bourgeoisie and the semi-proletariat, against the white agrarian faction of monopoly capital. This radicalisation resulted neither in a revolution, nor in a generic 'passive revolution', for the white agrarian establishment was essentially liquidated both economically and politically. The role of the petty-bourgeoisie and the nationalist leadership, their use of the state and their relationship with the movement must be interrogated further.

The character and function of the 'radicalised state' underwent a peculiar transformation: it suffered a suspension of its bureaucratic coherence (its 'bureaucratism'), just as its personnel was being mobilised in the interest of Fast Track Land Reform (Moyo and Yeros 2007a, 2011b). The reconstitution of Land Committees and Committees of Seven overrode local bureaucratic structures – something that the 'chaos' theorists have seen as the 'destruction of the state' (Hammar et al 2003) – but it also established *fast-track* procedures and new capacities for the expropriation and redistribution of land, while also reforming laws and amending the constitution to underpin the action and defend land occupiers against eviction. The breaking of ranks with monopoly capital is also exemplified in the fact that the state expropriated nearly 5,000 properties and redistributed them, going far beyond the estimated 1,000 properties that were actually occupied by the land movement. Moreover such acquisitions persisted beyond the immediate election contests.

From a left perspective, one may rightly fault the ruling party for streamlining the land movement and creating space for the petty-bourgeoisie. But it is not the case that it fulfilled a reactionary role, for it did not defend the *status quo ante*. Empirically, this was not the case and in our view, this formulation does not adequately recognise the existence of *real* intra-class conflict, between petty-bourgeois and monopoly capital, black and white elites and among black elites.

This conflict suggests that the third distinctive characteristic of Zimbabwe's radicalisation is the emergence of petty-bourgeois radicalism. This radicalism is another rare phenomenon – although, incidentally, this is also gaining ground in South Africa. he petty-bourgeoisie itself was radicalised, mainly by the land movement, but also by the nature of the external 'regime change' interventions. Certainly, it did so largely on its own terms, but there is a problem in attributing radicalisation solely to certain local-level war veterans, against all the rest that vied for land. Instead of one 'genuine' category of radicalism, there are different radicalisms, each with its own *class* project.

That the petty-bourgeoisie also became an agent of change surely presents very difficult political questions, as previous debates among African scholars have shown (Fanon 1967; Cabral 1978; Shivji 1976). Those who have opposed the petty-bourgeoisie outright, in the case of Zimbabwe, to the point of closing ranks with the 'regime change' agenda, have taken recourse to racialised discourses of 'corruption', 'patrimonialism' and 'orchestration'. This tendency vilified the whole of the land movement on the basis of the attendant use of force and the unfair advantages that some political elites sought.

Others have claimed to stand aloof of the difficult political questions, but have, nonetheless, deployed a liberal-populist 'people *versus* state' dichotomy. This approach renders the whole land reform process solely as a consequence of the agency of the landless against an indifferent state, at best, or a 'commandist' and 'clientelist' state, at worst (on this view, see Scoones et al 2010). Class analyses that reach similar conclusions can only do so by downplaying the radicalisation of the petty-bourgeoisie and treating it as if it never really broke ranks with monopoly capital (e.g. Sadomba, Chapter 3).

It would be more correct to say that the nationalist leadership in recent years has come to represent mainly *un-accommodated* bourgeois interests, which indeed have liberation convictions of their own, but which are under the illusion that they can reform monopoly capitalism so as to sustain a 'patriotic bourgeoisie' into the future. This situation explains the current pressures for 'indigenisation' programmes in strategic industries (to be discussed below), as opposed to more collectivist solutions (Moyo and Yeros 2011b: Moyo 2011b).

This situation also goes a long way to explain the violence that has accompanied land reform, mainly off the farms, as the nationalist leadership has, once again, proven unable to commit 'class suicide' and submit itself to the evolving and expanding popular demands on the ground (Moyo and Yeros 2009, 2011b). The bifurcation of the Fast Track Land Reform, the strategy of indigenisation of agro-estates and other industries and the recurrent violence are manifestations, not only of class conflict, but also of *intra*-class conflict between petty-bourgeois interests and monopoly capital.

But if we were to fault a radicalised nationalist leadership for an illusory petty-bourgeois project, a similar fault, albeit of a different order, may be attributed to the war veteran movement. Sadomba, for example, agrees that the war veteran movement became 'tactically sterile' and paid for this sterility dearly in the Murambatsvina assault on urban settlements in 2005. But, could a decentralised movement, even the one responsible for bringing radical land reform to fruition, overcome such tactical limitations? The uninterrupted escalation of the revolutionary situation would have required that the land movement undergo organisational and ideological innovation, one founded in proletarian consciousness and equipped with more sophisticated tactical and strategic thinking – which ultimately did not occur.

Masuko argues (in this volume) that, in this respect, the land movement did undergo innovation, beyond the single-issue platform of land reclamation, evident in the plethora of associational forms that have sprouted in the resettlement areas. Such associational forms are certainly the kernel of future progressive politics in the countryside, as Murisa (Chapter 7) also suggests. But their new issue-focus on service provision by the state (agricultural inputs, social infrastructure, markets, credit and subsidies) is far from articulating a new, radical mass movement; for now, this opportunity has dissipated. On the other hand, the liberal 'pro-democracy' movement, comprising the donor-funded MDC, NGOs and settler elements, continue to have no interest in the radical potential of such associations on the ground. Instead they persist with a limited 'pro-democracy' and market led agenda.

Structural reform and its new contradictions

Trajectories of accumulation: internal, from below and from above

The re-grouping of popular forces is all the more necessary given the new tendencies of class formation at the top. The land reform radically restructured land ownership, but it did not 'oust capital', which itself is now re-grouping (Moyo 2011b). This outcome leads us to the fourth distinctive characteristic of the radicalisation process: the deliberate design of competing trajectories of accumulation. A new tri-modal agrarian structure has been instituted through state policy, consisting of peasant, small-scale capitalist and large-scale estate farms, based on differential landownership regimes (state-sanctioned usufruct permits, non-tradable leases and freehold or state property, respectively), which in turn gives rise to different types of producers vying for different types of labour mobilisation and accumulation strategies (Moyo 2011b, 2011c). The evidence shows that Zimbabwe has unravelled the settler-dominated 'labour reserve' economy of the past, by amplifying the smallholder sector and incorporating a significant 'merchant' path, while retaining elements (albeit downsized) of the 'junker' and 'state' paths (for the general characteristics of these paths, see Moyo and Yeros 2005a.).

It is important to note that the diverse elements of this structure are not entirely unique to the continent, but their clear demarcation in state policy and the dynamic by which they have been established, do make this case unique. It has been argued before that, during the 1990s, in Africa as a whole, a new land concentration process was set off by neoliberal land reforms, a process led by domestic capital in association with foreign interests (Moyo 2008). This process installed a 'merchant' path generally, although it never became clearly articulated in state policy. Recently, there are signs of policy interest in this path, as noted in Mozambique, Namibia, South Africa and Ghana, for example.

From 2000 onwards, under the degenerating world-systemic conditions, this path has been overtaken by a larger process of large-scale land alienation by foreign capital itself, often with domestic allies. This alienation is now installing a new 'junker' path on the continent, most notably in the historical macro-regions of the colonial trade and concessionary economies, which had never shared the settler labour-reserve traits of Southern Africa (Amin 1972). What is unique in the case of Zimbabwe is that it has rowed against the current to meet the rest of Africa halfway, by breaking up the large-scale farming established in the course of the nineteenth-century scramble, broadening the small-scale capitalist sector, which had also been introduced by the colonial regime and preserving some agro-industrial estates (Moyo 2011b).

Any genuine class analysis of the new Zimbabwe must come to grips with the tendencies and contradictions of this tri-modal structure and avoid regime-change theories of 'rentier economy' (Davies 2005) or 'crony capitalism' (Bond 2009), or notions of 'passive revolution', which are based on nebulous assessments of the new class relations (e.g., Raftopoulos 2010). The fundamental question is whether Zimbabwe will be able to sustain, via this tri-model structure, an introverted process of accumulation 'from below'.

The details with regards to the socio-economic characteristics of the new land beneficiaries have been reported already (see Moyo et al 2009; Scoones et al 2010; Moyo, Chapter 2). What is important is to outline the tendencies and contradictions of the new agrarian structure. Three issues should concern us: the new type of labour reserve that has emerged, its attendant processes of class formation and the contest over accumulation strategies.

The structure of the new labour reserve

Re-peasantisation and the break-up of the settler agrarian monopoly has diminished the labour reserve of the past and undermined the functioning of the colonial cheap-labour system. As Walter Chambati argues in this volume (see also Chambati 2011), land reform has absorbed surplus labour into petty-commodity production for own consumption and for the domestic market and pried open access to natural resources and use values that previously were enclosed in the properties monopolised by white farmers. The immediate manifestation of this has been a shortage of labour, which has deprived especially the small-scale capitalist sector of the prior abundant workforce willing to work for wages below the cost of social reproduction.

Previously, the Large-Scale Commercial Farming sector (LSCF) owed much of its productivity to its reliance on the super-exploitation of semi-proletarianised labour. Indeed, by the late 1990s, 50 per cent of its workforce had come to consist of non-permanent, casual labour, which in turn reproduced itself precariously between the LSCF and the Communal Areas. Meanwhile, real wages on the farms had, under the weight of structural adjustment, collapsed to 24 per cent of the Poverty Datum Line, alongside sharp reductions of yields and incomes in the adjacent Communal Areas (see Chambati, Chapter 5). The intensification of super-exploitation all around was further facilitated by a racialised, quasi-feudal labour-tenancy system, together with a patriarchal system of customary authority, which continued to undermine the bargaining power of the semi-proletariat as a whole.

That the labour reserve diminished and the bargaining power of labour altered does not, of course, mean that the labour reserve economy has been extinguished. The persistence of simple reproduction among smallholders and the reconstitution of the small- and large-scale capitalist sectors, under the weight of Western sanctions, continue to re-create the structural conditions of super-exploitation, even among the new self-exploited peasantry. Super-exploitation is further abetted by residual labour-tenancy on the new farms, as well as intra-family and gender-based labour relations. Yet, the unravelling of racialised relations of personal dependence and the expansion of the smallholder sector have altered the balance of power among the three modes of farming. It is here that the new political struggle is now being fought.

Both small- and large-scale capitalist farmers have a structural interest in policy measures that will oblige small producers to work for wages below the cost of social reproduction. This structural interest would be reinforced should an export-oriented accumulation strategy come to pass (see Moyo and Nyoni, Chapter 6). But these two types of farmers are not identical, given that small-scale capitalist farmers, many with significant resource vulnerabilities, may also be co-opted by the state into production for domestic markets and industries. In fact, this objective has largely been their principal orientation to date. At the same time, smallholder farmers will themselves undergo differentiation, thereby adding to the labour pool. Yet, this may also be mitigated by inward-looking policy measures that both reinforce the conditions of smallholder production and induce the growth of cooperativism and rural industries capable of re-organising the labour process. The political struggle between the three modes of farming and the attendant disputes over labour, remains unequal and will be determined by a number of factors.

State interventionism and new 'developmentalism'

The dominant factor in shaping the accumulation trajectory is, of course, the structural power of monopoly capital, which has opposed the radicalisation process and undermined progressive agrarian change by imposing severe limits on Zimbabwe's economic recovery. From the beginning of the Fast Track, financial isolation and a capital strike had led to a severe shortage economy, leading the state towards an interventionist economic strategy. This interventionism under contemporary neoliberalism is the fifth distinctive characteristic of the Zimbabwe model.

We have argued elsewhere that the state initially had no comprehensive plan to defend against sanctions (Moyo and Yeros 2007a). A plan emerged as the internal and external contradictions escalated, taking the form of controls over prices, trade, capital and agricultural markets, the monopolisation of grain purchases by the Grain Marketing Board and the setting of food production targets. The plan also targeted subsidies to agriculture and industries, including for the production of ethanol, thereby reviving an erstwhile Import Substitution Industrialisation (ISI) that had been undermined by structural adjustment. State-owned agro-estates, together with state interests in mining, banking and other firms, are in the forefront of this strategy, especially for the production of local agro-fuels against a rising fuel-import bill. Furthermore, the plan includes an agricultural mechanisation policy to enhance motorised draught power, the bulk of which have been allocated to small- and large-scale capitalist farmers to compensate for the labour shortages produced by Fast Track Land Reform (Moyo 2011a; Moyo and Nyoni, Chapter 6). Other broad-based state investments include irrigation, electricity and transport facilities, although these have remained low given the fiscal constraints. This plan reflects both the class bias of the state and its reaction to the generalised strike by private banks and bilateral and multilateral donors.

Eventually, hyperinflation, political confrontations and informalisation of economic activity compelled the state back to an attempted normalisation with international capital. It is through this process that the state 'interrupted' the momentum of the revolutionary situation, culminating in the assault on urban land movements in 2005 (Moyo and Yeros 2007a, 2009, 2011b). Indeed, the heterodox plan lacked the foresight to defend against the ensuing capital strike, which could have been better resisted by a policy of immediate nationalisation of banks and strategic industries. Thus, the state became susceptible to carrot-and-stick strategies by foreign capital, including its refusal to fully default on debt.

Normalisation has led to cooptation back towards an extroverted strategy through various mechanisms (Moyo 2011a, 2011b). One has been the eventual shift of the land redistribution policy on agro-estates towards an essentially comprador 'indigenisation' strategy, by which black capitalists are to become majority shareholders in agro-estates, thereby succumbing to the logic of plantation agriculture and its associated financial circuit. Another has been the expansion of contract farming, linked to a similar external financial circuit, locking small-scale capitalists into agro-estates for the production of sugarcane for the European market (under the ACP-EU Lomé Convention), as well as for tobacco and cotton for the Chinese market. But the cooptation has been most evident in the adoption, in 2008, in the midst of peak hyperinflation, of a neoliberal policy on currency, capital, trade and agricultural markets. Thus, dependence on external finance, inputs and markets has exercised overriding power in tilting, once again, the internal balance between social classes, while Western sanctions against Zimbabwe, including those against the parastatals spearheading the economic recovery, have been retained.

Yet, the countertendencies are also notable. For the above policy of normalisation has not totally extinguished the dirigisme of the state: the new black bourgeoisie, still acutely vulnerable to a monopolistic world market, remains in conflict with international capital, as do, most obviously, the popular classes from which the nationalist leadership must still claim legitimacy. Despite the neoliberal turn, the state has not abandoned the policy of ISI, or its intention to mediate pro-actively in favour of black capital and, secondarily, smallholder farmers. The class character of state power, the strategies of the black bourgeoisie and the re-grouping of social forces are the three further factors that will co-determine the balance of forces.

Contrary to the trends on the rest of the continent, marked by a new wave of externally-driven land alienation for the production and export of foods and bio-fuels, the Zimbabwean state has persisted with its policy of seeking to build national food self-sufficiency and to substitute for imported petrol by expanding the cultivation of sugarcane on agro-estates owned by the state and public trusts. Producing ethanol for domestic transport and other industrial requirements has various local industrial spin-offs (Moyo 2011b). Such investments are being made via joint ventures with foreign capital, from the East, West and South, under the 'Look East Policy' inaugurated in 2004 (see more below).

The indigenisation strategy has also re-escalated, going beyond agriculture to secondary industries, banking and especially mining. Generally, indigenisation has been a multi-class strategy, whose class character has oscillated in accordance with the correlation of forces. In the 1980s, it shifted from a popular land reform policy to one geared towards the creation of a black bourgeoisie via affirmative action with respect to land. The latter continued throughout the 1990s, under structural adjustment, without much success, until its radicalisation in the Fast Track Land Reform Programme. Then, under the subsequent normalisation, the strategy shifted back to a bourgeois strategy, geared towards creating majority shareholding amongst black capitalists. Yet, a further elaboration of the policy has envisioned joint ventures between state-owned enterprises and foreign firms. This policy is reflected not only in the support for state-owned agro-estates; it has also turned on the mining sector, which has now become the principal target and which has enormous potential to fill the foreign-exchange gap.

Upon the discovery of massive diamond deposits, a struggle ensued, especially from 2007 onwards, for the control of the industry, against both small miners who entered the fray, as well as corporate capital of South African and Western origin. The strategy on diamonds and the possibility of circumventing sanctions, led to a confrontation with foreign capital and small miners, which has entailed the repression of the latter. In the event, the West, ostensibly in solidarity with the repressed small miners, resolved to broaden its sanctions tactics by invoking the 'Kimberly Certification Process' with regard to 'blood diamonds'. Then, as Zimbabwe won the certification battle, the United States proceeded unilaterally to impose new sanctions on two mining firms in partnership with the mining parastatals. Nonetheless, state policy on minerals now seems to be stabilising and is positioning the state to reap future profits, via joint ventures looking both East and West. The accommodation of Chinese capital has been central to this strategy, which has already begun exploration and production. Similarly, the expansion in the production of platinum by Western multinationals was compelled by the threat of losing concessions to the East. Meanwhile, high-ranking state personnel have positioned themselves in the state-owned Zimbabwe Mining

Development Corporation driving the joint ventures, which has undermined the legitimacy and transparency of the strategy. For some 'pro-democracy' forces (e.g., Cross 2011), this critique has become opportunistic, calling for the nationalisation of black capital but not Western capital!

It is important to add that a further elaboration of the indigenisation policy, beyond the re-distribution of majority shareholding and joint ventures, towards a higher degree of social access, has recently been emerging in the wake of popular agitation. This transformation involves the imposition of conditions on foreign firms to undertake investments in physical and social infrastructure, such as roads, schools and clinics, as well as the allocation of shares to 'community and employee trusts'. This strategy may soon be complemented by evolving plans to create institutional markets among smallholders, so as to strengthen local markets. The strategy reflects a renewed attempt, in response to more general criticisms of class bias, to broaden the benefits of indigenisation, especially of mining, to the rural areas. It also reflects the continued need of political elites (combining both ZANU-PF and MDC leaders) to respond to the reaction by capital and to meet popular demands for state support in the light of forthcoming elections.

Overall, these policies reflect the persistence of a specifically *nationalist* accumulation strategy promoted by black capitalists with connections to the state. For, despite having sunk roots of their own in the means of production, they remain vulnerable to both monopolistic forces and the need to maintain legitimacy vis-à-vis popular forces. In other words, black capital continues to seek to consolidate its position by recourse to a pro-active state, against what it considers to be its main obstacle, Western monopoly capital.

Yet, there are other tendencies at play among the black bourgeoisie, which could undermine its nationalist economic posture. For instance, the Fast Track Land Reform obtained a significant ethno-regional structure (Moyo 2011a), as aspiring capitalists, lacking other means to bid for land, mobilised sub-national, ethno-regional claims to land 'rights' to exclude non-local competitors. This tendency continues and could escalate as land bidding is re-focused on the enlargement of existing landholdings, at the expense of smallholders and as bidding spreads to the retained private and public agro-estates. These are essentially the ongoing petty-bourgeois tendencies of a class which remains profoundly insecure. Should the main 'enemy' come to be seen once again as 'internal' (and 'ethno-regional'), there would certainly be regression to a neocolonial type of politics and this would ultimately be more malleable to foreign interests. The immediate manifestation of such a tendency would be the escalation of factional politics, both within the ruling party and within the MDC (see Moyo and Yeros 2007b). Yet, this should not be seen as a foregone conclusion – or a perennial and de-contextualised 'ethnic' possibility in African politics – but as shifting strategies of accumulation, subject to pressures from above and from below.

Rural cooperativism and democratisation

This analysis takes us to a consideration of the economics and politics of the popular classes after land reform. While the larger farmers have been gravitating towards production for export markets (albeit still in minority numbers), the basic pillar of food sovereignty will remain the smallholder farmers, together with a significant portion of small-scale capitalists. There has been a clear shift in the orientation of production towards food grains, to which the new land beneficiaries have dedicated 78 per cent of their cropped land. And while national maize yields per hectare have suffered severe setbacks under conditions of drought and sanctions, beneficiaries in wetter agro-ecological regions have performed much better (Moyo 2011c; Moyo and Nyoni, Chapter 6). But the economic potential remains enormous, considering that land utilisation rates are already at 40 per cent – that is, the land utilisation level of the extroverted LSCF sector prior to Fast Track. But notable in this regard is that, on average, the A2 farmers with larger landholdings crop below 20 per cent of their land, while a few surpass the 50 per cent mark. In the absence of broad-based investments in infrastructure, fertilizer and machinery, fulfilment of the agricultural potential will be delayed and differentiation across regions will deepen, with adverse consequences for national cohesion.

One of the immediate consequences of Fast Track is the re-emergence of informal land rental markets between the better performers and the weaker ones, often between A2 and A1 farmers, respectively (Moyo, Chapter 2). Both macro-economic conditions and labour shortages, on both A1 and A2 farms, have contributed to this tendency. Land sharing is also common, although this often occurs among A1 farmers and kinship networks, as well as between all resettled farmers and farm workers, gold-panners and 'squatters' who have yet to be settled formally (Moyo et al 2000). Such tendencies represent local class differentiation across all agro-ecological regions and herald future conflicts over access to land and natural resources.

Although land tenure is generally seen to be secure, boundary and access disputes could intensify (Moyo, Chapter 2). One of the terrains of struggle that could intensify is the status of leasehold on A2 farms, which is being challenged by domestic and foreign elements which advocate the conversion of the current leasehold land rights into freehold tenures. In this case, small-scale capitalist farmers would find allies in private banks, which typically justify their refusal to finance resettlement farmers on the supposed absence of 'collateral'. Another terrain of struggle is the land tenure status of the remaining farm workers, who have been re-inserted into labour-tenancy relations (Chambati, Chapter 5). Yet, state policy still remains committed to both leasehold tenure and the protection of farm workers against eviction from A2 lands.

These struggles over production, land access, tenure and labour, as well as over the much-needed social services in general, require organised social forces capable of tilting the balance towards smallholders and farm workers. The most promising development is the local emergence of new cooperative movements to pool labour, savings and infrastructure, procure seeds and fertilizers, channel extension services, bid for producer prices and negotiate labour contracts (Murisa, Chapter 7). Some of these groups are orchestrated by state extension agents and by private contract farming firms, while others are led by the war veteran groups which grew out of the previous Committees of Seven in the land occupations. Yet others draw on kinship relations and existing former farming associations in the Communal Areas (see Murisa, Chapter 7; Moyo 2011c). Thus, among resettled farmers, approximately 40 per cent now belong to farmers' groups. Among the farm workers, there are cases of group negotiations for access to land and conditions of work, in the absence of a national agricultural labour union representation, whose credentials have not been in favour of agrarian reform (Chambati, Chapter 5).

This dynamic social development may shape the future of rural and national politics, depending on the ability of rural cooperativism to deepen its scope and branch out to form wider political alliances. The resurrection of mass politics requires building up the new producer associations into an advanced, united and autonomous cooperative movement of rural workers, capable, not only of obtaining ad hoc services, but also of dislocating the new black bourgeoisie from its political pedestal (Moyo and Yeros 2007a).

Rural cooperativism also holds the unique potential to transform gender relations and customary authority. These are the social and political pillars of historic super-exploitation, particularly that of women. Fast Track Land Reform tripled the proportion of rural women holding land in their own right, yet women remain greatly under-represented, with less than 20 per cent of the total farm units. The land movement also opened political space for women, which was filled in mass numbers, yet women seldom held leadership positions in land committees and local farmer associations (see Murisa, Chapter 7). The new cooperativism is the best possible vehicle for broadening the participation of women with respect to land rights, agrarian change and political leadership.

Moreover, cooperativism is the only realistic vehicle for withering away the retrogressive patriarchal aspects of customary authority. Contradictory tendencies have been evident here as well. It is true that the state extended customary authority to resettlement areas, both as a cooptation tactic and a low-cost dispute-resolution mechanism. The state also co-opted chiefs through their inclusion into the A2 farming scheme and mechanisation. Yet, the state has excluded chiefs from exercising authority over A1 land permits and A2 leases and has also maintained their subordination (in some power relations) to elected authorities in local government. Meanwhile, their cooptation into a new class position, where this has occurred, raised new questions regarding the trajectory of this institution. Furthermore, while the ethno-regional structure of Fast Track has also extended the kinship basis of customary authority, it has nonetheless been observed that beneficiaries from non-contiguous areas have not always embraced their new chiefs (Murisa, Chapter 7). Finally, the state has also been active in supporting farmers' groups via agricultural extension officers, contrary to suggestions that new farmers have not received state support (e.g. Scoones et al 2010; Cliffe et al 2011) or that they have been re-tribalised (Worby 2003).

Overall, it is clear that intervention into this fluid field by a new social agent based on cooperative and democratic principles can further erode customary authority, empower women, integrate farm workers and smallholders in agro-industrial production units and expand the potential for the formation of alliances among cooperative producers nationwide. This type of social agent may fulfil the aspirations for popular agrarian change which are necessary after Fast Track. But this transformation should go much further than welfarism would permit, by creating efficient worker-controlled cooperatives to sustain the struggle against monopoly capitalism and retain pressure on the reconfigured state.

Resistance through non-alignment

The changing security context

Internal dynamics, including the class character of the indigenisation strategy and the ongoing social struggles, will determine the ability of the state to sustain an inward-looking accumulation process and its legitimacy. However, the foreign policy of the state will also be crucial in circumventing Western sanctions, towards creating the external conditions for sustaining an inward-looking strategy, as well as in defending Zimbabwe from external intervention, ostensibly on 'humanitarian' grounds. In this regard, the sixth and final distinctive characteristic of Zimbabwe's radicalisation is its 'Look East' policy, which, despite its name, amounts to a vanguard redefinition of 'positive non-alignment' in the post-Cold War period.

The strategic context on the continent has been changing since the 1990s (Yeros forthcoming). As such, the current scramble for Africa has definite antecedents in the recent past. A new phase of land alienation was already underway under structural adjustment (Moyo 2008). To this situation was added a renewed interest in oil, gas and minerals at the turn of the century, until the most recent surge in land alienation for the production of food and bio-fuels. The determinants of the new scramble are to be found, not only in system-level changes, but also in the evolving geo-strategic facts on the ground in Africa.

In relation to energy resources, the 9/11 attacks on US targets was a turning point. The attacks raised the prospect of prolonged instability in Western Asia, setting off a policy debate on the possibility of expanding oil production in Africa, as proposed by the Cheney Report on energy (NEPDG 2001). In turn, this raised obvious concerns in China as to its possible exclusion from key sources of oil and shipping lanes, thereby compelling Beijing to fine-tune and upgrade its own Africa strategy over the following years (GoC 2006). The re-militarisation of US strategy has been most closely associated with this dynamic.

But the less acknowledged source of the scramble has been the changing security context on the continent. And here, several inter-related events shook the foundations of the US geo-strategy. The first event was the political transition in South Africa. A controlled transition though it may have been, it nonetheless deprived the Western alliance of a staunch ally in Southern Africa. The second was the state fracture and war in the DRC, by which the United States lost its main pillar in Central Africa. Thus, the two Cold War pillars of US strategy in these regions (the apartheid state and the Mobutu regime) collapsed in the space of a few years. The third event has been precisely the re-radicalisation of the liberation movement in Zimbabwe, which challenged outright the controlled character of the transitions to majority rule.

These events have been compounded by escalating disputes over the control of Somalia and Sudan in the East, over Ivory Coast in the West and over

North Africa, which again have threatened the control over critical sources of energy. All these have thrust collective imperialism back into crisis and raised the stakes of the scramble. It is in this light that the establishment of AFRICOM – which was deemed unnecessary even in the height of the Cold War – must be seen. AFRICOM's most immediate target may be China, but it is the loss of firm control over large swathes of the continent that has made it necessary.

In the wake of the Libya intervention and the 'Arab Spring' in general, external intervention has taken more complex forms, although it is clear that Zimbabwe has been a critical laboratory for the combined use of 'soft power' (via NGOs and supporting opposition parties) and direct economic pressure through sanctions. In this regard, Mahmood Mamdani (2011) has argued that internal democratic reforms, against the privileges of internal elites, are essential to prevent future Libya-style interventions. But we know that interventions are selective, typically to support extroverted economic interests and to conserve client and corrupt political systems. The North African revolts have themselves led to 'popular' elections which have propelled to power economically and socially conservative forces.

The Zimbabwe case shows us that progressive internal reforms themselves invite aggressive external interventions, which polarise politics towards regime change. That Western sabre-rattling against Zimbabwe has not resulted in a Libya-style intervention, or the external interventions that have been so common since the early 1990s in West and East Africa, has to do, in large measure, with the new SADC security framework which, despite all its prevarications, is now anchored in a mutual defence pact, which has been uniquely effective in preventing the further militarisation of the Zimbabwe question (Moyo and Yeros 2011b). Indeed, the SADC mutual defence pact, which grew out of the 1998 intervention in the DRC by Zimbabwe, Angola and Namibia against the US-sponsored invasion by Rwanda and Uganda (and then broadened to the rest of SADC in 2003), can be seen as a pioneering security structure not only in Africa, but also in the rest of the South.

Finally, the defence against Western aggression has also to do with Zimbabwe's deft foreign policy which has quite effectively used the emerging East against the West.

Zimbabwe's 'Look East' policy

Zimbabwe's current Look East policy (LEP), launched in 2004, is not as new as it appears. Since independence, Zimbabwe has abided consistently by most of its five founding foreign policy principles, which can be summarised as (a) national sovereignty and equality among nations, (b) attainment of a socialist, egalitarian and democratic society, (c) right of all peoples to self-determination and independence, (d) non-racialism at home and abroad and (e) positive non-alignment and peaceful co-existence among nations (Patel 1985; Patel and Chan 2006). The last principle is what concerns us here.

In the aftermath of the Cold War, non-alignment fell into disuse, often discarded, as by the British academic establishment, as 'antique' (Chan 2006: 180). Yet, African scholars have recognised an urgent need to reclaim this principle under the new world-systemic conditions. Thus, Issa Shivji has argued that Africa must 'define its solidarity with the oppressed people against both established and developing imperial hegemonies' (2009: 9). In so doing, the positing of an equivalence between Western imperialism and the emerging semi-peripheries must be avoided.

'Positive' non-alignment is precisely the principle at stake, which should not be confused with 'neutrality' or 'isolationism', but with (a) non-participation in the military projects of the great powers, which de facto means NATO, given that the emerging semi-peripheries have not embarked on militarisation; and (b) the freedom to judge each foreign policy issue on its own merits, based on national sovereignty and interests, which de facto means preserving the right and capacity to impose conditions on foreign economic interests regardless of their origins.

As Patel and Chan have argued, Zimbabwe's LEP must be seen as 'complementary, rather than as an alternative, to engaging with the West' (2006: 182). Indeed, Zimbabwe has neither turned its back on Western capital, nor has it accepted investment from China and the rest of the East or South without conditions. Nor, indeed, has it rejected military assistance from any single source, including the West, even though it has stubbornly confronted NATO strategy and, consequently, suffered an arms embargo since the DRC intervention in 1998.

Zimbabwe's LEP, in effect, has been pursued as a method of circumventing Western sanctions and, by engaging with China, as an instrument to force the West back into investing in Zimbabwe on conditions consistent with its indigenisation and empowerment policy. This strategy is now beginning to bear fruit. On the other hand, its arms procurement policy has not had the same effect. In the period 1980–1999, China accounted for 35 per cent of Zimbabwean imports of major conventional weapons, followed by the UK (26%), Brazil (11%), Italy (9%) and Spain (8%. In 2000–2009, China accounted for 39 per cent, followed by the Ukraine (35%) and Libya (27%) (see SIPRI 2011). Thus, while the West previously had participated in arms sales to Zimbabwe, together with China, from 2000 onwards, Zimbabwe has purchased arms only from non-Western countries, mainly China, but also the Ukraine and Libya.

Zimbabwe's LEP thus appears as a vanguard way of re-defining positive non-alignment in the post-Cold War world. Evidently, the only other African states that have effectively upheld a similar policy, although without the radical restructuring of their internal relations, have been Angola and Sudan.

Conclusion: lessons from Zimbabwe

There are several lessons to learn from Zimbabwe. The first and most obvious, is the need to rebuild autonomous research and intellectual capacity in Africa and the South more generally, a capacity which would be organic to local political struggles. One cannot fail to notice how markedly different the debates in this book are from those led by researchers in Northern institutions, which continue to deploy concepts that reproduce a colonial mindset, not least via the ubiquitous organising concept of 'neopatrimonialism'. This concept only serves to obscure the structural power of monopoly-finance capital and reduces all social relations to localised and ethnicised categories of domination and resistance. It also disables our understanding of the economic geography of Africa, which is now evolving rapidly beyond the structures inherited at independence (Amin 1972). Especially, the new tendencies towards 'tri-modalism' in Africa require urgent research. Clearly, no autonomous development is possible, unless we continue to produce adequate concepts and undertake systematic empirical research.

The second lesson is that radical change is possible. Zimbabwe may have particularities of its own, but the historical-structural and social sources of radical change are firmly rooted in the societies of the South everywhere (see Moyo and Yeros 2005a, 2011a). This is not to say that radical change depends on mere 'will'. Political resignation should not be answered by naïve voluntarism. It is necessary that the correlation of forces in every situation be assessed properly, with the intention of changing it, not preserving it. This also means that a clear understanding of the state apparatus and state power must be developed, not towards a blanket anti-statist policy of 'changing the world without taking power', which remains so hegemonic among social movements, but towards a strategy and tactics which seek to alter state power and unravel the state apparatus in the interest of the oppressed.

The third lesson is that mass mobilisations, in order to endure the countervailing forces that will inevitably align against them, must take seriously the agrarian component of society. This objective should not be merely to accumulate forces for change, but also to initiate a longer-term process of structural change and national resistance, of which the agrarian question is a fundamental component. All societies in recent years that have entered a process of radicalisation have discovered that their food dependence and their domestic disjunctures between agriculture, industry and energy are crucial sources of vulnerability. This potential weakness means that mass mobilisation must also take seriously the project of 're-peasantisation' as an explicitly modern project and as the only alternative in conquering autonomous development in the South (Amin 2012; Patnaik 2012).

Finally, it is crucial that a multi-disciplinary approach is encouraged systematically, as the challenges that are presented by radical change go beyond narrowly-focused disciplines and sub-disciplines which are incapable of seeing the whole. In the case of Zimbabwe, it is clear, for example, that radical change in the countryside became part and parcel of a regional security question, which went largely unnoticed by the dominant analyses. And in this case, a pioneering regional security framework succeeded in confirming the land reform, which otherwise would probably have been reversed by Western military intervention. Radical change and autonomous development require regional strategic autonomy and this also needs to be properly understood.

References

Amin, S., 1972, 'Underdevelopment and Dependence in Black Africa: Origins and Contemporary Forms', *Journal of Modern African Studies,* Vol. 10, No. 4: pp. 503–24.

Amin, S., 2012 (forthcoming), 'Contemporary Imperialism and the Agrarian Question', *Agrarian South: Journal of Political Economy,* Vol. 1, No. 1.

Bond, P., 2009, 'Mamdani on Zimbabwe Sets Back Civil Society', *ACAS Bulletin*, Vol. 82 (March). (http://concernedafricascholars.org/bulletin/82/bond).

Cabral, A., 1978, *A Arma da Teoria: Unidade e Luta I*, Second Edition, Seara Nova.

Cliffe, L., Alexander, J., Cousins B., and Gaidzanwa, R., 2011, 'An Overview of Fast Track Land Reform in Zimbabwe: Editorial introduction', *Journal of Peasant Studies*, Vol. 38, No.5, pp. 907-938.

Cross, E., 2011, 'Nationalise Zimbabwe's Diamond mines', *Africa Legal Brief*, 08 October 2011. (http://www.africalegalbrief.com/index.php/component/content/article/367-nationalise-zimbabwes-diamond-mines-eddie-cross.html).

Chambati, W., 2011, 'Restructuring of Agrarian Labour Relations after Fast Track Land Reform in Zimbabwe', *Journal of Peasant Studies,* Vol. 38, No. 5, pp. 1047–68.

Davies, R., 2005, 'Memories of Underdevelopment: A Personal Interpretation of Zimbabwe's Decline', in B. Raftopoulos and T. Savage, eds., *Zimbabwe: Injustice and Political Reconciliation*, Cape Town and Harare: IJR and Weaver Press.

Fanon, F., 1967, *The Wretched of the Earth*, trans. C. Farrington, London: Penguin Books.

GoC [Government of China], 2006, *China's Africa Policy.* (www.gov.cn/misc/2006-01/12/content_156509.htm).

Gwisai, M., 2002, 'MDC Go Back to the People'.

Hammar, A., B. Raftopoulos and S. Jensen, eds., 2003, *Zimbabwe's Unfinished Business: Rethinking Land, State and Nation in the Context of Crisis*, Harare: Weaver Press.

Mamdani, M., 2011, 'What Does Gaddafi's Fall Mean for Africa?'. (http://www.pambazuka.org/en/category/features/76055).

Mandaza, I., 1985, 'The State and Politics in the Post-White Settler Colonial Situation', in I. Mandaza, ed., *Zimbabwe: The Political Economy of Transition, 1980–1986*, Dakar: CODESRIA.

Moyo, S., 2001, 'The Land Occupation Movement and Democratization in Zimbabwe: Contradictions of neoliberalism', *Millennium Journal of International Studies*, Vol. 30, No. 2, pp. 311-330.

Moyo, S., 2008, *African Land Questions, Agrarian Transitions and the State: Contradictions of Neo-liberal Land Reforms*, Dakar: CODESRIA.

Moyo, S., 2011a, 'Three Decades of Agrarian Reform in Zimbabwe', *Journal of Peasant Studies*, Vol. 38, No. 3: pp. 493–531.

Moyo, S., 2011b, 'Land Concentration and Accumulation after Redistributive Reform in Post-Settler Zimbabwe', *Review of African Political Economy*, Vol. 38, No. 128: pp. 257–76.

Moyo, S., 2011c, 'Changing Agrarian Relations after Redistributive Land Reform in Zimbabwe', *Journal of Peasant Studies*, Vol. 38, No. 5: pp. 939–66.

Moyo, S., Chambati, W., Murisa, T., iziba, D., Dangwa, C., Mujeyi, K., and Nyoni, N. 2009, *Fast Track Land Reform Baseline Survey in Zimbabwe: Trends and Tendencies*, 2005/06, Harare: AIAS.

Moyo, S. and Yeros, P., 2005a, 'The Resurgence of Rural Movements under Neoliberalism', in S. Moyo and P. Yeros, eds., *Reclaiming the Land: The Resurgence of Rural Movements in Africa, Asia and Latin* America, London and Cape Town: Zed Books and David Philip.

Moyo, S. and Yeros, P., 2005b, 'Land Occupations and Land Reform in Zimbabwe: Towards the National Democratic Revolution?', in S. Moyo and P. Yeros, eds., *Reclaiming the Land: The Resurgence of Rural Movements in Africa, Asia and Latin America*, London and Cape Town: Zed Books and David Philip.

Moyo, S. and Yeros, P., 2007a, 'The Radicalised State: Zimbabwe's Interrupted Revolution', *Review of African Political Economy*, Vol. 111: pp. 103–21.

Moyo, S. and Yeros, P., 2007b, 'The Zimbabwe Question and the Two Lefts', *Historical Materialism*, Vol. 15, No. 3: pp. 171–204.

Moyo, S. and Yeros, P., 2009, 'Zimbabwe Ten Years On: Results and Prospects', *MR-zine*, 10 February. (http://mrzine.monthlyreview.org/2009/my100209.html).

Moyo, S. and Yeros, P., 2011a, 'The Fall and Rise of the National Question', in S. Moyo and Yeros, P., eds., *Reclaiming the Nation: The Return of the National Question in Africa, Asia and Latin America*, London: Pluto Press.

Moyo, S. and Yeros, P., 2011b, 'After Zimbabwe: State, Nation and Region in Africa', in S. Moyo and Yeros, P., eds., *Reclaiming the Nation: The Return of the National Question in Africa, Asia and Latin America*, London: Pluto Press.

NEPDG [National Energy Policy Development Group], 2001, *National Energy Policy*, May. (http://www.whitehouse.gov).

Patel, H., 1985, 'No Master, No Mortgage, No Sale', in T. Shaw and Y. Tandon, eds., *Regional Development at the International Level*, Vol. II, Lanham, MD: University Press of America.

Patel, H. and Chan, S., 2006, 'Zimbabwe's Foreign Policy: A Conversation', *The Round Table*, Vol. 95, No. 384: pp. 175–90.

Patnaik, P., 2012 (forthcoming), 'The Peasant Question and Contemporary Capitalism: Some Reflections with Reference to India', *Agrarian South: Journal of Political Economy*, Vol. 1, No. 1.

Raftopoulos, B., 2009, 'The Crisis in Zimbabwe, 1998–2008', in B. Raftopoulos and A.S. Mlambo, eds., *Becoming Zimbabwe: A History from the Pre-colonial Period to 2008*, Harare and Johannesburg: Weaver Press and Jacana Media.

Raftopoulos, B., 2010, 'The Global Political Agreement as a "Passive Revolution": Notes on Contemporary Politics in Zimbabwe', *The Round Table*, Vol. 99, No. 411: pp. 705–718.

Raftopoulos, B. and A.S. Mlambo, 2009, *Becoming Zimbabwe: A History from the Pre-colonial Period to 2008*, Harare and Johannesburg: Weaver Press and Jacana Media.

Ranger, T., 2004, 'Historiography, Patriotic History and the History of the Nation: The Struggle over the Past in Zimbabwe', *Journal of Southern African Studies*, Vol. 30, No. 2: pp. 215–34.

Sadomba, W., 2008, 'War veterans in Zimbabwe: complexities of a liberation movement in an African post-colonial settler society', PhD. Thesis, Wageningin University.

Scoones, I., Marongwe, N., Mavedzenge, B. Mahenehene, J., Murimbarimba, F., and Sukume, C., 2010 *Zimbabwe's Land Reform: Myths and Realities*, Suffolk, Harare and Johannesburg: James Currey, Weaver Press and Jacana Media.

Shivji, I.G., 1976, *Class Struggles in Tanzania*, New York, NY and London: Monthly Review Press.

Shivji, I.G., 2009, 'Mwalimu's Non-Alignment and Pan-Africanism in Relation to the Tasks of the Post-Neoliberal Generation', University of Dar es Salaam, mimeo.

SIPRI, 2011, Arms Transfers to Zimbabwe: Implications for an Arms Trade Treaty, Background Paper, March.

Skalnes, T., 1995, *The Politics of Economic Reform in Zimbabwe: Continuity and Change in Development*, New York: St. Martin's Press.

Worby, E., 2003, 'The end of modernity in Zimbabwe? Passages from Development to Sovereignty'. In A. Hammar, B. Raftopoulos and S. Jensen, eds., *Zimbabwe's Unfinished Business: Rethinking Land, State and Nation in the Context of Crisis*. Harare, Weaver Press.

Yeros, P., 2002, 'The Political Economy of Civilisation: Peasant-Workers in Zimbabwe and the Neo-colonial World', PhD Thesis, University of London.

CARDIFF UNIVERSITY * PRIFYSGOL CAERDYDD *

Lightning Source UK Ltd.
Milton Keynes UK
UKOW05f2355060114

224086UK00004B/342/P

9 782869 785533